Mathematical Modelling and Computational Methods for Innovation

Series editors: *Praveen Agarwal, Michael Ruzhansky, Martin Bohner*

Innovation is strongly entwined with mathematics. Mathematics plays a role in every innovation journey, from ideas to proof of concept, implementation, and real-world use. Mathematics provides the fundamental key to creating new technologies, products, and services in medicine, energy, quantum, AI, transport, and Agri-tech. The series offers a list of titles on the latest development and research in areas of mathematical modeling and computational mathematics for Innovation. It offers new reference works, including authored and edited volumes, on mathematical modeling and computational mathematics in the areas of science, engineering, education, industry, biology, agriculture, nature, environment, social life, IT, and other such mathematical innovation solutions within the scope of this series.

I-Function and Its Applications
P. Saxena, Praveen Agarwal, and Altaf Ahmad Bhat

https://www.routledge.com/Mathematical-Modeling-and-Computational-Methods-for-Innovation/book-series/MCMI

I-Function and Its Applications

Vinod Prakash Saxena, Praveen Agarwal, and
Altaf Ahmad Bhat

CRC Press
Taylor & Francis Group
Boca Raton London New York

CRC Press is an imprint of the
Taylor & Francis Group, an **informa** business
A CHAPMAN & HALL BOOK

Front cover image: is1003/ShutterStock

First edition published 2025
by CRC Press
2385 NW Executive Center Drive, Suite 320, Boca Raton FL 33431

and by CRC Press
4 Park Square, Milton Park, Abingdon, Oxon, OX14 4RN

CRC Press is an imprint of Taylor & Francis Group, LLC

ISBN: 978-1-032-84792-4 (hbk)
ISBN: 978-1-032-85355-0 (pbk)
ISBN: 978-1-003-51775-7 (ebk)

DOI: 10.1201/9781003517757

Typeset in Times
by codeMantra

Contents

Preface

I-function introduced by the first author is the most generalized form of special functions in general and generalized hypergeometric functions in particular. This function evolved while solving dual integral equations in general form. These equations arise in basic form while solving mixed boundary problems of potential theory, energy diffusion, and population dynamics. The I-function may provide a comprehensive representation of the solutions to all such problems.

At the same time, the I-function (already listed in AMS Mathematical Reviews) provides a fascinating field for teaching and research allied to complex, real and functional analysis. Lie Algebra of this function has also attracted several mathematicians.

Thus, I-function expands the area of special functions that have been developed and applied in a variety of fields, such as combinatory, astronomy, applied mathematics, physics, and engineering due mainly to their remarkable properties. The main purpose of this book is to be a forum of recently developed theories and formulas of I-function with its possible applications to some other research areas. This book provides readers with an opportunity to develop an understanding of recent trends of this function and the skills needed to apply advanced mathematical techniques to solve complex problems. Subject matters are normally related to properties of I-function involving the mathematical analysis and its numerous applications. The main objective of this book is to highlight the importance of fundamental results and techniques based on the theory of complex analysis and emphasize articles devoted to the mathematical aspect and for possible applications.

The authors are indebted to Dr. Renu Jain and Dr. Shilpi Jain for providing some of their research materials.

About the Authors

Altaf Ahmad Bhat earned his Ph.D. from the School of Mathematics and Allied Sciences, Jiwaji University Gwalior in 2019. He qualified CSIR NET in Mathematics in December 2015. His fields of specialization are Special Functions, Basic Hypergeometric Series, and Fractional Calculus. He has published more than 40 research papers in journals of international repute. He has 9 years of Teaching experience as Assistant Professor of Mathematics at Department of Computer Science and Engineering, Islamic University of Science and Technology – Awantipora – Pul wama – Kashmir – India from March 2011 to December 2015 and January 2018 to February 2022. Since November 2022 he has been Lecturer at the Department of Mathematics and Computing Skills Unit, Preparatory Studies Centre, University of Technology and Applied Sciences – Salalah – Sultanate of Oman.

Praveen Agarwal is Vice-Principal and Professor at Anand International College of Engineering, Jaipur, India. He was listed among Stanford University's ranking of the World's Top 2% of Scientists in 2020, 2021, 2022, and 2023. In the 2023 ranking of best scientists worldwide announced by Research.com, he ranked 21st at the India level and 2436th worldwide in Mathematics. He is the Editor of Book series *Mathematical Modelling & Computational Method for Innovation*, Taylor & Francis Group.

Vinod Prasad Saxena completed his M.Sc. in 1966 with distinction from Jiwaji University Gwalior, Madhya Pradesh. Four years afterwards, he earned a Ph.D. for his research contribution entitled "Integral Transforms and Their Technical Application". At the postdoctoral level, he worked as CSIR senior Research Fellow and postdoctoral Research Fellow at SATI, Vidisha and M.A. College of Technology Bhopal from May 1969 to July 1971. Besides, Dr. V. P. Saxena was a visiting scientist at the University of Cambridge, England. He was also a visiting scientist in 1978 under the British Council UGC (INDIA) Exchange of Young Scientists Programme and worked under Sri James Hill, Professor of LUC Asian University. Besides being a professor, he was Dean, Faculty of Science, Jiwaji University, Gwalior from 1980 to 1986 and again from 1990 to 1992. Dr. V. P. Saxena took over as emergency Vice Chancellor, Jiwaji University on August 9, 2000. Dr. V. P. Saxena visited England, China, Singapore, Argentina, Turkey, Cyprus, and Germany where he delivered a series of lectures and presented scores of research papers on Mathematics and Engineering.

About the Authors

1 Introduction

Special functions were introduced in the eighteenth century to define the solution of differential equations that come into view of mathematical models of certain problem in engineering, classical quantum physics, and other disciplines. In mechanics and other branches of classical modern physics mathematical modelling led to differential equations which had no solution in terms of known functions. This situation gave rise to the introduction of many of the special functions like Bessel function and Legendre polynomial. Afterwards, the special functions were unified in the form of hypergeometric functions and hypergeometric series.

In large growth of research work on special functions is a mathematical function that provides special needs in physics, engineering and recently in biology. It was seen that special functions can be used in the form of hypergeometric functions.

1.1 HYPERGEOMETRIC FUNCTIONS

Hypergeometric series is a significant generalization of geometric series. This is also known as hypergeometric function, and its notation as $F(a,b;c;x)$ was introduced by (Gauss, C. F. (1812) [1]) while delivering his famous thesis "Disquistiones generals circa seriem infintam" before the Royal Society in Gottingen.

The infinite series defined by him is

$$1+\frac{a.b}{c}\frac{x}{1!}+\frac{a(a+1)b(b+1)}{c(c+1)}\frac{x^2}{2!}+\frac{a(a+1)(a+2)b(b+1)(b+2)}{c(c+1)(c+2)}\frac{x^3}{3!}+\cdots \quad (1.1)$$

He also proved the famous summation theorem

$$_2F_1[a,b;c;1]=\frac{\Gamma(c)\Gamma(c-a-b)}{\Gamma(c-a)\Gamma(c-b)} \quad (1.2)$$

The quantities a, b and c are independent from x. This type of function is also represented by $_2F_1\ [a,b;c;x]$ and a,b,c, are parameters. Here x is argument of the function, where a, b are numerator parameters and c is denominator parameter, all these quantities may be real or complex. Equation (1.1) can be written as (Gaussian series).

$$_2F_1[a,b;c;x]=\sum_{n=0}^{\infty}\frac{(a)_n(b)_n x^n}{(c)_n n!} \quad (1.3)$$

where

$$(a)_n=\frac{\Gamma(a+n)}{\Gamma(a)}=(a)(a+1)(a+2)\ldots(a+n-1);n\geq 1$$
$$(a)_0=1,a\neq 0. \quad (1.4)$$

The above notations are called the Pochhammer symbols [2].

DOI: 10.1201/9781003517757-1

We are concerned with various properties of the special functions under consideration. Equation (1.3) satisfies a certain differential equation is, for us, only one among many facts of interest.

Convergence conditions for hypergeometric function are derived as follows:

$$\lim_{n\to\infty}\left|\frac{(a)_{n+1}(b)_{n+1}z^{n+1}}{(c)_{n+1}(n+1)!}\cdot\frac{(c)_n n!}{(a)_n(b)_n z^n}\right|$$

$$=\lim_{n\to\infty}\left|\frac{(a+n)(b+n)z}{(c+n)(n+1)}\right|=|z|$$

So long as none of a,b,c is zero or a negative integer, the series in (1.3) has the circle the series, $|z|<1$ as its circle of convergence. If either or both a and b is zero or a negative integer, the series terminates and convergence does not enter the discussion.

On the boundary $|z|=1$, of the region of convergence, a sufficient condition for absolute convergence of the series is Re $(c-a-b)>0$.

To prove this, let

$$b = 1/2 \text{ Re } (c-a-b) > 0,$$

Then compare the terms of the series

$$1+\sum_{n=1}^{\infty}\left|\frac{(a)_n(b)_n z^n}{(c)_n n!}\right|$$

with corresponding terms of the series

$$\sum_{n=1}^{\infty}\frac{1}{n^{1+\delta}}$$

This is to be convergent series, since $|z|=1$ and

$$\lim_{n\to\infty}\left|\frac{n^{1+\delta}(a)_n(b)_n}{(c)_n n!}\right|$$

$$=\lim_{n\to\infty}\left|\frac{(a)_n}{(n-1)!n^a}\cdot\frac{(b)_n}{(n-1)!n^b}\cdot\frac{(c)_n}{(n-1)!n^c}\cdot\frac{(n-1)!n^{1+\delta}}{(n)!n^{c-a-b}}\right| \tag{1.5}$$

$$=\left|\frac{1}{\Gamma(a)}\cdot\frac{1}{\Gamma(b)}\cdot\frac{\Gamma(c)}{1}\right|\lim_{n\to\infty}\left|\frac{1}{n^{(c-a-b-\delta)}}\right|=0$$

because Re $(c-a-b-\delta)=2\delta-\delta>0$, the series (1.3) is absolutely convergent on $|z|=1$ when Re $(c-a-b)>0$.

Wallis [3] is the first who uses the term 'hypergeometric' to denote any series which was beyond the ordinary geometric series. During the next 150 years, many other mathematicians studied similar series, like Euler [4] who gave amongst many other results, the famous relation

$$ {}_2F_1[-n,b:c:x]=(1-x)^{c+n-b}\,{}_2F_1(c+n,c-b:c:x) \tag{1.6}$$

This equation is also called a modern infinite hypergeometric series The Gauss hypergeometric function $F(a, b{:}c{:}x)$ has played a significant role in the development of analysis on account of the classical work of Euler [4], Gauss [1] and Riemann [5].

In the definition of Gauss hypergeometric function, it was discovered by Kummer [6], when replaced x by x/b and subsequently letting $b\rightarrow\infty$ leads to another interesting series.

$$ {}_1F_1\left[\begin{matrix} a; \\ & x \\ c; \end{matrix}\right]$$

The above notation is known as confluent hypergeometric function.

A similar technique led to the hypergeometric functions ${}_0F_1$. This result proved to be the expansion of ${}_0F_1$ turned out to be Bessel function.

Gauss hypergeometric function ${}_2F_1$ and confluent hypergeometric function ${}_1F_1$ from the core of special functions and include, as a special case, these functions are commonly used functions. Thus ${}_2F_1$ includes, and most of the orthogonal polynomials like Jacobi. The confluent hypergeometric functions include, as their special cases, and modified forms of confluent hypergeometric functions are Whittaker function and Bessel function.

${}_pF_q$ is a natural generalization of ${}_2F_1$, which is defined as a generalized hypergeometric function, i.e.

$$ \begin{aligned} {}_pF_q\left[(a_p);(b_q):x\right] &= {}_pF_q\left[\begin{matrix} (a_p); \\ & x \\ (b_q); \end{matrix}\right] \\ &= \sum_{n=0}^{\infty}\frac{\prod_{j=1}^{p}(a_j)_n x^n}{\prod_{j=1}^{q}(b_j)_n n!} \end{aligned} \tag{1.7}$$

Here,

$$ {}^kC_n=\binom{k}{n}=\frac{\Gamma(k+1)}{n!\,\Gamma(k+1-n)} \tag{1.8}$$

The number $a_1, a_2,\ldots, a_p$ and $b_1, b_2,\ldots, b_q$ are referred to as numerator and denominator parameters respectively. In Equation (1.7), no denominator parameter should be zero or a negative integer and if any parameter is zero or negative integer, then series reduces to a polynomial

The series ${}_pF_q$ convergent for all complex values of x when $p \le q$.

Also when $p=q+1$ the series is convergent, if $|x|<1$. It converges when $x=1$ if

$$\operatorname{Re}\left[\sum_{j=1}^{q} b_j - \sum_{j=1}^{p} a_j\right] > 0$$

If $p > q + 1$ the series never converges except when $x = 0$ and the function is defined only when the series terminates.

The generalized hypergeometric functions were studied systematically by Thomae [7].

Equation (1.7) has a fundamental system of solutions represented by hypergeometric series multiplied by a power of x.

For complete integration of the hypergeometric equation Mellin–Barnes integrals were used by Barne [8]. Dixon and Farrar [27] have given the asymptotic expansion of general Mellin–Barnes-type integrals.

Mellin has considered a similar more general class of trigonometric functions, Laplace integrals extended over a product of Gamma function in several important papers.

Almost all account of confluent, Gauss and generalized hypergeometric functions have been given in standard works by Slater [29], Exton [9], and Rainville [10].

Further Generalizations

1.2 MAC-ROBERT'S E-FUNCTION

Mac-Robert's E-function [28] arose from an attempt to give a meaning to the symbol

${}_pF_q$ when $p > q + 1$. For $p \le q + 1$, we have

$$E\left[p,\alpha_r : q : \rho_s : x\right] = \frac{\Gamma(\alpha_1)\Gamma(\alpha_2)\ldots\Gamma(\alpha_p)}{\Gamma(\rho_1)\Gamma(\rho_2)\ldots\Gamma(\rho_q)} \times {}_pF_q\left[\alpha_1,\alpha_2,\ldots,\alpha_p : \rho_1,\rho_2,\ldots,\rho_q : -\frac{1}{x}\right] \tag{1.9}$$

where $x \ne 0$, if $p < q$ and $|x|>1$, $p = q + 1$.

For $p \ge q + 1$, we have

$$E\left(p;\alpha_r : q;\rho_s;x\right) = \sum_{r=1}^{p} \frac{\prod_{j=1}^{p}\Gamma(\alpha_j-\alpha_r)}{\prod_{j=1}^{q}\Gamma(\rho_j-\alpha_r)}\Gamma(\alpha_r)x^{\alpha_r} \times {}_{q+1}F_{p+1}\left(\alpha_r,\alpha_r-b_1+1,\ldots\alpha_r-\rho_s+1; \alpha_r-\alpha_1,\ldots *\ldots,\alpha_r-\alpha_p+1;(-1)^{p+q}x\right). \tag{1.10}$$

where $|x| < 1$, if $p = q + 1$. The prime $\prod$ indicates the omission of the factor $\Gamma(\alpha_r - \alpha_r)$. The asterisk indicates the omission of parameter $(\alpha_r - \alpha_r + 1)$; an empty product is to be interpreted as 1; and zero or negative value of the α_r are tacitly excluded.

The asymptotic expansion, for $x \to \infty$, $1/2(p - q + 1)\pi$, (1.10) is given by the right-hand side of (1.9). E-function is also defined in terms of Mellin–Barnes integral as given below:

$$E\left[x\left|\begin{array}{c}\beta_1,\beta_2\ldots,\beta_q\\ \alpha_1,\alpha_2\ldots,\alpha_p\end{array}\right.\right]=\frac{1}{2\pi i}\int_L\frac{\prod_{j=1}^{p}\Gamma(\alpha_j-\xi)\Gamma(\xi)}{\prod_{j=1}^{q}\Gamma(\beta_j-\xi)}x^{\xi}\,d\xi$$

where an empty product is interpreted as 1 and the parameters are such that no pole $\Gamma(\alpha_j-\xi)$ $(j = 1,\ldots,p)$ coincides with $\xi=-N$, where N is a +ve integer. These assumptions will be retained throughout.

The three different paths L of integration are as follows:

i. L runs from $-\iota\infty$ to $+\iota\infty$ so that all the poles of $\Gamma(\alpha_j-\xi)$ $(j = 1,\ldots,p)$ are to the right and the poles of $\Gamma(\xi)$ to the left of L. The integral converges if $q<p+1$ and $|\arg x|<\left(m+n-\frac{1p}{2}-\frac{1q}{2}\pi\right)$.

ii. L is a loop starting and ending at $+\infty$ and encircling all the poles of $\Gamma(\alpha_j-\xi)(j = 1,\ldots, p)$ once in the negative direction, but none of the poles of $\Gamma(\xi)$. The integral converges if $p\geq 1$ and either $q+1<p$ or $p=q+1$ and $|\arg x|<1$.

iii. L is a loop starting and ending at $-\infty$ and encircling all poles of $\Gamma(\xi)$ once in the positive direction, but none of the poles of $\Gamma(\alpha_j-\xi)(j = 1,\ldots, p)$. The integral converges if $q\geq 0$ and either $q+1>p$ or $p=q+1$ and $|x|>1$.

1.3 MEIJER'S G-FUNCTION

Meijer's G-function also provides an interpretation of the symbol $_pF_q$ when $p > q + 1$. This interpretation is incomplete agreement with the one given by Mac-Robert's E-function.

The G-function was defined by Meijer [11] in terms of Mellin–Barnes type integral, it allows a greater freedom in the values of p and q without replacing any non-trivial restriction on m and n, which is given as

$$G_{p,q}^{m,n}\left[x\left|\begin{array}{c}a_1,a_2,\ldots,a_p\\ b_1,b_2,\ldots,b_q\end{array}\right.\right]$$
$$=\frac{1}{2\pi i}\int_L\frac{\prod_{j=1}^{m}\Gamma(b_j-\xi)\prod_{j=1}^{n}\Gamma(1-a_j+\xi)}{\prod_{j=m+1}^{q}\Gamma(1-b_j+\xi)\prod_{j=n+1}^{p}\Gamma(a_j-\xi)}x^{\xi}\,d\xi \tag{1.11}$$

where an empty product is interpreted as 1; $0 \le m \le q$ and the parameters are such that no poles of $\Gamma(b_j - \xi)$, $(j = 1,2,\ldots, m)$ coincide with any poles of $\Gamma(1 - a_j + \xi)$, $(j = 1,2,\ldots, n)$. these assumptions with be retained throughout.

We can write G-function more briefly as

$$G_{p,q}^{m,n}\left[x\left|\begin{matrix}a_j\\ b_j\end{matrix}\right.\right] \text{ or } G_{p,q}^{m,n}[x] \tag{1.12}$$

i. L is a loop from $-i\infty$ to $+i\infty$ so all poles of $\Gamma(b_j - \xi)$, $(j = 1, 2,\ldots, m)$ are to the right and all the poles of $\Gamma(1 - a_j + \xi)$, $(j = 1, 2,\ldots, n)$ to the left of L. The integral converges if $p + q < 2(m + n)$ and
$|x| < (m + n - 1/2p - 1/2\, q)\,\pi$.

ii. L is a loop starting and ending at $+i\infty$ and enriching all poles of $\Gamma(b_j - \xi)$, $(j = 1,2,\ldots, m)$ once in the negative direction, but none of the poles of $\Gamma(1 - a_j + \xi)$, $(j = 1,2,\ldots, n)$. The integral converges if $q \ge 1$ and either $p < q$ or $p = q$ and $|x| < 1$.

iii. L is a loop starting and ending at $-i\infty$ and enriching all poles of $\Gamma(1 - a_j+\xi)$, $(j = 1,2,\ldots, n)$ once in the positive direction, but none of the poles of $\Gamma(b_j - \xi)$, $(j = 1,2,\ldots, m)$. The integral converges if $p \ge 1$ and either $p < q$ or $p = q$ and $|x| > 1$.

${}_pF_q$ and E-function are contained by the G-function as its particular cases. They are expressed by the relation

$${}_pF_q\left[\begin{matrix}(a_j)_{1,p}\\ \\ (b_j)_{1,q}\end{matrix};x\right] = \frac{\prod_{j=1}^{q}\Gamma(b_j)}{\prod_{j=1}^{p}\Gamma(a_j)}G_{p,q+1}^{1,p}\left[x\left|\begin{matrix}(1-a_j)_{1,p}\\ 0,(1-b_j)_{1,q}\end{matrix}\right.\right]$$

$$= \frac{\prod_{j=1}^{q}\Gamma(b_j)}{\prod_{j=1}^{p}\Gamma(a_j)}G_{p+1,q}^{1,p}\left[-\frac{1}{x}\left|\begin{matrix}1,b_1,\ldots,b_q\\ a_1,\ldots,a_p\end{matrix}\right.\right] \tag{1.13}$$

$$E[p,\alpha_r:q:\beta_s:x] = G_{q+1,p}^{1,p}\left[x\left|\begin{matrix}1,(\beta_s)_{1,q}\\ (\alpha_r)_{1,p}\end{matrix}\right.\right] \tag{1.14}$$

Other significant uses of G-function are given by Mathai and Saxena [12], such as uses of G-function in statistical theory of distributions and series expansion of general G-function.

Compared with Meijer and Mac-Robert have contributed so much research in their respective fields, a lot of other researchers have also made wealthier literature on these functions. In the monumental work by Luke [13] is the availability of a detailed account of G-function.

G-function and E-function are quite general in character, a number of special functions like Wright's generalized Bessel function [14], Wright's, generalized parabolic cylinder function, a function due to Mittage-Leffler [30], and several other functions do not form their special cases.

By considering Meijer's G-function [15], extended the field of symmetrical Fourier kernels.

A more general function was introduced by Fox [15], which was named as, Fox's H-function.

1.4 FOX'S H-FUNCTION

The Fox's H-function is a generalized form of Meijer's G-function [15]. This occurred in the study of functional equations considered by Chandrasekharan and Narsimhan [16], and Bochner [17].

The H-function is defined in terms of the following Mellin–Barnes type of integral

$$H[x] = H_{p,q}^{m,n}[z] = H\left[x \left| \begin{matrix} (a_j,\alpha_j)_{1,p} \\ (b_j,\beta_j)_{1,q} \end{matrix} \right. \right]$$

$$= H\left[x \left| \begin{matrix} (a_1,\alpha_1),\ldots,(a_p,\alpha_p) \\ (b_1,\beta_1),\ldots,(b_q,\beta_q) \end{matrix} \right. \right] \quad (1.15)$$

$$= \frac{1}{2\pi i}\int_L \phi(\xi) x^{\xi} d\xi \quad (1.16)$$

where z is a complex variable and

$$x^{\xi} = \exp\,[\xi log\left|x\right| + i \arg z].$$

Here $\log\left|x\right|$ represents the natural logarithm of $\left|x\right|$ and ($\arg x$) is not necessarily the principle values. An empty product is interpreted as unity. Also

$$\phi(\xi) = \frac{\prod_{j=1}^{m}\Gamma(b_j - \beta_j\xi)\prod_{j=1}^{n}\Gamma(1-a_j+\alpha_j\xi)}{\prod_{j=m+1}^{q}\Gamma(1-b_j+\beta_j\xi)\prod_{j=n+1}^{p}\Gamma(a_j-\alpha_j\xi)} \quad (1.17)$$

m, n, p, and q are negative integers satisfying $0 \le n \le p$, $0 \le m \le q$ and α_j ($j = 1,2, \ldots q$) are assumed to be positive quantities for standardization purposes. a_j ($j = 1,2, \ldots n$). b_j ($j =1,2, \ldots m$) are complex numbers such that none of the points.

$$\xi = \left\{\frac{(b_h+v)}{\beta_h}\right\}(h=1,2\ldots\ldots m, v=0,\ 1,\ 2\ldots\ldots n)$$

which are the poles of $\Gamma\left(b_h - \beta_h\xi\right)$, $(h = 1, 2, \ldots m)$ and points

$$\xi = \{a_j - \eta - 1)/\alpha_j\}, (j = 1, 2, \ldots n, \eta = 0, 1, 2, \ldots)$$

are the poles of $\Gamma\left(1 - a_j + \alpha_j\xi\right)$ coincide with one another, i.e.

$$\alpha_j\,(b_h + v) \neq b_h\,(a_j - \eta - i)\;(\,v, n = 0, 1, 2, \ldots, h = 0, 1, 2, \ldots m, i = 1,2 \ldots n.)$$

The contour L runs from $(\sigma - i\infty)$ to $(\sigma + i\infty)$ such that the poles of $\Gamma\left(b_h - \beta_h\xi\right)$, $h = 1, 2 \ldots m$ lie to the right of L and the poles of $\Gamma\left(1 - a_j + \alpha_j\xi\right)$, $j = 1, 2\ldots n$ lie to the left of L.

The behaviour of the H-function for large and small values of the argument was discussed by Braaksma [18].

Here,

$$H[x] = H_{p,q}^{m,n}\left[x\left|\begin{matrix}(a_1,\alpha_1),\ldots,(a_p,\alpha_p)\\ \\ (b_1,\beta_1),\ldots,(b_q,\beta_q)\end{matrix}\right.\right] \tag{1.18}$$

where

$$\alpha = \min\left\{\mathrm{Re}\left(\frac{b_j}{\beta_j}\right)\right\}, (j{=}1, 2\ldots m)$$

$$A = \sum_{j=1}^{q}\beta_j - \sum_{j=1}^{p}\alpha_j > 0$$

$$B = \sum_{j=1}^{n}\alpha_j - \sum_{j=n+1}^{p}\alpha_j + \sum_{j=1}^{m}\beta_j - \sum_{j=m+1}^{q}\beta_j > 0$$

$|\arg z| < ½\, B\pi$, and $\beta = \max\,[Re\,(a_j - 1/\alpha_j)].(j = 1,2, \ldots\ldots n)$.

The relation between H-function and G-function is denoted by

$$H_{p,q}^{m,n}\left[x\left|\begin{matrix}(a_j,v)_{1,p}\\(b_j,v)_{1,q}\end{matrix}\right.\right] = \frac{1}{v}G_{p,q}^{m,n}\left[x^{\frac{1}{v}}\left|\begin{matrix}(a_j)_{1,p}\\(b_j)_{1,q}\end{matrix}\right.\right] \tag{1.19}$$

where v is real and positive, H-function contains many other special functions as its particular cases.

Braaksma [18] has done a detailed analytical study of the H-function. In general, the transformation involving integral in Lebesgue or improper Riemann form is defined as

$$f(x) = \int_a^b K(u,x)h(u)\,du \tag{1.20}$$

where $K(u, x)$ and $h(u)$ are both integrable with respect to u here $K(u, x)$ is known as kernel of the transformation, x may be real or complex and u is real any of the limits a and b may be finite or infinite. In the above equation, $f(x)$ is the image of $h(u)$ in the transform and $h(u)$ the original of $f(x)$. If the integral equation can be expressed as

$$h(u) = \int_a^b H(u,x) f(x)\,dx \tag{1.21}$$

where $H(u,x)$ is a known function then the above equation is called the inversion formula of the above equation. If

$$K(u,x) = H(u,x)$$

$K(u,x)$ is known as symmetric Fourier kernel and, if

$$K(u,x) \neq H(u,x) \in C[a,b]$$

then $K(u,x)$ and $H(u,x)$ are called unsymmetrical Fourier kernels. $K(u,x)$ and $H(u,x)$ are also known as reciprocal kernels of each other. The theory of integral transform is very useful in solving various types of boundary value problems.

Fourier kernel is also associated with H-function and obtained the asymptotic behaviour of the kernel for a large value of the argument. (Keserwani, R.N (1965) [19]), obtained analytic continuations and asymptotic expansions for H- function, and several pairs of symmetric and unsymmetric Fourier kernels involving H-function.

1.5 SAXENA'S I-FUNCTION

Due to motivated by the importance of Fox's H-function and G-function in mathematical analysis in 1982, Saxena [20] has been introduced a new function which is called an I-function or Saxena's I-function.

The definition of the I-function is as follows:

$$I(z) = I^{m,n}_{p_i,q_i;r}\left[x \,\middle|\, \begin{matrix} (a_j,a_j)_{j,n};(a_{ji},\alpha_{ji})_{n+1,p_i} \\ (b_j,\beta_j)_{1,m};(a_{ji},\beta_{ji})_{m+1,q_i} \end{matrix}\right] \tag{1.22}$$

$$= \frac{1}{2\pi i}\int_L \phi(\xi) x^{\xi}\,d\xi$$

$$\phi(\xi) = \frac{\prod_{j=1}^{m}\Gamma(b_j - \beta_j\xi)\prod_{j=1}^{n}\Gamma(1 - a_j + \alpha_j\xi)}{\sum_{i=1}^{r}\left\{\prod_{j=m+1}^{q_i}\Gamma(1 - b_{ji} + \beta_{ji}\xi)\prod_{j=n+1}^{p_i}\Gamma(a_{ji} - \alpha_{ji}\xi)\right\}} \tag{1.23}$$

i. L is a contour which running from σ – i ∞ to σ + i ∞, so that all the poles of $\Gamma(b_j - \beta_j\xi)$, $j = 1,2, \ldots n$, are to the right and all the poles of $\Gamma(1-a_j-\alpha_j\xi)$, $j = 1,2, \ldots, m$, to the left of L, the integral convergences if $p+q<2<(m+n)$, $|\arg x| < \left(m+n-\frac{p}{2}-\frac{q}{2}\right)\pi$

ii. L is a loop starting and ending at $\sigma + i\infty$ and encircling all poles of $\Gamma(b_j - \beta_j\xi)$, $j = 1,2, \ldots\ldots n$, once in the negative direction, but none of the poles of $\Gamma(1-a_j+\alpha_j\xi)$, $j = 1, 2, \ldots\ldots, m$, the integral converges of $q \geq 1$ and $p < q$ or $p = q$ and $|x| < 1$

iii. L is a loop beginning and ending at $\sigma - i\infty$ and encircling all poles of $\Gamma(1 - a_j + \alpha_j\xi)$, $j = 1,2, \ldots\ldots, m$, once in the positive direction, but none of the poles of, $\Gamma(b_j - \beta_j\xi)$, $j = 1,2, \ldots\ldots n$ the integral converges of $p \geq 1$ and either $p > q$ or $p = q$ and $|x| < 1$.

This function is defined under the existing conditions:

i. $\lambda > 0$, $|\arg x| < \pi\,\lambda/2$
ii. $\lambda > 0$, $\mathrm{Re}\,(\mu + 1) < 0$, $|\arg x| > \pi_\lambda/2$

λ and μ denote the quantities

$$\lambda = \sum_{j=1}^{p}\alpha_j + \sum_{j=1}^{q}\beta_j = \max_{1\le i\le r}\left[\sum_{j=q+1}^{S_i}\alpha_{ji} + \sum_{j=q+1}^{t_i}\beta_{ji}\right] \tag{1.24}$$

$$\mu = \sum_{j=1}^{m}\left(b_j-\beta_j\sigma\right) - \sum_{j=1}^{n}\left(a_j-\alpha_j\sigma\right)$$

$$-\max_{1\le i\le r}\left[\sum_{j=n+1}^{p_i}\left(a_j-\alpha_j\sigma\right) - \sum_{j=m+1}^{q_i}\left(b_j-\beta_j\sigma\right) + \frac{p_i}{2} - \frac{q_i}{2}\right] \tag{1.25}$$

Here,

$$\xi = \sigma + it$$

1.6 FRACTIONAL INTEGRAL OPERATORS

For the solutions of various integral equations, the theory of fraction integral operators is very useful. Various definitions of fractional integrations have been given by many authors, like Kober [21], Erdelyi [22], Saxena [23], etc.

The following generalized operators have been used by Fox [24] in his analysis.

$$\Im[\gamma,\varepsilon:m]w(x) = \frac{m}{\Gamma\gamma}x^{-\varepsilon-\gamma m+m-1}\int_0^x\left(x^m - v^m\right)^{\gamma-1}v^{\varepsilon}w(v)dv \tag{1.26}$$

$$\Re[\gamma,\varepsilon:m]w(x) = \frac{m}{\Gamma\gamma}x^{\varepsilon}\int_0^x\left(v^m - x^m\right)^{\gamma-1}v^{-\varepsilon-\gamma\varepsilon+\varepsilon m-1}w(v)dv \tag{1.27}$$

The operator $\Im$ exists if $w(x) \in L_p(0,\infty)$

$$P \geq 1, \gamma > 0, \varepsilon > (1-p)/p$$

If, in addition, $w(x)$ can be differentiated sufficiently often then the operator $\Im$ exists for both negative and positive values of γ.

Also for $\Re$, $w(x) \in L_p(0,\infty)$, $p \geq 1, m > 0, \varepsilon > \frac{(-1)}{p}$ while γ can take any negative or positive value.

In the same paper, Fox has shown

$$M[\Im(\gamma,\varepsilon,m)]w(x) = \frac{[(\varepsilon+1-\xi)/m]}{[\gamma+(\varepsilon+1-\xi)/m]} M[w(x)]$$

$$M[\Re(\gamma,\varepsilon,m)]w(x) = \frac{[(\varepsilon+\xi)/m]}{[\gamma+(\varepsilon+\xi)/m]} M[w(x)] \tag{1.28}$$

Kober [21] has introduced operator $\Im, \Re$ if $m = 1$, and the above form generalized by Erdelyi [22].

Beta functions in the integral forms are very useful for the application of the above operators

$\Re$ and $\Im$, which are defined as:

$$\int_0^x \left(x^{\frac{1}{\alpha}} - v^{\frac{1}{\alpha}}\right)^{d-e-1} v^{\frac{e}{\alpha}-1-\xi} dv = \frac{\alpha\Gamma(d-e)\Gamma(e-\alpha\xi)}{\Gamma(d-\alpha\xi)} x^{\frac{d}{\alpha}-\frac{1}{\alpha}-\xi} \tag{1.29}$$

if $d > e$ and $\frac{e}{\alpha} > \sigma, (\xi = \sigma + it)$

$$\int_x^\infty \left(v^{\frac{1}{\alpha}} - x^{\frac{1}{\alpha}}\right)^{d-e-1} v^{\frac{1}{\alpha}-\frac{d}{\alpha}-1-\xi} dv = \frac{\alpha\Gamma(d-e)\Gamma(e+\alpha\xi)}{\Gamma(d+\alpha\xi)} x^{-\frac{e}{\alpha}-\xi} \tag{1.30}$$

given $d > e$ at the lower limit and, $\alpha\,\mathrm{Re}(\xi) + e > 0$ at the upper limit.

These operators were used to solve dual integral equations by Saxena [25]. Kesarwani [26] used the operators $\Im$ and $\Re$in slightly different forms. In the next section, we discuss the existence conditions of I-function with its special cases. This will help in developing analytic aspects of the function based on complex integration and Gamma function properties.

1.7 THE I-FUNCTION AND EXISTENCE CONDITIONS

$$I(z) = I^{m,n}_{p_i,q_i;r}\left[z \left| \begin{matrix} (a_j,\alpha_j)_{j,n};(a_{ji},\alpha_{ji})_{n+1,p_i} \\ (b_j,\beta_j)_{1};(a_{ji},\beta_{ji})_{m+1,q_i} \end{matrix} \right.\right] = \frac{1}{2\pi i}\int_L \phi(\xi) z^{\xi} d\xi$$

$$\phi(\xi) = \frac{\prod_{j=1}^{m} \Gamma(b_j - \beta_j \xi) \prod_{j=1}^{n} \Gamma(1 - a_j + \alpha_j \xi)}{\sum_{i=1}^{r} \left\{ \prod_{j=m+1}^{q_i} \Gamma(1 - b_{ji} + \beta_{ji} \xi) \prod_{j=n+1}^{p_i} \Gamma(a_{ji} - \alpha_{ji} \xi) \right\}} \tag{1.31}$$

under the existence conditions of I-function defined in Section 1.5.

The special cases are derived from the asymptotic behaviour of the Gamma function these special cases deal with dependence on function parameters like

The I-function and the Fox's H-function both are important in mathematical analysis. We use the existence convergence condition of this function with the asymptotic behaviour of the Gamma function. Here we discuss the convergence conditions of the integral using the Gamma function and derive some special cases in terms of Gamma functions [31].

We use the formula of the Gamma function [32].

$$\lim_{|y| \to \infty} |\Gamma(x + iy)| = \lim_{|y| \to \infty} (2\pi)^{\frac{1}{2}} e^{\frac{-\pi}{2}|y|} |y|^{x - \frac{1}{2}} \tag{1.32}$$

In (1.31) by putting $z = \mathrm{Re}^{i\theta}$ ($\xi = \sigma + \mathrm{it}$) and taking limit as $t \to \infty$ in the right-hand side of (1.31) putting the value of Equation (1.32), we obtain

$$I^{m,n}_{p_i,\, q_i : r} |z|$$

$$= \lim_{t \to \infty} \frac{1}{2\pi i} \int_L \frac{\prod_{j=1}^{m} \Gamma(b_j - \beta_j \xi) \prod_{j=1}^{n} \Gamma(1 - a_j + \alpha_j \xi)}{\sum_{i=1}^{r} \left\{ \prod_{j=m+1}^{q_i} \Gamma(1 - b_{ji} + \beta_{ji} \xi) \prod_{j=n+1}^{p_i} \Gamma(a_{ji} - \alpha_{ji} \xi) \right\}} z^{\xi} \, d\xi \tag{1.33}$$

$$= \lim_{t \to \infty} \frac{1}{2\pi i} \int_L \frac{\prod_{j=1}^{m} (2\pi)^{\frac{1}{2}} e^{\frac{-\pi}{2}|-\beta_j t|} |-\beta_j t|^{\left(b_j - \beta_j \sigma - \frac{1}{2}\right)}}{\left[\sum_{i=1}^{r} \prod_{j=m+1}^{q_i} (2\pi)^{\frac{1}{2}} e^{\frac{-\pi}{2}|\beta_{ji} t|} |\beta_{ji} t|^{\left(\frac{1}{2} - b_{ji} + \beta_{ji} \sigma\right)} \right]}$$

$$\times \frac{\prod_{j=1}^{n} (2\pi)^{\frac{1}{2}} e^{\frac{-\pi}{2}|\alpha_j t|} |-\alpha_j t|^{\left(\frac{1}{2} - a_j + \alpha_j \sigma\right)}}{\left[\prod_{j=n+1}^{p_i} (2\pi)^{\frac{1}{2}} e^{\frac{-\pi}{2}|-\alpha_{ji} t|} |-\alpha_{ji} t|^{\left(a_{ji} - \alpha_{ji} \sigma - \frac{1}{2}\right)} \right]} \left(\mathrm{Re}^{i\theta}\right)^{\sigma + it} d\xi \tag{1.34}$$

$$= \lim_{t\to\infty} \frac{e^{-\frac{\pi}{2}\left(\sum_{j=1}^{m}\beta_j+\sum_{j=1}^{n}\alpha_j\right)t}}{\left[\sum_{i=1}^{r} e^{-\frac{\pi}{2}\left(\sum_{j=m+1}^{q_i}\beta_{ji}+\sum_{j=n+1}^{p_i}\alpha_{ji}\right)t}\right]} \tag{1.35}$$

$$\times \frac{t^{\left\{\sum_{j=1}^{m}\left(b_j-\beta_j\sigma\right)+\sum_{j=1}^{n}\left(-a_j+\alpha_j\sigma\right)+\frac{n}{2}-\frac{m}{2}\right\}} e^{|-\theta t|}}{\left[t^{\left\{\sum_{j=m+1}^{q_i}(-b_{ji}+\beta_{ji}\sigma)+\sum_{j=n+1}^{p_i}\left(a_{ji}-\alpha_{ji}\sigma\right)+\frac{q_i}{2}-\frac{p_i}{2}+\frac{n}{2}-\frac{m}{2}\right\}}\right]}$$

Let

$$A = \max\left[\sum_{j=m+1}^{q_i}\beta_{ji}+\sum_{j=n+1}^{p_i}\alpha_{ji}\right], \forall i = 1,2,3\ldots,r$$

$$B = \min\left[\sum_{j=m+1}^{q_i}\left(-b_{ji}+\beta_{ji}\sigma\right)+\sum_{j=n+1}^{p_i}\left(a_{ji}-\alpha_{ji}\sigma\right)+\left(\frac{q_i}{2}-\frac{p_i}{2}+\frac{n}{2}-\frac{m}{2}\right)\right], \forall i = 1,2,3\ldots,r$$

Then we have from Equation (1.35),

$$I_{p,q:r}^{m,n}|z| \sim \lim_{t\to\infty} \frac{e^{-\frac{\pi}{2}\left(\sum_{j=1}^{m}\beta_j+\sum_{j=1}^{n}\alpha_j-A\right)t}\, t^{\left\{\sum_{j=1}^{m}\left(b_j-\beta_j\sigma\right)+\sum_{j=1}^{n}\left(-a_j+\alpha_j\sigma\right)-B+\frac{n}{2}-\frac{m}{2}\right\}}}{\left[\sum_{i=1}^{r} e^{-\frac{\pi}{2}\left(\sum_{j=m+1}^{q_i}\beta_{ji}+\sum_{j=n+1}^{p_i}\alpha_{ji}-A\right)t}\right]} \tag{1.36}$$

$$\times \frac{e^{|-\theta t|}}{t^{\left\{\sum_{j=m+1}^{q_i}\left(-b_{ji}+\beta_{ji}\sigma\right)+\sum_{j=n+1}^{p_i}\left(a_{ji}-\alpha_{ji}\sigma\right)-B+\frac{q_i}{2}\frac{p_i}{2}+\frac{n}{2}-\frac{m}{2}\right\}}}$$

1.8 EXISTENCE CONDITIONS – SPECIAL CASE [33]

Let $\alpha_j = \eta\alpha_{j+1}$ then we get $\alpha_1 = \eta\alpha_2, \alpha_2 = \eta\alpha_3\ldots$, according to this condition, we can write

$$\alpha_1 = \eta^{n-1}\alpha_n$$

Similarly, we assume
same as $\beta_j = \delta\beta_{j+1}$ then we get

$$\beta_1 = \delta\beta_2, \beta_2 = \delta\beta_3 \ldots,$$

$$\beta_1 = \delta^{m-1}\beta_m$$

substituting the above in (2.6) we arrive at

$$I_{p,q:r}^{m,n}|z| \sim \lim_{t\to\infty} \frac{e^{-\frac{\pi}{2}\left((\beta_1+\beta_2+\beta_3,\cdots,+\beta_m)+(\alpha_1+\alpha_2+\alpha_3+\cdots+\alpha_n)-A\right)t}}{\sum_{i=1}^{r}\left[e^{-\frac{\pi}{2}\left(\sum_{j=m+1}^{q_i}\beta_{ji}-\sum_{j=n+1}^{p_i}\alpha_{ji}-A\right)t}\right]}$$

$$\times \frac{t^{\left\{\sum_{j=1}^{m}\left(b_j-(\beta_1+\beta_2+\beta_3+,\cdots+\beta_m)\sigma+\sum_{j=1}^{n}\left(-a_j+(\alpha_1+\alpha_2+\alpha_3+\cdots+\alpha_n)\sigma\right)-B+\frac{n}{2}-\frac{m}{2}\right)\right\}}e^{|-\theta t|}}{\sum_{i=1}^{r}\left[t^{\left\{\sum_{j=m+1}^{q_i}(-b_{ji}+\beta_{ji}\sigma)+\sum_{j=n+1}^{p_i}(a_{ji}-\alpha_{ji}\sigma)-B+\frac{q_i}{2}-\frac{p_i}{2}+\frac{n}{2}-\frac{m}{2}\right\}}\right]} \tag{1.37}$$

Hence above series will be,

$$\begin{aligned}
&\left((\beta_1+\beta_2+\beta_3,\ldots,+\beta_m)+(\alpha_1+\alpha_2+\alpha_3+\cdots+\alpha_n)-A\right)\\
&=\left[\beta_1\left\{1+\delta+\delta^2+\delta^3+\cdots+\delta^{m-1}\right\}\right.\\
&\left.+\alpha_1\left\{1+\eta+\eta^2+\eta^3+\cdots+\eta^{n-1}\right\}-A\right]\\
&=\left[\beta_1\left\{\frac{1}{1-\delta^{m-1}}\right\}+\alpha_1\left\{\frac{1}{1-\eta^{n-1}}\right\}-A\right]
\end{aligned} \tag{1.38}$$

Therefore,

$$I_{p,q:r}^{m,n}|z| \sim \lim_{t\to\infty} \frac{e^{-\frac{\pi}{2}\left(\frac{\beta_1}{1-\delta^{m-1}}+\frac{\alpha_1}{1-\eta^{n-1}}-A\right)t}\, t^{\left\{\sum_{j=1}^{m}(b_j-\beta_j\sigma)+\sum_{j=1}^{n}(-a_j+\alpha_j\sigma)-B+\frac{n}{2}-\frac{m}{2}\right\}}}{\sum_{i=1}^{r}\left[e^{-\frac{\pi}{2}\left(\sum_{j=m+1}^{q_i}\beta_{ji}-\sum_{j=n+1}^{p_i}\alpha_{ji}A\right)t}\right]}$$

$$\times \frac{e^{|-\theta t|}}{\sum_{i=1}^{r}\left[t^{\left\{\sum_{j=m+1}^{q_i}(-b_{ji}+\beta_{ji}\sigma)+\sum_{j=n+1}^{p_i}(a_{ji}\,\alpha_{ji}\sigma)B+\frac{q_i}{2}-\frac{p_i}{2}+\frac{n}{2}-\frac{m}{2}\right\}}\right]} \tag{1.39}$$

for

$$\mathrm{Re}^{\left(\frac{\beta_1}{1-\delta^{m-1}}+\frac{\alpha_1}{1-\eta^{n-1}}\right)} > \mathrm{Re}(A)$$

1.9 PARTICULAR CASES

i. If we put $\alpha_{ji} = \alpha_j, \beta_{ji} = \beta_j, r = 1$, in Equations (1.38) and (1.39) reduce to

$$H_{p,q}^{m,n}\left|z\right| \sim \lim_{t\to\infty} \frac{e^{-\frac{\pi}{2}\left(\left(\beta_1+\beta_2+\beta_{3,\dots,}+\beta_m\right)+\left(\alpha_1+\alpha_2+\alpha_3+\dots+\alpha_n\right)-A\right)t}}{\left[e^{-\frac{\pi}{2}\left(\sum_{j=m+1}^{q}\beta_j-\sum_{j=n+1}^{p}\alpha_j-A\right)t}\right]}$$

$$\times \frac{t^{\left\{\sum_{j=1}^{m}\left(b_j-\left(\beta_1+\beta_2+\beta_3+,\dots+\beta_m\right)\sigma\right)+\sum_{j=1}^{n}\left(-a_j+\left(\alpha_1+\alpha_2+\alpha_3+\dots+\alpha_n\right)\sigma\right)-B+\frac{n}{2}-\frac{m}{2}\right\}}e^{\left|-\theta t\right|}}{\left[t^{\left\{\sum_{j=m+1}^{q}\left(-b_j+\beta_j\sigma\right)+\sum_{j=n+1}^{p}\left(a_j-\alpha_j\sigma\right)-B+\frac{q}{2}-\frac{p}{2}+\frac{n}{2}-\frac{m}{2}\right\}}\right]} \tag{1.40}$$

$$H_{p,q}^{m,n}\left|z\right| \sim \lim_{t\to\infty} \frac{e^{-\frac{\pi}{2}\left(\frac{\beta_1}{1-\delta^{m-1}}+\frac{\alpha_1}{1-\eta^{n-1}}-A\right)t}\, t^{\left\{\sum_{j=1}^{m}\left(b_j-\beta_j\sigma\right)+\sum_{j=1}^{n}\left(-a_j+\alpha_j\sigma\right)-B+\frac{n}{2}-\frac{m}{2}\right\}}}{\left[e^{-\frac{\pi}{2}\left(\sum_{j=m+1}^{q}\beta_j-\sum_{j=n+1}^{p}\alpha_j A\right)t}\right]} \tag{1.41}$$

$$\times \frac{e^{\left|-\theta t\right|}}{\left[t^{\left\{\sum_{j=m+1}^{q}\left(-b_j+\beta_j\sigma\right)+\sum_{j=n+1}^{p}\left(a_j-\alpha_j\sigma\right)B+\frac{q}{2}\frac{p}{2}+\frac{n}{2}-\frac{m}{2}\right\}}\right]}$$

ii. Again we put in Equation (2.10) and (2.11) $\alpha_j = \beta_j = 1$, then it will reduce to,

$$G^{m,n}_{p,q}|z| \sim \lim_{t\to\infty} \frac{t^{\left\{\sum_{j=1}^{m}(b_j-\sigma)+\sum_{j=1}^{n}(-a_j+\sigma)-B+\frac{n}{2}-\frac{m}{2}\right\}} e^{|-\theta t|}}{\left[t^{\left\{\sum_{j=m+1}^{q}(-b_j+\sigma)+\sum_{j=n+1}^{p}(a_j-\sigma)B+\frac{q}{2}-\frac{p}{2}+\frac{n}{2}-\frac{m}{2}\right\}}\right]} \tag{1.42}$$

where

$$B=\min\left[\sum_{j=m+1}^{q}(-b_j+\sigma)+\sum_{j=n+1}^{p}(a_j-\sigma)+\left(\frac{q}{2}-\frac{p}{2}+\frac{n}{2}-\frac{m}{2}\right)\right].$$

Following are some selected identities of I-function

1.10 SIMPLE IDENTITIES OF THE I-FUNCTION

This section gives some useful identities of the I-function. Proofs are avoided in parallel cases.

i.

$$S_1 \equiv z^{\sigma} I^{m,n}_{p_i,q_i:r}\left[z\left|\begin{matrix}(a_j,\alpha_j)_{1,n}; & (a_{ji},\alpha_{ji})_{n+1,p_i}\\ (b_j,\beta_j)_{1,m}; & (b_{ji},\beta_{ji})_{m+1,q_i}\end{matrix}\right.\right]$$

$$=I^{m,n}_{p_i,q_i:r}\left[z\left|\begin{matrix}(a_h-1,1),(a_j,1)_{j,n}; & (a_{ji},1)_{n+1,p_i}\\ (b_j,1)_{1,m}; & (b_{ji},1)_{m+1,q_i}\end{matrix}\right.\right] \tag{1.43}$$

$$+I^{m,n}_{p_i,q_i:r}\left[z\left|\begin{matrix}(a_j,1); & (a_{ji},1)\\ (b_j+1,1),(b_j,1)_{j\neq k}; & (b_{ji},1)\end{matrix}\right.\right]$$

where $m,n\geq 1$, provided $\sigma \geq 0$.

We have

$$S_1=\frac{1}{2\pi i}\int_L \frac{\prod_{j=1}^{m}\Gamma(\mathrm{b_j}-\beta_j\xi)\prod_{j=1}^{n}\Gamma(1-\mathrm{a_j}+\alpha_j\xi)z^{\xi+\sigma}d\xi}{\sum_{i=1}^{r}\left\{\prod_{j=m+1}^{q_i}\Gamma(1-\mathrm{b_{ij}}+\beta_{ji}\xi)\prod_{j=n+1}^{p_i}\Gamma(\mathrm{a_{ji}}-\alpha_{ji}\xi)\right\}}$$

Put $\sigma+\xi=t \Rightarrow \xi=t-\sigma,\ d\xi=dt$.

Thus, we get

$$S_1=\frac{1}{2\pi i}\int_L \frac{\prod_{j=1}^{m}\Gamma(\mathrm{b_j}-\beta_j\{t-\sigma\})\prod_{j=1}^{n}\Gamma(1-\mathrm{a_j}+\alpha_j\{t-\sigma\})z^{t}dt}{\sum_{i=1}^{r}\left\{\prod_{j=m+1}^{q_i}\Gamma(1-\mathrm{b_{ij}}+\beta_{ji}\{t-\sigma\})\prod_{j=n+1}^{p_i}\Gamma(\mathrm{a_{ji}}-\alpha_{ji}\{t-\sigma\})\right\}}$$

Hence the result.

$$z^{\sigma} H_{p,q}^{m,n}\left[z \left| \begin{matrix} (a_1,\alpha_1),\ldots\ldots & (a_p,\alpha_p) \\ (b_1,\beta_1),\ldots\ldots & (b_q,\beta_q) \end{matrix} \right. \right]$$

$$= H_{p,q}^{m,n}\left[z \left| \begin{matrix} (a_1+\sigma\alpha_1,\alpha_1),\ldots\ldots & (a_p+\alpha_p\sigma,\alpha_p) \\ (b_1+\beta_1\sigma,\beta_1),\ldots\ldots & (b_q+\beta_q\sigma,\beta_q) \end{matrix} \right. \right] \tag{1.44}$$

ii.

$$S_2 \equiv (1-a_h+b_k) I_{p_i,q_i:r}^{m,n}\left[z \left| \begin{matrix} (a_j,1); & (a_{ji},1) \\ (b_j,1); & (b_{ji},1) \end{matrix} \right. \right]$$

$$= I_{p_i,q_i:r}^{m,n}\left[z \left| \begin{matrix} (a_h-1,1),(a_j,1)_{j\neq h}; & (a_{ji},1) \\ (b_j,1); & (b_{ji},1) \end{matrix} \right. \right] \tag{1.45}$$

$$+ I_{p_i,q_i:r}^{m,n}\left[z \left| \begin{matrix} (a_j,1); & (a_{ji},1) \\ (b_j+1,1),(b_j,1)_{j\neq k}; & (b_{ji},1) \end{matrix} \right. \right]$$

We have

$$S_2 = \frac{1}{2\pi i}\int_L \frac{(1-a_h+b_k)\prod_{j=1}^{m}\Gamma(b_j-\xi)\prod_{j=1}^{n}\Gamma(1-a_j+\xi)z^{\xi}d\xi}{\sum_{i=1}^{r}\left\{\prod_{j=m+1}^{q_i}\Gamma(1-b_{ij}+\xi)\prod_{j=n+1}^{p_i}\Gamma(a_{ji}-\xi)\right\}}$$

$$= \frac{1}{2\pi i}\int_L \frac{\prod_{j=1}^{m}\Gamma(b_j-\xi)(1-a_h+\xi)\prod_{j=1}^{n}\Gamma(1-a_j+\xi)z^{\xi}d\xi}{\sum_{i=1}^{r}\left\{\prod_{j=m+1}^{q_i}\Gamma(1-b_{ij}+\xi)\prod_{j=n+1}^{p_i}\Gamma(a_{ji}-\xi)\right\}}$$

$$+ \frac{1}{2\pi i}\int_L \frac{(b_k-\xi)\prod_{j=1}^{m}\Gamma(b_j-\xi)\prod_{j=1}^{n}\Gamma(1-a_j+\xi)z^{\xi}d\xi}{\sum_{i=1}^{r}\left\{\prod_{j=m+1}^{q_i}\Gamma(1-b_{ij}+\xi)\prod_{j=n+1}^{p_i}\Gamma(a_{ji}-\xi)\right\}}$$

$$= \frac{1}{2\pi i}\int_L \frac{\prod_{j=1}^{m}\Gamma(b_j-\xi)\Gamma(1-\{a_h-1\}+\xi)\prod_{j=1}^{n}\Gamma(1-a_j+\xi)z^{\xi}d\xi}{\sum_{i=1}^{r}\left\{\prod_{j=m+1}^{q_i}\Gamma(1-b_{ij}+\xi)\prod_{j=n+1}^{p_i}\Gamma(a_{ji}-\xi)\right\}}$$

$$+ \frac{1}{2\pi i}\int_L \frac{\Gamma(b_j-1+\xi)\prod_{j=1}^{m}\Gamma(b_j-\xi)\prod_{j=1}^{n}\Gamma(1-a_j+\xi)z^{\xi}d\xi}{\sum_{i=1}^{r}\left\{\prod_{j=m+1}^{q_i}\Gamma(1-b_{ij}+\xi)\prod_{j=n+1}^{p_i}\Gamma(a_{ji}-\xi)\right\}}$$

This leads to the result (1.45)

iii.

$$S_3 \equiv \left(b_k\alpha_h - a_h\beta_k + \beta_k\right) I^{m,n}_{p_i,q_i:r}\left[z \left| \begin{matrix} \left(a_j,\alpha_j\right)_{1,n}; & \left(a_{ji},\alpha_{ji}\right)_{n+1,p_i} \\ \left(b_j,\beta_j\right)_{1,m}; & \left(b_{ji},\beta_{ji}\right)_{m+1,q_i} \end{matrix} \right. \right]$$

$$= \alpha_h I^{m,n}_{p_i,q_i:r}\left[z \left| \begin{matrix} \left(a_j,\alpha_j\right); & \left(a_{ji},\alpha_{ji}\right) \\ \left(b_k+1,\beta_k\right), \left(b_j,\beta_j\right)_{j\neq ki}; & \left(b_{ji},\beta_{ji}\right) \end{matrix} \right. \right] \tag{1.46}$$

$$+\beta_k I^{m,n}_{p_i,q_i:r}\left[z \left| \begin{matrix} \left(a_h-1,\alpha_h\right),\left(a_j,\alpha_j\right)_{j\neq hi}; & \left(a_{ji},\alpha_{ji}\right) \\ \left(b_j,\beta_j\right); & \left(b_{ji},\beta_{ji}\right) \end{matrix} \right. \right]$$

where $m,n \geq 1$ and $1 \leq h \leq n$, $1 \leq k \leq m$.

For $r = 1$, the above identity takes the following form:

$$\left(b_k\alpha_h - a_h\beta_k + \beta_k\right) H^{m,n}_{p,q}\left[z \left| \begin{matrix} \left(a_j,\alpha_j\right)_{1,p} \\ \left(b_j,\beta_j\right)_{1,q} \end{matrix} \right. \right]$$

$$= \alpha_h H^{m,n}_{p,q}\left[z \left| \begin{matrix} \left(a_j,\alpha_j\right)_{1,p} \\ \left(b_k+1,\beta_k\right),\left(b_j,\beta_j\right)_{1,q \text{ but } j\neq k} \end{matrix} \right. \right] \tag{1.47}$$

$$+\beta_k H^{m,n}_{p,q}\left[z \left| \begin{matrix} \left(a_h-1,\alpha_h\right),\left(a_j,\alpha_j\right)_{1,p \text{ but } j\neq h} \\ \left(b_j,\beta_j\right)_{1,q} \end{matrix} \right. \right]$$

Provided $m,n \geq 1$ and $1 \leq h \leq n$, $1 \leq k \leq m$

iv.

$$S_4 \equiv I^{m,n+1}_{p_i+1,q_i+1:r}\left[z \left| \begin{matrix} (1-k,v),\left(a_j,\alpha_j\right)_{1,n}; & \left(a_{ji},\alpha_{ji}\right)_{n+1,p_i} \\ \left(b_j,\beta_j\right)_{1,m}; & \left(b_{ji},\beta_{ji}\right)_{m+1,q_i},(1,h),(2-k,v) \end{matrix} \right. \right]$$

$$= (k-1) I^{m,n}_{p_i,q_i+1:r}\left[z \left| \begin{matrix} \left(a_j,\alpha_j\right)_{1,n}; & \left(a_{ji},\alpha_{ji}\right)_{n+1,p_i} \\ \left(b_j,\beta_j\right)_{1,m}; & \left(b_{ji},\beta_{ji}\right)_{m+1,q_i},(1,h) \end{matrix} \right. \right]$$

$$+ I^{m,n+1}_{p_i+1,q_i+2:r}\left[z \left| \begin{matrix} (0,v),\left(a_j,\alpha_j\right)_{1,n}; & \left(a_{ji},\alpha_{ji}\right)_{n+1,p_i} \\ \left(b_j,\beta_j\right)_{1,m}; & \left(b_{ji},\beta_{ji}\right)_{m+1,q_i},(1,v),(1,v) \end{matrix} \right. \right] \tag{1.48}$$

This is proved by using the formula

$$\frac{\Gamma(k+vs)}{\Gamma(hs)\Gamma(k-1+vs)}=\frac{\Gamma(k-1)}{\Gamma(hs)}+\frac{vs}{\Gamma(hs)}$$

$$S_5 \equiv I^{m+1,n+1}_{p_i+1,q_i+2:r}\left[z\left|\begin{matrix}(a_j,\alpha_j)_{1,n}; & (a_{ji},\alpha_{ji})_{n+1,p_i},(k-1,v)\\ (k,v),(b_j,\beta_j)_{1,m}; & (b_{ji},\beta_{ji})_{m+1,q_i},(1,h)\end{matrix}\right.\right]$$

$$=(k-1)I^{m,n+1}_{p_i,q_i+1:r}\left[z\left|\begin{matrix}(a_j,\alpha_j)_{1,n}; & (a_{ji},\alpha_{ji})_{n+1,p_i}\\ (b_j,\beta_j)_{1,m}; & (b_{ji},\beta_{ji})_{m+1,q_i},(1,h)\end{matrix}\right.\right] \tag{1.49}$$

$$-I^{m,n+1}_{p_i+1,q_i+2:r}\left[z\left|\begin{matrix}(0,v),(a_j,\alpha_j)_{1,n}; & (a_{ji},\alpha_{ji})_{n+1,p_i}\\ (b_j,\beta_j)_{1,m}; & (b_{ji},\beta_{ji})_{m+1,q_i},(1,v),(1,v)\end{matrix}\right.\right]$$

$$S_6 \equiv I^{m,n+1}_{p_i+1,q_i+2:r}\left[z\left|\begin{matrix}(1-k,\alpha),(a_j,\alpha_j)_{1,n}; & (a_{ji},\alpha_{ji})_{n+1,p_i}\\ (b_j,\beta_j)_{1,m}; & (b_{ji},\beta_{ji})_{m+1,q_i},(-k,\alpha),(1,h)\end{matrix}\right.\right]$$

$$=I^{m,n}_{p_i,q_i+1:r}\left[z\left|\begin{matrix}(a_j,\alpha_j)_{1,n}; & (a_{ji},\alpha_{ji})_{n+1,p_i}\\ (b_j,\beta_j)_{1,m}; & (b_{ji},\beta_{ji})_{m+1,q_i},(1,h)\end{matrix}\right.\right]$$

$$-I^{m,n+2}_{p_i+1,q_i+3:r}\left[z\left|\begin{matrix}(1-k,\alpha),(0,\alpha),(a_j,\alpha_j)_{1,n}; & (a_{ji},\alpha_{ji})_{n+1,p_i}\\ (b_j,\beta_j)_{1,m}; & (b_{ji},\beta_{ji})_{m+1,q_i},(-k,\alpha),(1,\alpha),(1,h)\end{matrix}\right.\right] \tag{1.50}$$

$$S_7 \equiv kI^{m+1,n+1}_{p_i+1,q_i+2:r}\left[z\left|\begin{matrix}(a_j,\alpha_j)_{1,n}; & (a_{ji},\alpha_{ji})_{n+1,p_i},(k+1,\alpha)\\ (k,\alpha),(b_j,\beta_j)_{1,m}; & (b_{ji},\beta_{ji})_{m+1,q_i},(1,h)\end{matrix}\right.\right]$$

$$=I^{m,n}_{p_i,q_i+1:r}\left[z\left|\begin{matrix}(a_j,\alpha_j)_{1,n}; & (a_{ji},\alpha_{ji})_{n+1,p_i}\\ (b_j,\beta_j)_{1,m}; & (b_{ji},\beta_{ji})_{m+1,q_i},(1,h)\end{matrix}\right.\right] \tag{1.51}$$

$$+I^{m+1,n+1}_{p_i+2,q_i+3:r}\left[z\left|\begin{matrix}(0,\alpha),(a_j,\alpha_j)_{1,n}; & (a_{ji},\alpha_{ji})_{n+1,p_i},(k+1,\alpha)\\ (k,\alpha),(b_j,\beta_j)_{1,m}; & (b_{ji},\beta_{ji})_{m+1,q_i},(1,\alpha),(1,h)\end{matrix}\right.\right]$$

1.11 DERIVATIVE OF THE I-FUNCTION

In this section, we derive certain differentiation formulas of the I-function.

i.

$$D_1 \equiv z\frac{d}{dz}I^{m,n}_{p_i,q_i:r}\left[z\left|\begin{matrix}(a_j,1)_{1,n};&(a_{ji},1)_{n+1,p_i}\\(b_j,1)_{1,m};&(b_{ji},1)_{m+1,q_i}\end{matrix}\right.\right]$$

$$= b_1 I^{m,n}_{p_i,q_i:r}\left[z\left|\begin{matrix}(a_j,1)_{1,n};&(a_{ji},1)_{n+1,p_i}\\(b_j,1)_{1,m};&(b_{ji},1)_{m+1,q_i}\end{matrix}\right.\right] \tag{1.52}$$

$$-I^{m,n}_{p_i,q_i:r}\left[z\left|\begin{matrix}(a_j,1)_{1,n};&(a_{ji},1)_{n+1,p_i}\\(b_1+1,1)(b_j,1)_{2,m};&(b_{ji},1)_{m+1,q_i}\end{matrix}\right.\right]$$

Provided $m \geq 1$.

We have

$$D_1 = \frac{z\frac{d}{dz}}{2\pi i}\int_L \frac{\prod_{j=1}^{m}\Gamma(b_j-\xi)\prod_{j=1}^{n}\Gamma(1-a_j+\xi)z^{\xi}d\xi}{\sum_{i=1}^{r}\left\{\prod_{j=m+1}^{q_i}\Gamma(1-b_{ij}+\xi)\prod_{j=n+1}^{p_i}\Gamma(a_{ji}-\xi)\right\}}$$

$$= \frac{1}{2\pi i}\int_L \frac{\prod_{j=1}^{m}\Gamma(b_j-\xi)\prod_{j=1}^{n}\Gamma(1-a_j+\xi)(b_1+\xi-b_1)z^{\xi}d\xi}{\sum_{i=1}^{r}\left\{\prod_{j=m+1}^{q_i}\Gamma(1-b_{ij}+\xi)\prod_{j=n+1}^{p_i}\Gamma(a_{ji}-\xi)\right\}}$$

$$= \frac{b_1}{2\pi i}\int_L \frac{\prod_{j=1}^{m}\Gamma(b_j-\xi)\prod_{j=1}^{n}\Gamma(1-a_j+\xi)z^{\xi}d\xi}{\sum_{i=1}^{r}\left\{\prod_{j=m+1}^{q_i}\Gamma(1-b_{ij}+\xi)\prod_{j=n+1}^{p_i}\Gamma(a_{ji}-\xi)\right\}}$$

$$-\frac{1}{2\pi i}\int_L \frac{(b_1-\xi)\prod_{j=1}^{m}\Gamma(b_j-\xi)\prod_{j=1}^{n}\Gamma(1-a_j+\xi)z^{\xi}d\xi}{\sum_{i=1}^{r}\left\{\prod_{j=m+1}^{q_i}\Gamma(1-b_{ij}+\xi)\prod_{j=n+1}^{p_i}\Gamma(a_{ji}-\xi)\right\}}$$

Hence the result.

ii.

$$D_2 \equiv z\frac{d}{dz}I^{m,n}_{p_i,q_i:r}\left[z\left|\begin{matrix}(a_j,1)_{1,n}; & (a_{ji},1)_{n+1,p_i}\\ (b_j,1)_{1,m}; & (b_{ji},1)_{m+1,q_i}\end{matrix}\right.\right]$$

$$= I^{m,n}_{p_i,q_i:r}\left[z\left|\begin{matrix}(a_1-1,1),(a_j,1)_{2,n}; & (a_{ji},1)_{n+1,p_i}\\ (b_j,1)_{1,m}; & (b_{ji},1)_{m+1,q_i}\end{matrix}\right.\right] \tag{1.53}$$

$$+(a_1-1)I^{m,n}_{p_i,q_i:r}\left[z\left|\begin{matrix}(a_j,1)_{1,n}; & (a_{ji},1)_{n+1,p_i}\\ (b_1+1,1)(b_j,1)_{2,m}; & (b_{ji},1)_{m+1,q_i}\end{matrix}\right.\right]$$

iii.

$$D_3 \equiv D_x^r\left\{x^{\lambda}I^{m,n}_{p_i,q_i:r}\left[zx^{\sigma}\left|\begin{matrix}(a_j,\alpha_j)_{1,n}; & (a_{ji},\alpha_{ji})_{n+1,p_i}\\ (b_j,\beta_j)_{1,m}; & (b_{ji},\beta_{ji})_{m+1,q_i}\end{matrix}\right.\right]\right\}$$

$$\times x^{\lambda-r}I^{m,n+1}_{p_i+1,q_i+1:r}\left[zx^{\sigma}\left|\begin{matrix}(-\lambda,\sigma),(a_j,\alpha_j)_{1,n}; & (a_{ji},\alpha_{ji})_{n+1,p_i}\\ (b_j,\beta_j)_{1,m}; & (b_{ji},\beta_{ji})_{m+1,q_i};(-\lambda+r,\sigma)\end{matrix}\right.\right] \tag{1.54}$$

where $D_x = \frac{d}{dx}$

We have

$$D_3 = \frac{1}{2\pi i}\int_L \frac{\prod_{j=1}^m \Gamma(b_j-\beta_j\xi)\prod_{j=1}^n \Gamma(1-a_j+\alpha_j\xi)z^{\xi}D_x^r x^{\lambda+\sigma\xi}}{\sum_{i=1}^r\left\{\prod_{j=m+1}^{q_i}\Gamma(1-b_{ij}+\beta_{ji}\xi)\prod_{j=n+1}^{p_i}\Gamma(a_{ji}-\alpha_{ji}\xi)\right\}}d\xi$$

$$= \frac{1}{2\pi i}\int_L \frac{\prod_{j=1}^m \Gamma(b_j-\beta_j\xi)\prod_{j=1}^n \Gamma(1-a_j+\alpha_j\xi)(\lambda+\sigma\xi)(\lambda+\sigma\xi-1)\ldots.(\lambda+\sigma\xi-r+1))}{\sum_{i=1}^r\left\{\prod_{j=m+1}^{q_i}\Gamma(1-b_{ij}+\beta_{ji}\xi)\prod_{j=n+1}^{p_i}\Gamma(a_{ji}-\alpha_{ji}\xi)\right\}}$$

$$\times z^{\xi}x^{\lambda+\sigma\xi-r}d\xi$$

$$= \frac{1}{2\pi i}\int_L \frac{\prod_{j=1}^m \Gamma(b_j-\beta_j\xi)\prod_{j=1}^n \Gamma(1-a_j+\alpha_j\xi)\Gamma(\lambda+\sigma\xi+1)}{\sum_{i=1}^r\left\{\prod_{j=m+1}^{q_i}\Gamma(1-b_{ij}+\beta_{ji}\xi)\prod_{j=n+1}^{p_i}\Gamma(a_{ji}-\alpha_{ji}\xi)\right\}\Gamma(\lambda+\sigma\xi+1+r)}$$

$$\times z^{\xi}x^{\lambda+\sigma\xi-r}d\xi$$

$$= \frac{x^{\lambda-r}}{2\pi i}\int_L \frac{\prod_{j=1}^{m}\Gamma(b_j-\beta_j\xi)\prod_{j=1}^{n}\Gamma(1-a_j+\alpha_j\xi)\Gamma(1-(-\lambda)+\sigma\xi)}{\sum_{i=1}^{r}\left\{\prod_{j=m+1}^{q_i}\Gamma(1-b_{ij}+\beta_{ji}\xi)\prod_{j=n+1}^{p_i}\Gamma(a_{ji}-\alpha_{ji}\xi)\right\}\Gamma(1-(-\lambda+\gamma)+\sigma\xi)}$$

$$\times z^{\xi}x^{\lambda+\sigma\xi-r}d\xi$$

Hence the result.

iv.

$$D_4 \equiv D_x^r\left\{x^{\lambda}I_{p_i,q_i:r}^{m,n}\left[(cx+d)^{\sigma}\left|\begin{matrix}(a_j,\alpha_j)_{1,n}; & (a_{ji},\alpha_{ji})_{n+1,p_i}\\ (b_j,\beta_j)_{1,m}; & (b_{ji},\beta_{ji})_{m+1,q_i}\end{matrix}\right.\right]\right\}$$

$$=\left(\frac{c}{cx+d}\right)^r I_{p_i+1,q_i+1:r}^{m,n+1}\left[(cx+d)^{\sigma}\left|\begin{matrix}(0,\sigma),(a_j,\alpha_j)_{1,n}; & (a_{ji},\alpha_{ji})_{n+1,p_i}\\ (b_j,\beta_j)_{1,m}; & (b_{ji},\beta_{ji})_{m+1,q_i};(\gamma,\sigma)\end{matrix}\right.\right] \tag{1.55}$$

where c and d are complex numbers and σ is real and positive.

v.

$$D_5 \equiv (D_x.x-c_1)\ldots(D_x.x-c_1)\left\{x^{\lambda}I_{p_i,q_i:r}^{m,n}\left[zx^{\sigma}\left|\begin{matrix}(a_j,\alpha_j)_{1,n}; & (a_{ji},\alpha_{ji})_{n+1,p_i}\\ (b_j,\beta_j)_{1,m}; & (b_{ji},\beta_{ji})_{m+1,q_i}\end{matrix}\right.\right]\right\}$$

$$= x^{\lambda}I_{p_i+1,q_i+1:r}^{m,n+1}\left[zx^{\sigma}\left|\begin{matrix}(c_j-\lambda-1,\sigma)_{1,n},(a_j,\alpha_j)_{1,n}; & (a_{ji},\alpha_{ji})_{n+1,p_i}\\ (b_j,\beta_j)_{1,m}; & (b_{ji},\beta_{ji})_{m+1,q_i};(c_j-\lambda,\sigma)_{1,l}\end{matrix}\right.\right] \tag{1.56}$$

where $\sigma > 0$ and symbol $D_x(x)$ indicates that the function of x in front of it is first multiplied by x and then product is differentiated with respect to x.

We have

$$D_5 = \frac{1}{2\pi i}\int_L \frac{\prod_{j=1}^{m}\Gamma(b_j-\beta_j\xi)\prod_{j=1}^{n}\Gamma(1-a_j+\alpha_j\xi)z^{\xi}}{\sum_{i=1}^{r}\left\{\prod_{j=m+1}^{q_i}\Gamma(1-b_{ij}+\beta_{ji}\xi)\prod_{j=n+1}^{p_i}\Gamma(a_{ji}-\alpha_{ji}\xi)\right\}}$$

$$\times(D_x.x-c_1)\ldots(D_x.x-c_l)x^{\lambda+\sigma\xi}d\xi$$

$$= \frac{1}{2\pi i}\int_L \frac{\prod_{j=1}^{m}\Gamma(b_j-\beta_j\xi)\prod_{j=1}^{n}\Gamma(1-a_j+\alpha_j\xi)z^{\xi}}{\sum_{i=1}^{r}\left\{\prod_{j=m+1}^{q_i}\Gamma(1-b_{ij}+\beta_{ji}\xi)\prod_{j=n+1}^{p_i}\Gamma(a_{ji}-\alpha_{ji}\xi)\right\}}$$

$$\times(1+\lambda-c_1+\sigma\xi)\ldots(1+\lambda-c_l+\sigma\xi)x^{\lambda+\sigma\xi}d\xi$$

$$= \frac{1}{2\pi i}\int_L \frac{\prod_{j=1}^{m}\Gamma(b_j-\beta_j\xi)\prod_{j=1}^{n}\Gamma(1-a_j+\alpha_j\xi)\prod_{j=1}^{l}(1+\lambda-c_j+\sigma\xi)x^{\lambda+\sigma\xi}z^{\xi}}{\sum_{i=1}^{r}\{\prod_{j=m+1}^{q_i}\Gamma(1-b_{ij}+\beta_{ji}\xi)\prod_{j=n+1}^{p_i}\Gamma(a_{ji}-\alpha_{ji}\xi)\}}d\xi$$

$$= \frac{x^{\lambda}}{2\pi i}\int_L \frac{\prod_{j=1}^{m}\Gamma(b_j-\beta_j\xi)\prod_{j=1}^{n}\Gamma(1-a_j+\alpha_j\xi)}{\sum_{i=1}^{r}\left\{\prod_{j=m+1}^{q_i}\Gamma(1-b_{ij}+\beta_{ji}\xi)\prod_{j=n+1}^{p_i}\Gamma(a_{ji}-\alpha_{ji}\xi)\right\}}$$

$$\times\frac{\prod_{j=1}^{l}(1-\{c_j-\lambda-1\}+\sigma\xi)x^{\sigma\xi}z^{\xi}}{\prod_{j=1}^{l}(1-\{c_j-\lambda\}+\sigma\xi)}d\xi$$

Hence the result.

$$D_6 \equiv (xD_x)^h\left\{x^{\lambda}I^{m,n}_{p_i,q_i:r}\left[zx^{\sigma}\left|\begin{matrix}(a_j,\alpha_j)_{1,n}; & (a_{ji},\alpha_{ji})_{n+1,p_i}\\ (b_j,\beta_j)_{1,m}; & (b_{ji},\beta_{ji})_{m+1,q_i}\end{matrix}\right.\right]\right\}$$

$$= x^{\lambda}I^{m,n+h}_{p_i+h,q_i+h:r}\left[zx^{\sigma}\left|\begin{matrix}(-\lambda,\sigma)_h,(a_j,\alpha_j)_{1,n}; & (a_{ji},\alpha_{ji})_{n+1,p_i}\\ (b_j,\beta_j)_{1,m}; & (b_{ji},\beta_{ji})_{m+1,q_i};(1-\lambda,\sigma)_h\end{matrix}\right.\right] \quad (1.57)$$

Here $(a,\alpha)_h$ abbreviates the h-parameters sequence $(a,\alpha),(a,\alpha),\ldots\ldots,(a,\alpha)$.

For $r = 1$ the above relation takes the form

$$(xD_x)^h\left\{x^{\lambda}H^{m,n}_{p,q}\left[zx^{\sigma}\left|\begin{matrix}(a_1,\alpha_1),(a_2,\alpha_2),\ldots & (a_p,\alpha_p)\\ (b_1,\beta_1),(b_q,\beta_q),\ldots & (b_q,\beta_q)\end{matrix}\right.\right]\right\}$$

$$= x^{\lambda}H^{m,n+h}_{p+h,q+h}\left[zx^{\sigma}\left|\begin{matrix}(-\lambda,\sigma)_h & (a_j,\alpha_j)_{1,p}\\ (b_j,\beta_j)_{1,q}; & (1-\lambda,\sigma)_h\end{matrix}\right.\right] \quad (1.58)$$

2 Transformation of I-Function and H-Function for Rational Parameters

In this chapter, the special cases of I-function when the coefficients of '*s*' in respective gamma functions in the integrands are rational numbers are given. The formulas investigated in this chapter will unify the higher cadre of generalized hypergeometric functions [34].

2.1 TRANSFORMATION OF H-FUNCTION

We know that

$$H_{p,q}^{m,n}[z] = H_{p,q}^{m,n}\left[z \left| \begin{matrix} (a_1,\alpha_1)\ldots\ldots(a_p,\alpha_p) \\ (b_1,\beta_1)\ldots\ldots(b_q,\beta_q) \end{matrix} \right. \right] \tag{2.1}$$

$$= \frac{1}{2\pi i}\int_L t\,(s) z^s\,ds$$

where

$$t(s) = \frac{\prod_{j=1}^{m}\Gamma(b_j - \beta_j s)\prod_{j=1}^{n}\Gamma(1 - a_j + \alpha_j s)}{\prod_{j=m+1}^{q}\Gamma(1 - b_j + \beta_j s)\prod_{j=n+1}^{p}\Gamma(a_j - \alpha_j s)} \tag{2.2}$$

Now,

$$H_{p,q}^{m,n}[z] = \frac{1}{2\pi i}\int_L \frac{\prod_{j=1}^{m}\Gamma(b_j - \beta_j s)\prod_{j=1}^{n}\Gamma(1 - a_j + \alpha_j s)}{\prod_{j=m+1}^{q}\Gamma(1 - b_j - \beta_j s)\prod_{j=n+1}^{p}\Gamma(a_j - \alpha_j s)} z^s\,ds \tag{2.3}$$

Put $\alpha_j = M_j$, $\beta_j = N_j$ in (3.3), we get

DOI: 10.1201/9781003517757-2

$$= \frac{1}{2\pi i}\int_L \frac{\prod_{j=1}^{m}\Gamma\left(b_j - N_j s\right)\prod_{j=1}^{n}\Gamma\left(1-a_j+M_j s\right)}{\prod_{j=m+1}^{q}\Gamma\left(1-b_j+N_j s\right)\prod_{j=n+1}^{n}\Gamma\left(a_j - M_j s\right)} Z^s\, ds$$

$$= \frac{1}{2\pi i}\int_L \frac{\prod_{j=1}^{m}\Gamma\left(N_j\left(\frac{b_j}{N_j}-s\right)\right)\prod_{j=1}^{n}\Gamma\left(M_j\left(\frac{1-q_j}{M_j}+s\right)\right)}{\prod_{j=m+1}^{q}\Gamma\left(N_j\left(\frac{1-b_j}{N_j}+s\right)\right)\prod_{j=n+1}^{p}\Gamma\left(M_j\left(\frac{q_j}{M_j}-s\right)\right)} z^s\, ds \tag{2.4}$$

Put $\frac{b_j}{N_j} - s = Z_j, \frac{1-a_j}{M_j} + s = \gamma_j, \frac{1-b_j}{N_j} + s = Y_j, \frac{a_j}{M_j} - s = X_j$ in (3.4),

we get

$$= \frac{1}{2\pi i}\int_L \frac{\prod_{j=1}^{m}\Gamma\left(N_j Z_j\right)\prod_{j=1}^{n}\Gamma\left(M_j Y_j\right)}{\prod_{j=m+1}^{q}\Gamma\left(N_j Y_j\right)\prod_{j=n+1}^{p}\Gamma\left(M_j X_j\right)} Z^s ds \tag{2.5}$$

By using formula,

$$\Gamma(mZ) = (2\pi)^{\frac{1-m}{2}}\, m^{z-1/2}\prod_{j=1}^{m}\left(Z + \frac{j-1}{m}\right),$$
$$Z \neq 0, -1/m, 2/m \ldots\ldots m = 1, 2, 3. \tag{2.6}$$

Then,

$$\prod_{j=1}^{m}\Gamma\left(N_j Z_j\right) = \Gamma(N_1 Z_1)\Gamma(N_2 Z_2)\Gamma(N_3 Z_3)\ldots\ldots\ldots\Gamma(N_m Z_m)$$

$$= \left[(2\pi)^{\frac{1-N_1}{2}}(N_1)^{Z_1-1/2}\prod_{j=1}^{N_1}\Gamma\left(Z_1 + \frac{j-1}{N_1}\right)\right]$$

$$\times\left[(2\pi)^{\frac{1-N_2}{2}}(N_2)^{Z_2-1/2}\prod_{j=1}^{N_2}\Gamma\left(Z_2 + \frac{j-1}{N_2}\right)\right]$$

$$\times\left[(2\pi)^{\frac{1-N_3}{2}}(N_3)^{Z_3-1/2}\prod_{j=1}^{N_3}\Gamma\left(Z_3 + \frac{j-1}{N_3}\right)\right]$$

$$\times\left[(2\pi)^{\frac{1-N_m}{2}}(N_m)^{Z_m-1/2}\prod_{j=1}^{N_m}\Gamma\left(Z_m + \frac{j-1}{N_m}\right)\right]$$

$$= (2\pi)^{\left(\frac{1}{2}+\frac{1}{2}+\frac{1}{2}\ldots\ldots m\text{ times}\right)} . (2\pi)^{-\left[\frac{N_1+N_2+N_3+\ldots\ldots+N_m}{2}\right]} . (N_{.1}.N_2.N_3\ldots N_m)^{-\frac{1}{2}}$$

$$\prod_{j=1}^{m}(N_j)^{Z_j} . \prod_{j=1}^{N_1}\Gamma\left(\frac{b_1}{N_1} - s + \frac{j-1}{N_1}\right) . \prod_{j=1}^{N_2}\Gamma\left(\frac{b_2}{N_2} - s + \frac{j-1}{N_2}\right).$$

$$\prod_{j=1}^{N_3}\Gamma\left(\frac{b_3}{N_3} - s + \frac{j-1}{N_3}\right)\ldots\prod_{j=1}^{N_m}\Gamma\left(\frac{b_m}{N_m} - s + \frac{j-1}{N_m}\right)$$

$$= (2\pi)^{m/2} . (2\pi)^{-\left[\frac{N_1+N_2+N_3+\ldots+N_m}{2}\right]} . (N_1.N_2\ldots N_m)^{-\frac{1}{2}} \times N_1^{\left(\frac{b_1}{N_1}-s\right)} . N_2^{\left(\frac{b_2}{N_2}-s\right)} . N_3^{\left(\frac{b_3}{N_3}-s\right)}$$

$$\ldots N_m^{\left(\frac{b_m}{N_m}-s\right)} . \prod_{j=1}^{N_1}\Gamma\left(\frac{b_1+j-1}{N_1} - s\right).$$

$$\times\prod_{j=1}^{N_2}\Gamma\left(\frac{b_2+j-1}{N_2} - s\right) . \prod_{j=1}^{N_3}\Gamma\left(\frac{b_3+j-1}{N_3} - s\right)\ldots\prod_{j=1}^{N_m}\left(\frac{b_m+j-1}{N_m} - s\right).$$

$$= (2\pi)^{m/2} . (2\pi)^{\frac{-(N_1+N_2+N_3+\ldots\ldots+N_m)}{2}} . (N_1.N_2.N_3\ldots N_m)^{-\frac{1}{2}} (N_1.N_2.N_3\ldots N_m)^{-s}$$

$$N_1^{\frac{b_1}{N_1}} . N_2^{\frac{b_2}{N_2}} N_3^{\frac{b_3}{N_3}} \ldots\ldots N_m^{\frac{b_m}{N_m}} . \prod_{j=1}^{N_1}\Gamma\left(\frac{b_1+j-1}{N_1} - s\right)$$

$$\times\prod_{j=1}^{N_2}\Gamma\left(\frac{b_2+j-1}{N_2} - s\right) . \prod_{j=1}^{N_3}\Gamma\left(\frac{b_3+j-1}{N_3} - s\right)\ldots\prod_{j=1}^{N_m}\left(\frac{b_m+j-1}{N_m} - s\right)$$

$$= (2\pi)^{m/2} . (2\pi)^{\frac{-(N_1+N_2+N_3+\ldots\ldots+N_m)}{2}} (N_1.N_2.N_3\ldots N_m)^{-\frac{1}{2}}$$

$$\times\prod_{j=1}^{m}(N_j)^{-s} N_j^{\frac{b_j}{N_j}} \prod_{i=1}^{N}\Gamma(\beta_i - s) \tag{2.7}$$

where $\beta_i = \dfrac{b_i + j - 1}{N_i}$

$N = N_1 + N_2 + N_3 + \ldots\ldots + N_m$, $j = 1$ to N.

$$\prod_{j=1}^{n}\Gamma\left(M_j\gamma_j\right)=\Gamma\left(M_1\gamma_1\right)\Gamma\left(M_2\gamma_2\right).\Gamma\left(M_3\gamma_3\right).....\Gamma\left(M_n\gamma_n\right)$$

$$=\left[(2\pi)^{\frac{1-M_1}{2}}(M_1)^{\gamma_1-1/2}\prod_{j=1}^{M_1}\Gamma\left(\gamma_1+\frac{j-1}{M_1}\right)\right]$$

$$\times\left[(2\pi)^{\frac{1-M_2}{2}}M_2{}^{\gamma_2-1/2}\prod_{j=1}^{M_2}\Gamma\left(\gamma_2+\frac{j-1}{M_2}\right)\right].$$

$$\times\left[(2\pi)^{\frac{1-M_3}{2}}(M_3)^{\gamma_3-1/2}\prod_{j=1}^{M_3}\left(\gamma_3+\frac{j-1}{M_3}\right)\right]$$

$$\left[(2\pi)^{\frac{1-M_n}{2}}(M_n)^{\gamma_n-1/2}\prod_{j=1}^{M_n}\left(\gamma_n+\frac{j-1}{M_n}\right)\right]$$

$$=(2\pi)^{\left(1/2+1/2+1/2\ldots..n\text{ times}\right)}(2\pi)^{-\left[\frac{M_1+M_2+M_3+\ldots+M_n}{2}\right]}\left(M_1.M_2.M_3\ldots..M_n\right)^{-1/2}$$

$$\left[M_1^{\frac{1-a_1}{M_1}}.\,M_2^{\frac{1-a_2}{M_2}}\ldots.M_n^{\frac{1-a_n}{M_n}}\right]\left(M_1.M_2.M_3\ldots..M_n\right)^{+s}.$$

$$\prod_{j=1}^{M_1}\Gamma\left(\frac{1-a_1}{M_1}+s+\frac{j-1}{M_1}\right).\prod_{j=1}^{M_2}\Gamma\left(\frac{1-a_2}{M_2}+s+\frac{j-1}{M_2}\right).$$

$$\prod_{j=1}^{M_3}\Gamma\left(\frac{1-a_3}{M_3}+s+\frac{j-1}{M_3}\right)\ldots\prod_{j=1}^{M_n}\Gamma\left(\frac{1-a_n}{M_n}+s+\frac{j-1}{M_n}\right).$$

$$=(2\pi)^{n/2}(2\pi)^{-\left[\frac{M_1+M_2+M_3+\ldots+M_n}{2}\right]}\left(M_1.M_2\ldots M_n\right)^{-1/2}\prod_{j=1}^{n}\left(M_j\right)^{\frac{1-a_j}{M_j}}\left(M_j\right)^{s}.$$

$$\times\prod_{j=1}^{M_1}\Gamma\left(\frac{j-a_1}{M_1}+s\right).\prod_{j=1}^{M_2}\Gamma\left(\frac{j-a_2}{M_2}+s\right).\prod_{j=1}^{M3}\Gamma\left(\frac{j-a_3}{M_3}+s\right)$$

$$\ldots\prod_{j=1}^{M_n}\Gamma\left(\frac{j-a_n}{M_n}+s\right)$$

$$=(2\pi)^{n/2}(2\pi)^{-\frac{[M_1+M_2+\ldots M_n]}{2}}.\left(M_1.M_2\ldots M_n\right)^{-1/2}\prod_{j=1}^{n}\left(M_j\right)^{\frac{1-a_j}{M_j}}\left(M_j\right)^{s}$$

$$\prod_{i=1}^{M}\Gamma(\alpha_i+s) \tag{2.8}$$

where $\alpha_i = \dfrac{j-a_i}{M_i} + s$

$$M = M_1 + M_2 + M_3 + \ldots. + M_n$$

$$\prod_{j=M+1}^{q} \Gamma\left(N_j Y_j\right) = \Gamma\left(N_{m+1} Y_{m+1}\right).\ \Gamma\left(N_{m+2} Y_{m+2}\right).\ \Gamma\left(N_{m+3} Y_{m+3}\right) \ldots\ldots. \Gamma\left(N_q Y_q\right)$$

$$= \left[(2\pi)^{\frac{1-N_{m+1}}{2}} \left(N_{m+1}\right)^{Y_{m+1}-\frac{1}{2}} \prod_{j=1}^{N_{m+1}} \Gamma\left(Y_{m+1} + \frac{j-1}{N_{m+1}} \right) \right].$$

$$\times \left[(2\pi)^{\frac{1-N_{m+2}}{2}} \left(N_{m+2}\right)^{Y_{m+2}-\frac{1}{2}} \prod_{j=1}^{N_{m+2}} \Gamma\left(Y_{m+2} + \frac{j-1}{N_{m+2}} \right) \right].$$

$$\left[(2\pi)^{\frac{1-N_q}{2}} \left(N_q\right)^{Y_q-\frac{1}{2}} \prod_{j=1}^{N_q} \Gamma\left(Y_q + \frac{j-1}{N_q} \right) \right]$$

$$= (2\pi)^{\frac{q-m}{2}} . (2\pi)^{-\frac{\left[N_{m+1}+N_{m+2}+\ldots+N_q\right]}{2}} . \prod_{j=m+1}^{q} \left(N_j\right)^{\frac{1-b_j}{N_j}} \left(N_j\right)^s \prod_{j=m+1}^{q} \left(N_j\right)^{-\frac{1}{2}}$$

$$\prod_{j=1}^{N_{m+1}} \Gamma\left(\frac{1-b_{m+1}}{N_{m+1}} + s + \frac{j-1}{N_{m+1}} \right). \prod_{j=1}^{N_{m+2}} \Gamma\left(\frac{1-b_{m+2}}{N_{m+2}} + s + \frac{j-1}{N_{m+2}} \right)$$

$$\prod_{j=1}^{N_{m+3}} \Gamma\left(\frac{1-b_{m+3}}{N_{m+3}} + s + \frac{j-1}{N_{m+3}} \right) \ldots\ldots \prod_{j=1}^{N_q} \Gamma\left(\frac{1-b_q}{N_q} + s + \frac{j-1}{N_q} \right)$$

$$= (2\pi)^{\frac{q-m}{2}} (2\pi)^{-\frac{\left[N_{m+1}+N_{m+2}+\ldots.+N_q\right]}{2}} . \prod_{j=m+1}^{q} \left(N_j\right)^{\frac{1-b_j}{N_j}} \left(N_j\right)^s \prod_{j=m+1}^{q} \left(N_j\right)^{-\frac{1}{2}} .$$

$$\prod_{j=1}^{N_{m+1}} \Gamma\left(\frac{j-b_{m+1}}{N_{m+1}} + s \right) \prod_{j=1}^{N_{m+2}} \Gamma\left(\frac{j-b_{m+2}}{N_{m+2}} + s \right) \ldots\ldots \prod_{j=1}^{N_q} \Gamma\left(\frac{j-b_q}{N_q} + s \right)$$

$$= (2\pi)^{\frac{q-m}{2}} (2\pi)^{-\frac{\left[N_{m+1}+N_{m+2}+\ldots+N_q\right]}{2}} . \prod_{j=m+1}^{q} \left(N_j\right)^{\frac{1-b_j}{N_q}} \left(N_j\right)^s \prod_{j=m+1}^{q} \left(N_j\right)^{-\frac{1}{2}}$$

$$\prod_{i=m+1}^{Q} \Gamma\left(\frac{j-b_i}{N_i} + s \right)$$

$$= (2\pi)^{\frac{q-m}{2}} (2\pi)^{-\frac{\left[N_{m+1}+N_{m+2}+\ldots N_q\right]}{2}} . \prod_{j=m+1}^{q} \left(N_j\right)^{\frac{1-b_j}{N_a}} \left(N_j\right)^s \prod_{j=m+1}^{q} \left(N_j\right)^{-\frac{1}{2}} \prod_{i=m+1}^{Q} \Gamma\left(\lambda_i + s\right) \tag{2.9}$$

where

$$Q = N_{m+1} + N_{m+1} + N_{m+3} + \ldots + N_q$$

$$\lambda_i = \Gamma\left(\frac{j-b_i}{N_i} + s\right)$$

$$\prod_{j=n+1}^{P} \Gamma(M_j X_j) = \Gamma(M_{n+1}X_{n+1})\Gamma(M_{n+2}X_{n+2})\Gamma(M_{n+3}X_{n+3})\ldots\ldots\ \Gamma(M_P X_P)$$

$$= \left[(2\pi)^{\frac{1-M_{n+1}}{2}} (M_{n+1})^{X_{n+1}-1/2} \prod_{j=1}^{M_{n+1}} \Gamma\left(X_{n+1} + \frac{j-1}{M_{n+1}}\right)\right].$$

$$\times\left[(2\pi)^{\frac{1-M_{n+2}}{2}} (M_{n+2})^{X_{n+2}-1/2} \prod_{j=1}^{M_{n+2}} \Gamma\left(X_{n+2} + \frac{j-1}{M_{n+2}}\right)\right]$$

$$\times\left[(2\pi)^{\frac{1-M_{n+3}}{2}} (M_{n+3})^{X_{n+3}-1/2} \prod_{j=1}^{M_{n+3}} \Gamma\left(X_{n+3} + \frac{j-1}{M_{n+3}}\right)\right]$$

$$\left[(2\pi)^{\frac{1-M_P}{2}} (M_P)^{X_P-1/2} \prod_{j=1}^{M_P} \Gamma\left(X_P + \frac{j-1}{M_P}\right)\right]$$

$$= (2\pi)^{\frac{p-n}{2}} (2\pi)^{-\frac{[M_{n+1}+M_{n+2}+\ldots+M_p]}{2}} \prod_{j=n+1}^{p} (M_j)^{\frac{a_j}{M_j}} (M_j)^{-s} \prod_{j=n+1}^{R} (M_j)^{-1/2}$$

$$\prod_{i=n+1}^{p} \Gamma\left(\frac{a_i + j - 1}{M_i} - s\right)$$

where $R = M_{n+1} + M_{n+2} + \ldots + M_p$

$$\delta_i = \frac{a_i + j - 1}{M_i}$$

$$= (2\pi)^{\frac{p-n}{2}} (2\pi)^{-\frac{[M_{n+1}+M_{n+2}+\ldots+M_p]}{2}} . \prod_{j=n+1}^{p} (m_j)^{\frac{a_j}{M_j}} (M_j)^{-s} \prod_{j=m+1}^{p} (M_j)^{-1/2} \prod_{i=n+1}^{R} \Gamma(\delta_i - s) \tag{2.10}$$

Putting in Equation (2.3) we get

$$H_{p,q}^{m,n}[z]=\int_L (2\pi)^{m/2}(2\pi)^{-\frac{[N_1+N_2+\ldots+N_m]}{2}}$$

$$\times\prod_{j=1}^{m}(N_j)^{-1/2}.\prod_{j=1}^{m}(N_j)^{\frac{b_j}{N_j}}(2\pi)^{n/2}(2\pi)^{-\frac{[M_1+M_2+\ldots+M_n]}{2}}\prod_{j=1}^{n}(M_j)^{-1/2}$$

$$\times\frac{\prod_{j=1}^{n}(M_j)^{\frac{1-a_j}{n_j}}\prod_{i=1}^{N}\Gamma(\beta_i-s)\prod_{i=1}^{M}\Gamma(\alpha_i+s)\prod_{j=1}^{m}(N_j)^{-s}\prod_{j=1}^{n}(M_j)^{s}}{(2\pi)^{\frac{q-m}{2}}(2\pi)^{-\frac{[N_{m+1}+N_{m+2}+\ldots+N_q]}{2}}\prod_{j=m+1}^{q}(N_j)^{\frac{1-b_j}{N_j}}\prod_{j=m+1}^{q}(N_j)^{-1/2}(2\pi)^{\frac{p-n}{2}}}$$

$$\times\frac{1}{(2\pi)^{-\frac{[M_{n+1}+M_{n+2}+..+M_p]}{2}}\prod_{j=m+1}^{p}(M_j)^{\frac{a_j}{M_j}}\prod_{j=m+1}^{q}(M_j)^{-1/2}\prod_{i=m+1}^{Q}\Gamma(\lambda_i+s)}$$

$$\times\frac{1}{\prod_{i=n+1}^{R}\Gamma(\delta_i-s)\prod_{j=m+1}^{q}(N_j)^{s}\prod_{j=n+1}^{p}(M_j)^{-s}}z^s ds$$

$$=(2\pi)^{\frac{m+n-q+m-p+n}{2}}(2\pi)^{-\frac{[(N_1+N_2+\ldots+N_m)+(M_1+M_2+\ldots M_n)+(N_{m+1}+N_{m+2}+\ldots+N_q)+(M_{n+1}+M_{n+2}+\ldots+M_p)]}{2}}$$

$$\times\frac{\prod_{j=1}^{m}(N_j)^{-1/2}\prod_{j=1}^{m}(N_j)^{\frac{b_j}{N_j}}\prod_{j=1}^{n}(M_j)^{-1/2}\prod_{j=1}^{n}(M_j)^{\frac{1-a_j}{n_j}}}{\prod_{j=m+1}^{q}(N_j)^{\frac{1-b_j}{N_j}}\prod_{j=m+1}^{q}(N_j)^{-1/2}\prod_{j=m+1}^{p}(M_j)^{\frac{a_j}{M_j}}\prod_{j=m+1}^{q}(M_j)^{-1/2}}$$

$$\times\int_L\frac{\prod_{i=1}^{N}\Gamma(\beta_i-s)\prod_{i=1}^{M}\Gamma(\alpha_i+s)\prod_{j=1}^{m}(N_j)^{-s}\prod_{j=1}^{n}(M_j)^{s}}{\prod_{i=m+1}^{Q}\Gamma(\lambda_i+s)\prod_{i=n+1}^{R}\Gamma(\delta_i-s)\prod_{j=m+1}^{q}(N_j)^{s}\prod_{j=n+1}^{p}(M_j)^{-s}}z^s\,ds$$

where $\eta=\sum_{r=1}^{m}N_r,\mu=\sum_{r=1}^{n}M_r$

$$\xi = \sum_{r=m+1}^{q} Q_r , \psi = \sum_{r=n+1}^{p} R_r$$

$$= (2\pi)^{\frac{2m+2n-p-q}{2}} (2\pi)^{-\left(\frac{\eta+\mu+\xi+\psi}{2}\right)} \frac{\prod_{j=1}^{m}(N_j)^{-1/2} \prod_{j=1}^{m}(N_j)^{\frac{b_j}{N_j}} \prod_{j=1}^{n}(M_j)^{-1/2} \prod_{j=1}^{n}(M_j)^{\frac{1-a_j}{n_j}}}{\prod_{j=m+1}^{q}(N_j)^{\frac{1-b_j}{N_j}} \prod_{j=m+1}^{q}(N_j)^{-1/2} \prod_{j=n+1}^{p}(M_j)^{\frac{a_j}{M_j}} \prod_{j=m+1}^{q}(M_j)^{-1/2}}$$

$$\times \frac{1}{2\pi i} \int \frac{\prod_{i=1}^{\eta} \Gamma(\beta_{i-s}) \prod_{i=1}^{\mu} \Gamma(\alpha_i + s)}{\prod_{i=m+1}^{\xi} \Gamma(\lambda_i + s) \prod_{i=n+1}^{\psi} \Gamma(\delta_i - s)} \left[\frac{\prod_{j=1}^{m}(N_j)^{-1} \prod_{j=1}^{m} M_j}{\prod_{j=m+1}^{q}(N_j) \prod_{j=m+1}^{p}(M_j)^{-1}} .z \right]^s ds$$

$$= (2\pi)^{-\left(\frac{\eta+\mu+\xi+\psi}{2}\right)} \frac{\prod_{j=1}^{m}(N_j)^{-1/2} \prod_{j=1}^{m}(N_j)^{\frac{b_j}{N_j}} \prod_{j=1}^{n}(M_j)^{-1/2} \prod_{j=1}^{n}(M_j)^{\frac{1-a_j}{n_j}}}{\prod_{j=m+1}^{q}(N_j)^{\frac{1-b_j}{N_j}} \prod_{j=m+1}^{q}(N_j)^{-1/2} \prod_{j=n+1}^{p}(M_j)^{\frac{a_j}{M_j}} \prod_{j=m+1}^{q}(M_j)^{-1/2}}$$

$$G_{\xi,\psi}^{\eta,\mu}\left[\rho \left| \begin{matrix} \alpha_1 , \alpha_2 ,\ldots\ldots \alpha_\mu , \delta_{n+1} ,\ldots\ldots \delta_\psi , \\ \beta_1 , \beta_2 ,\ldots\ldots \beta_\eta , \lambda_{m+1} ,\ldots\ldots \lambda_\xi \end{matrix} \right. \right] \tag{2.11}$$

Special Cases

$$I_{p_i, q_i:r}^{m, n}[z] = I_{p_i, q_i:r}^{m, n}\left[z \left| \begin{matrix} (q_i, \alpha_i)_{1, n} \cdots & (q_{j1}, \alpha_{j1})_{n+1, p} \\ (b_j, \beta_j)_{1, m} \cdots & (b_{j1}, \beta_{j1})_{m+1, q} \end{matrix} \right. \right]$$

$$= \frac{1}{2\pi i} \int_L t(s) z^s \, ds$$

where

$$t = \frac{\prod_{j=1}^{m} \Gamma(b_j - \beta_j s) \, \Gamma(1 - a_j + \alpha_j s)}{\sum_{i=1}^{r} \left\{ \prod_{j=m+1}^{q_i} \Gamma(1 - b_{j1} + \beta_{j1} s) \prod_{j=n+1}^{p_i} (a_{j1} - \alpha_{j1} s) \right\}}$$

Putting $r = 1$in (2.12) we get

$$I_{p_i,q_i:1}^{m,n}:1\,[z]=I_{p_i,q_i:1}^{m,n}\left[z\left|\begin{matrix}(q_i,\alpha_i)_{1,n}\ldots\ldots & (q_j,\alpha_j)_{n+1,p}\\(b_j,\beta_j)_{1,m}\ldots\ldots & (b_j,\beta_j)_{m+1,q}\end{matrix}\right.\right]$$

$$:\quad=\frac{1}{2\pi i}\int_L\frac{\prod_{j=1}^{m}\Gamma(b_j-\beta_j s)\prod_{j=1}^{n}\Gamma(1-a_j+\alpha_j s)}{\sum_{i=1}^{1}\left\{\prod_{j=m+1}^{q_i}\Gamma(1-b_{j1}+\beta_{j1}s)\prod_{j=n+1}^{p_i}\Gamma(q_{j1}-\alpha_{j1}s)\right\}}z^s\,ds\,1\,[z]=$$

$$=\frac{1}{2\pi i}\int_L\frac{\prod_{j=1}^{m}\Gamma(b_j-\beta_j s)\prod_{j=1}^{n}\Gamma(1-a_j+\alpha_j s)}{\prod_{j=m+1}^{q_i}\Gamma(1-b_j+\beta_j s)\prod_{j=n+1}^{q_i}\Gamma(a_j-\alpha_j)s}z^s\,ds$$

$$I_{p_i,q_i:1}^{m,n}\left[z\left|\begin{matrix}(q_i,\alpha_i)_{1,n}\ldots\ldots & (q_j,\alpha_j)_{n+1,p}\\(b_j,\beta_j)_{1,m}\ldots\ldots & (b_j,\beta_j)_{m+1,q}\end{matrix}\right.\right]\tag{2.13}$$

Put

$\alpha_j=M_j,\ \beta_j=N_j$

$$=\frac{1}{2\pi i}\int_L\frac{\prod_{j=1}^{m}\Gamma\left(N_j\left(\frac{b_j}{N_j}-s\right)\right)\prod_{j=1}^{n}\Gamma\left(M_j\left(\frac{1-q_j}{M_j}+s\right)\right)}{\prod_{j=m+1}^{q}\Gamma\left(N_j\left(\frac{1-b_j}{N_j}+s\right)\right)\prod_{j=n+1}^{p}\Gamma\left(M_j\left(\frac{q_j}{M_j}-s\right)\right)}z^s\,ds\tag{2.14}$$

Putting $\frac{b_j}{N_j}-S=Z_j,\ \frac{1-a_j}{M_j}+s=\gamma_j,\ \frac{1-b_j}{N_j}+s=Y_j,\ \frac{a_j}{M_j}-s=X_j$ in (3.4),

we get

$$=\frac{1}{2\pi i}\int_L\frac{\prod_{j=1}^{m}\Gamma(N_jZ_j)\prod_{j=1}^{n}\Gamma(M_jY_j)}{\prod_{j=m+1}^{q}\Gamma(N_jY_j)\prod_{j=n+1}^{p}\Gamma(M_jX_j)}z^s\,ds\tag{2.15}$$

By using formula,

$$\Gamma(mZ)=(2\pi)^{\frac{1-m}{2}}\,m^{z-1/2}\prod_{j=1}^{m}\left(Z+\frac{j-1}{m}\right),\tag{2.16}$$

$$Z\neq 0,\ -1/m,\ 2/m\ldots\ldots m=1,2,3.$$

We arrive at

$$\prod_{j=1}^{m}\Gamma(N_jZ_j)=\Gamma(N_1Z_1)\Gamma(N_2Z_2)\Gamma(N_3Z_3)\ldots\ldots\ldots\Gamma(N_mZ_m)$$

$$=\left[(2\pi)^{\frac{1-N_1}{2}}(N_1)^{Z_1-\frac{1}{2}}\prod_{j=1}^{N_1}\Gamma\left(Z_1+\frac{j-1}{N_1}\right)\right]$$

$$\times\left[(2\pi)^{\frac{1-N_2}{2}}(N_2)^{Z_2-\frac{1}{2}}\prod_{j=1}^{N_2}\Gamma\left(Z_2+\frac{j-1}{N_2}\right)\right]$$

$$\times\left[(2\pi)^{\frac{1-N_3}{2}}(N_3)^{Z_3-\frac{1}{2}}\prod_{j=1}^{N_3}\Gamma\left(Z_3+\frac{j-1}{N_3}\right)\right]$$

$$\left[(2\pi)^{\frac{1-N_m}{2}}(N_m)^{Z_m-\frac{1}{2}}\prod_{j=1}^{N_m}\Gamma\left(Z_m+\frac{j-1}{N_m}\right)\right]$$

$$=(2\pi)^{\left(\frac{1}{2}+\frac{1}{2}+\frac{1}{2}\ldots m\text{ times}\right)}.(2\pi)^{-\left[\frac{N_1+N_2+N_3+\ldots\ldots+N_m}{2}\right]}.(N_1.N_2.N_3\ldots\ldots N_m)^{-\frac{1}{2}}$$

$$\times\prod_{j=1}^{m}(N_j)^{Z_j}.\prod_{j=1}^{N_1}\Gamma\left(\frac{b_1}{N_1}-s+\frac{j-1}{N_1}\right).\prod_{j=1}^{N_2}\Gamma\left(\frac{b_2}{N_2}-s+\frac{j-1}{N_2}\right).$$

$$\times\prod_{j=1}^{N_3}\Gamma\left(\frac{b_3}{N_3}-s+\frac{j-1}{N_3}\right)\ldots\prod_{j=1}^{N_m}\Gamma\left(\frac{b_m}{N_m}-s+\frac{j-1}{N_m}\right)$$

$$=(2\pi)^{\frac{m}{2}}.(2\pi)^{-\left[\frac{N_1+N_2+N_3+\ldots+N_m}{2}\right]}.(N_1.N_2\ldots N_m)^{-\frac{1}{2}}.$$

$$\times\ N_1^{\left(\frac{b_1}{N_1}-s\right)}.N_2^{\left(\frac{b_2}{N_2}-s\right)}.N_3^{\left(\frac{b_3}{N_3}-s\right)}\ldots N_m^{\left(\frac{b_m}{N_m}-s\right)}.\prod_{j=1}^{N_1}\Gamma\left(\frac{b_1+j-1}{N_1}-s\right).$$

$$\times\prod_{j=1}^{N_2}\Gamma\left(\frac{b_2+j-1}{N_2}-s\right).\prod_{j=1}^{N_3}\Gamma\left(\frac{b_3+j-1}{N_3}-s\right)\ldots\prod_{j=1}^{N_m}\Gamma\left(\frac{b_m+j-1}{N_m}-s\right).$$

$$=(2\pi)^{\frac{m}{2}}.(2\pi)^{\frac{-(N_1+N_2+N_3+\ldots\ldots+N_m)}{2}}.(N_1.N_2.N_3\ldots N_m)^{-\frac{1}{2}}(N_1.N_2.N_3\ldots\ldots N_m)^{-s}$$

$$\times N_1^{\frac{b_1}{N_1}}.N_2^{\frac{b_2}{N_2}}N_3^{\frac{b_3}{N_3}}\ldots\ldots\ldots\ldots N_m^{\frac{b_m}{N_m}}.\prod_{j=1}^{N_1}\Gamma\left(\frac{b_1+j-1}{N_1}-s\right)$$

$$\times\prod_{j=1}^{N_2}\Gamma\left(\frac{b_2+j-1}{N_2}-s\right).\prod_{j=1}^{N_3}\Gamma\left(\frac{b_3+j-1}{N_3}-s\right)\ldots\prod_{j=1}^{N_m}\left(\frac{b_m+j-1}{N_m}-s\right)$$

$$=(2\pi)^{m/2}.(2\pi)^{\frac{-(N_1+N_2+N_3+\ldots..+N_m)}{2}}.(N_1.N_2.N_3\ldots N_m)^{-1/2}$$

$$\times\prod_{j=1}^{m}(N_j)^{-s}\,N_j^{\frac{b_j}{N_j}}\prod_{i=1}^{N}\Gamma(\beta_i-s) \tag{2.17}$$

where $\beta_i=\dfrac{b_i+j-1}{N_i}$

$$N=N_1+N_2+N_3+\ldots\ldots+N_m,\ j=1\text{ to }N.$$

$$\prod_{j=1}^{n}\Gamma(M_j\gamma_j)=\Gamma(M_1\gamma_1)\Gamma(M_2\gamma_2).\Gamma(M_3\gamma_3)\ldots..\Gamma(M_n\gamma_n)$$

$$=\left[(2\pi)^{\frac{1-M_1}{2}}(M_1)^{\gamma_1-1/2}\prod_{j=1}^{M_1}\Gamma\left(\gamma_1+\frac{j-1}{M_1}\right)\right]$$

$$.\left[(2\pi)^{\frac{1-M_2}{2}}M_2^{\gamma_2-1/2}\prod_{j=1}^{M_2}\Gamma\left(\gamma_2+\frac{j-1}{M_2}\right)\right]$$

$$.\left[(2\pi)^{\frac{1-M_3}{2}}(M_3)^{\gamma_3-1/2}\prod_{j=1}^{M_3}\left(\gamma_3+\frac{j-1}{M_3}\right)\right]$$

$$\left[(2\pi)^{\frac{1-M_n}{2}}(M_n)^{\gamma_n-1/2}\prod_{j=1}^{M_n}\left(\gamma_n+\frac{j-1}{M_n}\right)\right]$$

$$=(2\pi)^{\left(1/2+1/2+1/2\ldots..n\ times\right)}.$$

$$(2\pi)^{-\left[\frac{M_1+M_2+M_3+\ldots+M_n}{2}\right]}(M_1.M_2.M_3\ldots..M_n)^{-1/2}$$

$$\left[M_1^{\frac{1-a_1}{M_1}}.M_2^{\frac{1-a_2}{M_2}}\ldots.M_n^{\frac{1-a_n}{M_n}}\right](M_1.M_2.M_3\ldots..M_n)^{+s}$$

$$\prod_{j=1}^{M_1}\Gamma\left(\frac{1-a_1}{M_1}+s+\frac{j-1}{M_1}\right).\prod_{j=1}^{M_2}\Gamma\left(\frac{1-a_2}{M_2}+s+\frac{j-1}{M_2}\right).$$

$$\prod_{j=1}^{M_3}\Gamma\left(\frac{1-a_3}{M_3}+s+\frac{j-1}{M_3}\right)\ldots\prod_{j=1}^{M_n}\Gamma\left(\frac{1-a_n}{M_n}+s+\frac{j-1}{M_n}\right).$$

$$=(2\pi)^{n/2}(2\pi)^{-\left[\frac{M_1+M_2+M_3+\ldots+M_n}{2}\right]}(M_1.M_2\ldots M_n)^{-1/2}\prod_{j=1}^{n}(M_j)^{\frac{1-a_j}{M_j}}(M_j)^s.$$

$$\times\prod_{j=1}^{M_1}\Gamma\left(\frac{j-a_1}{M_1}+s\right).\prod_{j=1}^{M_2}\Gamma\left(\frac{j-a_2}{M_2}+s\right).\prod_{j=1}^{M3}\Gamma\left(\frac{j-a_3}{M_3}+s\right)\ldots\prod_{j=1}^{M_n}\Gamma\left(\frac{j-a_n}{M_n}+s\right)$$

$$=(2\pi)^{n/2}(2\pi)^{-\frac{[M_1+M_2+\ldots M_n]}{2}}.(M_1.M_2\ldots M_n)^{-1/2}\prod_{j=1}^{n}(M_j)^{\frac{1-a_j}{M_j}}(M_j)^s$$

$$\prod_{i=1}^{M}\Gamma(\alpha_i+s)\tag{2.18}$$

where $\alpha_i=\dfrac{j-a_i}{M_i}+s$

$$M=M_1+M_2+M_3+\ldots+M_n$$

$$\prod_{j=M+1}^{q}\Gamma(N_jY_j)=\Gamma(N_{m+1}Y_{m+1}).\ \Gamma(N_{m+2}Y_{m+2}).\ \Gamma(N_{m+3}Y_{m+3})\ldots\ldots\Gamma(N_qY_q)$$

$$=\left[(2\pi)^{\frac{1-N_{m+1}}{2}}(N_{m+1})^{Y_{m+1}-1/2}\prod_{j=1}^{N_{m+1}}\Gamma\left(Y_{m+1}+\frac{j-1}{N_{m+1}}\right)\right].$$

$$\left[(2\pi)^{\frac{1-N_{m+2}}{2}}(N_{m+2})^{Y_{m+2}-1/2}\prod_{j=1}^{N_{m+2}}\Gamma\left(Y_{m+2}+\frac{j-1}{N_{m+2}}\right)\right].$$

$$\left[(2\pi)^{\frac{1-N_q}{2}}(N_q)^{Y_q-1/2}\prod_{j=1}^{N_q}\Gamma\left(Y_q+\frac{j-1}{N_q}\right)\right]$$

$$=(2\pi)^{\frac{q-m}{2}}.(2\pi)^{-\frac{[N_{m+1}+N_{m+2}+\ldots+N_q]}{2}}.\prod_{j=m+1}^{q}(N_j)^{\frac{1-b_j}{N_j}}(N_j)^s\prod_{j=m+1}^{q}(N_j)^{-1/2}$$

$$\prod_{j=1}^{N_{m+1}}\Gamma\left(\frac{1-b_{m+1}}{N_{m+1}}+s+\frac{j-1}{N_{m+1}}\right).\prod_{j=1}^{N_{m+2}}\Gamma\left(\frac{1-b_{m+2}}{N_{m+2}}+s+\frac{j-1}{N_{m+2}}\right).$$

$$\prod_{j=1}^{N_{m+3}}\Gamma\left(\frac{1-b_{m+3}}{N_{m+3}}+s+\frac{j-1}{N_{m+3}}\right)\ldots\prod_{j=1}^{N_q}\Gamma\left(\frac{1-b_q}{N_q}+s+\frac{j-1}{N_q}\right)$$

$$=(2\pi)^{\frac{q-m}{2}}(2\pi)^{-\frac{\left[N_{m+1}+N_{m+2}+\ldots+N_q\right]}{2}}.\prod_{j=m+1}^{q}\left(N_j\right)^{\frac{1-b_j}{N_j}}\left(N_j\right)^s\prod_{j=m+1}^{q}\left(N_j\right)^{-1/2}.$$

$$\prod_{j=1}^{N_{m+1}}\Gamma\left(\frac{j-b_{m+1}}{N_{m+1}}+s\right)\prod_{j=1}^{N_{m+2}}\Gamma\left(\frac{j-b_{m+2}}{N_{m+2}}+s\right)\ldots\prod_{j=1}^{N_q}\Gamma\left(\frac{j-b_q}{N_q}+s\right)$$

$$=(2\pi)^{\frac{q-m}{2}}(2\pi)^{-\frac{\left[N_{m+1}+N_{m+2}+\ldots+N_q\right]}{2}}.\prod_{j=m+1}^{q}\left(N_j\right)^{\frac{1-b_j}{N_q}}\left(N_j\right)^s\prod_{j=m+1}^{q}\left(N_j\right)^{-1/2}\prod_{i=m+1}^{Q}\Gamma\left(\frac{j-b_i}{N_i}+s\right)$$

$$=(2\pi)^{\frac{q-m}{2}}(2\pi)^{-\frac{\left[N_{m+1}+N_{m+2}+\ldots N_q\right]}{2}}.\prod_{j=m+1}^{q}\left(N_j\right)^{\frac{1-b_j}{N_a}}\left(N_j\right)^s\prod_{j=m+1}^{q}\left(N_j\right)^{-1/2}\prod_{i=m+1}^{Q}\Gamma\left(\lambda_i+s\right).$$

$$=\Gamma\left(\mathrm{N}_{m+1}\mathrm{Y}_{m+1}\right).\ \Gamma\left(\mathrm{N}_{m+2}\mathrm{Y}_{m+2}\right).\ \Gamma\left(\mathrm{N}_{m+3}\mathrm{Y}_{m+3}\right)\ldots\ldots\Gamma\left(\mathrm{N}_q\mathrm{Y}_q\right) \tag{2.19}$$

where

$$Q=N_{m+1}+N_{m+1}+N_{m+3}+\ldots+N_q$$

$$\lambda_i=\Gamma\left(\frac{j-b_i}{N_i}+s\right)$$

$$\prod_{j=n+1}^{P}\Gamma\left(M_jX_j\right)=\Gamma\left(M_{n+1}X_{n+1}\right)\Gamma\left(M_{n+2}X_{n+2}\right)\Gamma\left(M_{n+3}X_{n+3}\right)\ldots\ldots\ \Gamma\left(M_PX_P\right)$$

$$=\left[(2\pi)^{\frac{1-M_{n+1}}{2}}\left(M_{n+1}\right)^{X_{n+1}-1/2}\prod_{j=1}^{M_{n+1}}\Gamma\left(X_{n+1}+\frac{j-1}{M_{n+1}}\right)\right].$$

$$\left[(2\pi)^{\frac{1-M_{n+2}}{2}}\left(M_{n+2}\right)^{X_{n+2}-1/2}\prod_{j=1}^{M_{n+2}}\Gamma\left(X_{n+2}+\frac{j-1}{M_{n+2}}\right)\right].$$

$$\left[(2\pi)^{\frac{1-M_{n+3}}{2}}\left(M_{n+3}\right)^{X_{n+3}-1/2}\prod_{j=1}^{M_{n+3}}\Gamma\left(X_{n+3}+\frac{j-1}{M_{n+3}}\right)\right]$$

$$\ldots\ldots\left[(2\pi)^{\frac{1-M_P}{2}}\left(M_P\right)^{X_P-1/2}\prod_{j=1}^{M_P}\Gamma\left(X_P+\frac{j-1}{M_P}\right)\right]$$

$$= (2\pi)^{\frac{p-n}{2}} (2\pi)^{-\frac{[M_{n+1}+M_{n+2}+\ldots+M_p]}{2}} \prod_{j=n+1}^{p} (M_j)^{\frac{a_j}{M_j}} (M_j)^{-s}$$

$$\prod_{j=n+1}^{R} (M_j)^{-1/2} \prod_{i=n+1}^{p} \Gamma\left(\frac{a_i + j - 1}{M_i} - s\right)$$

where $R = M_{n+1} + M_{n+2} + \ldots + M_p$

$$\delta_i = \frac{a_i + j - 1}{M_i}$$

$$= (2\pi)^{\frac{p-n}{2}} (2\pi)^{-\frac{[M_{n+1}+M_{n+2}+\ldots+M_p]}{2}} . \tag{2.20}$$

$$\prod_{j=n+1}^{p} (m_j)^{\frac{a_j}{M_j}} (M_j)^{-s} \prod_{j=m+1}^{p} (M_j)^{-1/2} \prod_{i=n+1}^{R} \Gamma(\delta_i - s)$$

Putting in Equation (2.13) we get

$$I_{p,\,q;1}^{m,\,n}[z] = \int_L (2\pi)^{m/2} (2\pi)^{-\frac{[N_1+N_2+\ldots+N_m]}{2}}$$

$$\times \prod_{j=1}^{m} (N_j)^{-1/2} . \prod_{j=1}^{m} (N_j)^{\frac{b_j}{N_j}} (2\pi)^{n/2} (2\pi)^{-\frac{[M_1+M_2+\ldots+M_n]}{2}} \prod_{j=1}^{n} (M_j)^{1/2}$$

$$\times \frac{\prod_{j=1}^{n} (M_j)^{\frac{1-a_j}{n_j}} \prod_{i=1}^{N} \Gamma(\beta_i - s) \prod_{i=1}^{M} \Gamma(\alpha_i + s) \prod_{j=1}^{m} (N_j)^{-s} \prod_{j=1}^{n} (M_j)^{s}}{(2\pi)^{\frac{q-m}{2}} (2\pi)^{-\frac{[N_{m+1}+N_{m+2}+\ldots+N_q]}{2}} . \prod_{j=m+1}^{q} (N_j)^{\frac{1-b_j}{N_j}} \prod_{j=m+1}^{q} (N_j)^{-1/2} (2\pi)^{\frac{p-n}{2}}}$$

$$\times \frac{1}{(2\pi)^{-\frac{[M_{n+1}+M_{n+2}+..+M_p]}{2}} \prod_{j=m+1}^{p} (M_j)^{\frac{a_j}{M_j}} \prod_{j=m+1}^{q} (M_j)^{-1/2} \prod_{i=m+1}^{Q} \Gamma(\lambda_i + s)}$$

$$\times \frac{1}{\prod_{i=n+1}^{R} \Gamma(\delta_i - s) \prod_{j=m+1}^{q} (N_j)^{s} \prod_{j=n+1}^{p} (M_j)^{-s}} z^s ds$$

$$= (2\pi)^{\frac{m+n-q+m-p+n}{2}} (2\pi)^{-\frac{[(N_1+N_2+\ldots+N_m)+(M_1+M_2+\ldots M_n)+(N_{m+1}+N_{m+2}+\ldots+N_q)+(M_{n+1}+M_{n+2}+\ldots+M_p)]}{2}}$$

$$\times\frac{\prod_{j=1}^{m}(N_j)^{-1/2}\prod_{j=1}^{m}(N_j)^{\frac{b_j}{N_j}}\prod_{j=1}^{n}(M_j)^{-1/2}\prod_{j=1}^{n}(M_j)^{\frac{1-a_j}{n_j}}}{\prod_{j=m+1}^{q}(N_j)^{\frac{1-b_j}{N_j}}\prod_{j=m+1}^{q}(N_j)^{-1/2}\prod_{j=m+1}^{p}(M_j)^{\frac{a_j}{M_j}}\prod_{j=m+1}^{q}(M_j)^{-1/2}}$$

$$\times\int_L\frac{\prod_{i=1}^{N}\Gamma(\beta_i-s)\prod_{i=1}^{M}\Gamma(\alpha_i+s)\prod_{j=1}^{m}(N_j)^{-s}\prod_{j=1}^{n}(M_j)^{s}}{\prod_{i=m+1}^{Q}\Gamma(\lambda_i+s)\prod_{i=n+1}^{R}\Gamma(\delta_i-s)\prod_{j=m+1}^{q}(N_j)^{s}\prod_{j=n+1}^{p}(M_j)^{-s}}z^s ds$$

where $\eta=\sum_{r=1}^{m}N_r,\ \mu=\sum_{r=1}^{n}M_r$

$$\xi=\sum_{r=m+1}^{q}Q_r,\ \psi=\sum_{r=n+1}^{p}R_r$$

$$=(2\pi)^{\left(\frac{2m+2n-p-q}{2}\right)}(2\pi)^{-\left(\frac{\eta+\mu+\xi+\psi}{2}\right)}\frac{\prod_{j=1}^{m}(N_j)^{-1/2}\prod_{j=1}^{m}(N_j)^{\frac{b_j}{N_j}}\prod_{j=1}^{n}(M_j)^{-1/2}\prod_{j=1}^{n}(M_j)^{\frac{1-a_j}{n_j}}}{\prod_{j=m+1}^{q}(N_j)^{\frac{1-b_j}{N_j}}\prod_{j=m+1}^{q}(N_j)^{-1/2}\prod_{j=n+1}^{p}(M_j)^{\frac{a_j}{M_j}}\prod_{j=m+1}^{q}(M_j)^{-1/2}}$$

$$\times\frac{1}{2\pi i}\int\frac{\prod_{i=1}^{\eta}\Gamma(\beta_{i-s})\prod_{i=1}^{\mu}\Gamma(\alpha_i+s)}{\prod_{i=m+1}^{\xi}\Gamma(\lambda_i+s)\prod_{i=n+1}^{\psi}\Gamma(\delta_i-s)}\left[\frac{\prod_{j=1}^{m}(N_j)^{-1}\prod_{j=1}^{m}M_j}{\prod_{j=m+1}^{q}(N_j)\prod_{j=m+1}^{p}(M_j)^{-1}}.z\right]^s ds$$

$$=(2\pi)^{\frac{2m+2n-p-q}{2}}$$

$$(2\pi)^{-\left(\frac{\eta+\mu+\xi+\psi}{2}\right)}\frac{\prod_{j=1}^{m}(N_j)^{-1/2}\prod_{j=1}^{m}(N_j)^{\frac{b_j}{N_j}}\prod_{j=1}^{n}(M_j)^{-1/2}\prod_{j=1}^{n}(M_j)^{\frac{1-a_j}{n_j}}}{\prod_{j=m+1}^{q}(N_j)^{\frac{1-b_j}{N_j}}\prod_{j=m+1}^{q}(N_j)^{-1/2}\prod_{j=n+1}^{p}(M_j)^{\frac{a_j}{M_j}}\prod_{j=m+1}^{q}(M_j)^{-1/2}}$$

$$G_{\xi,\psi}^{\eta,\mu}\left[\rho\left|\begin{matrix}\alpha_1,\alpha_2,\ldots\ldots\alpha_\mu,\delta_{n+1},\ldots\ldots\delta_\psi,\\ \beta_1,\beta_2,\ldots\ldots\beta_\eta,\lambda_{m+1},\ldots\ldots\lambda_\xi\end{matrix}\right.\right]$$

(2.21)

The transformation relations derived above can generate new properties of the I-function and other higher order generalized hypergeometric functions.

2.2 SPECIAL CASES, PROPERTIES, AND TRANSFORMATION

We have

$$H_{p,q}^{m,n}[z] = H_{p,q}^{m,n}\left[z \,\middle|\, \begin{matrix} (a_1,\alpha_1)\ldots\ldots(a_p,\alpha_p) \\ (b_1,\beta_1)\ldots\ldots(b_q,\beta_q) \end{matrix} \right] \tag{2.22}$$

$$= \frac{1}{2\pi i}\int_L t\,(s) z^s\, ds$$

where

$$t(s)\frac{\prod_{j=1}^{m}\Gamma(b_j-\beta_j s)\prod_{j=1}^{n}\Gamma(1-a_j+\alpha_j s)}{\prod_{j=m+1}^{q}\Gamma(1-b_j+\beta_j s)\prod_{j=n+1}^{p}\Gamma(a_j-\alpha_j s)} \tag{2.23}$$

Now,

$$H_{p,q}^{m,n}[z] = \frac{1}{2\pi i}\int_L \frac{\prod_{j=1}^{m}\Gamma(b_j-\beta_j s)\prod_{j=1}^{n}\Gamma(1-a_j+\alpha_j s)}{\prod_{j=m+1}^{q}\Gamma(1-b_j-\beta_j s)\prod_{j=n+1}^{p}\Gamma(a_j-\alpha_j s)} z^s\, ds \tag{2.24}$$

Putting $\alpha_j = M_j$, $\beta_j = N_j$ in (2.24), we get

$$= \frac{1}{2\pi i}\int_L \frac{\prod_{j=1}^{m}\Gamma(b_j-N_j s)\prod_{j=1}^{n}\Gamma(1-a_j+M_j s)}{\prod_{j=m+1}^{q}\Gamma(1-b_j+N_j s)\prod_{j=n+1}^{n}\Gamma(a_j-M_j s)} Z^s ds$$

$$= \frac{1}{2\pi i}\int_L \frac{\prod_{j=1}^{m}\Gamma\left(N_j\left(\frac{b_j}{N_j}-s\right)\right)\prod_{j=1}^{n}\Gamma\left(M_j\left(\frac{1-q_j}{M_j}+s\right)\right)}{\prod_{j=m+1}^{q}\Gamma\left(N_j\left(\frac{1-b_j}{N_j}+s\right)\right)\prod_{j=n+1}^{p}\Gamma\left(M_j\left(\frac{q_j}{M_j}-s\right)\right)} z^s ds \tag{2.25}$$

$$= \frac{1}{2\pi i}\int_L \frac{\prod_{j=1}^{m}\Gamma(N_j Z_j)\prod_{j=1}^{n}\Gamma(M_j Y_j)}{\prod_{j=m+1}^{q}\Gamma(N_j Y_j)\prod_{j=n+1}^{p}\Gamma(M_j X_j)} z^s\, ds \tag{2.26}$$

For $\frac{b_j}{N_j} - s = Z_j$, $\frac{1-a_j}{M_j} + s = \gamma_j$, $\frac{1-b_j}{N_j} + s = Y_j$, $\frac{a_j}{M_j} - s = X_j$ in (2.25),

By using formula,

$$\Gamma(mZ) = (2\pi)^{\frac{1-m}{2}} m^{z-\frac{1}{2}} \prod_{j=1}^{m} \left(Z + \frac{j-1}{m} \right), \tag{2.27}$$

$$Z \neq 0, -\tfrac{1}{m}, \tfrac{2}{m} \ldots\ldots m = 1, 2, 3.$$

We have,

$$\prod_{j=1}^{m} \Gamma(N_j Z_j) = \Gamma(N_1 Z_1)\Gamma(N_2 Z_2)\Gamma(N_3 Z_3)\ldots\ldots\ldots\Gamma(N_m Z_m)$$

$$= \left[(2\pi)^{\frac{1-N_1}{2}} (N_1)^{Z_1-\frac{1}{2}} \prod_{j=1}^{N_1} \Gamma\left(Z_1 + \frac{j-1}{N_1} \right) \right]$$

$$\left[(2\pi)^{\frac{1-N_2}{2}} (N_2)^{Z_2-\frac{1}{2}} \prod_{j=1}^{N_2} \Gamma\left(Z_2 + \frac{j-1}{N_2} \right) \right]$$

$$\left[(2\pi)^{\frac{1-N_3}{2}} (N_3)^{Z_3-\frac{1}{2}} \prod_{j=1}^{N_3} \Gamma\left(Z_3 + \frac{j-1}{N_3} \right) \right]$$

$$\left[(2\pi)^{\frac{1-N_m}{2}} (N_m)^{Z_m-\frac{1}{2}} \prod_{j=1}^{N_m} \Gamma\left(Z_m + \frac{j-1}{N_m} \right) \right]$$

$$= (2\pi)^{\left(\frac{1}{2}+\frac{1}{2}+\frac{1}{2}\ldots\ldots m times\right)} . (2\pi)^{-\left[\frac{N_1+N_2+N_3+\ldots\ldots+N_m}{2}\right]} . (N_1 . N_2 . N_3 \ldots\ldots N_m)^{-\frac{1}{2}}$$

$$\prod_{j=1}^{m} (N_j)^{Z_j} . \prod_{j=1}^{N_1} \Gamma\left(\frac{b_1}{N_1} - s + \frac{j-1}{N_1} \right) . \prod_{j=1}^{N_2} \Gamma\left(\frac{b_2}{N_2} - s + \frac{j-1}{N_2} \right).$$

$$\prod_{j=1}^{N_3} \Gamma\left(\frac{b_3}{N_3} - s + \frac{j-1}{N_3} \right) \ldots \prod_{j=1}^{N_m} \Gamma\left(\frac{b_m}{N_m} - s + \frac{j-1}{N_m} \right)$$

$$= (2\pi)^{\frac{m}{2}} (2\pi)^{-\left[\frac{N_1+N_2+N_3+\ldots+N_m}{2}\right]} . (N_1 . N_2 \ldots N_m)^{-\frac{1}{2}}$$

$$N_1^{\left(\frac{b_1}{N_1}-s\right)} . N_2^{\left(\frac{b_2}{N_2}-s\right)} . N_3^{\left(\frac{b_3}{N_3}-s\right)} \ldots N_m^{\left(\frac{b_m}{N_m}-s\right)} . \prod_{j=1}^{N_1} \Gamma\left(\frac{b_1 + j - 1}{N_1} - s \right).$$

$$\prod_{j=1}^{N_2} \Gamma\left(\frac{b_2 + j - 1}{N_2} - s \right) . \prod_{j=1}^{N_3} \Gamma\left(\frac{b_3 + j - 1}{N_3} - s \right) \ldots \prod_{j=1}^{N_m} \left(\frac{b_m + j - 1}{N_m} - s \right).$$

$$= (2\pi)^{m/2} . (2\pi)^{\frac{-(N_1+N_2+N_3+.....+N_m)}{2}} .(N_1.N_2.N_3 \ldots N_m)^{-1/2} (N_1.N_2.N_3 .....N_m)^{-s}$$

$$N_1^{\frac{b_1}{N_1}}.N_2^{\frac{b_2}{N_2}} N_3^{\frac{b_3}{N_3}} \ldots\ldots\ldots\ldots N_m^{\frac{b_m}{N_m}}.\prod_{j=1}^{N_1}\Gamma\left(\frac{b_1+j-1}{N_1}-s\right)$$

$$\prod_{j=1}^{N_2}\Gamma\left(\frac{b_2+j-1}{N_2}-s\right).\prod_{j=1}^{N_3}\Gamma\left(\frac{b_3+j-1}{N_3}-s\right)\ldots\prod_{j=1}^{N_m}\left(\frac{b_m+j-1}{N_m}-s\right)$$

$$= (2\pi)^{m/2} . (2\pi)^{\frac{-(N_1+N_2+N_3+.....+N_m)}{2}} .(N_1.N_2.N_3 \ldots N_m)^{-1/2}$$

$$\prod_{j=1}^{m}(N_j)^{-s} N_j^{\frac{b_j}{N_j}} \prod_{i=1}^{N}\Gamma(\beta_i - s) \tag{2.28}$$

where $\beta_i = \beta_i = \dfrac{b_i + j - 1}{N_i}$

$$N = N_1 + N_2 + N_3 + \ldots\ldots + N_m,\ j = 1 \text{ to } N.$$

$$\prod_{j=1}^{n}\Gamma(M_j\gamma_j) = \Gamma(M_1\gamma_1)\Gamma(M_2\gamma_2).\Gamma(M_3\gamma_3).....\Gamma(M_n\gamma_n)$$

$$= \left[(2\pi)^{\frac{1-M_1}{2}} (M_1)^{\gamma_1 - 1/2} \prod_{j=1}^{M_1}\Gamma\left(\gamma_1 + \frac{j-1}{M_1}\right)\right]$$

$$\left[(2\pi)^{\frac{1-M_2}{2}} M_2^{\gamma_2 - 1/2} \prod_{j=1}^{M_2}\Gamma\left(\gamma_2 + \frac{j-1}{M_2}\right)\right].$$

$$\left[(2\pi)^{\frac{1-M_3}{2}} (M_3)^{\gamma_3 - 1/2} \prod_{j=1}^{M_3}\left(\gamma_3 + \frac{j-1}{M_3}\right)\right]$$

$$\left[(2\pi)^{\frac{1-M_3}{2}} (M_3)^{\gamma_3 - 1/2} \prod_{j=1}^{M_3}\left(\gamma_3 + \frac{j-1}{M_3}\right)\right]$$

$$= (2\pi)^{\left(1/2 + 1/2 + 1/2 \ldots n \text{ times}\right)} .(2\pi)^{-\left[\frac{M_1+M_2+M_3+\ldots+M_n}{2}\right]} (M_1.M_2.M_3.....M_n)^{-1/2}$$

$$\left[M_1^{\frac{1-a_1}{M_1}} . M_2^{\frac{1-a_2}{M_2}} \ldots . M_n^{\frac{1-a_n}{M_n}}\right] (M_1.M_2.M_3.....M_n)^{+s}.$$

$$\prod_{j=1}^{M_1}\Gamma\left(\frac{1-a_1}{M_1}+s+\frac{j-1}{M_1}\right).\prod_{j=1}^{M_2}\Gamma\left(\frac{1-a_2}{M_2}+s+\frac{j-1}{M_2}\right).$$

$$\prod_{j=1}^{M_3}\Gamma\left(\frac{1-a_3}{M_3}+s+\frac{j-1}{M_3}\right)....\prod_{j=1}^{M_n}\Gamma\left(\frac{1-a_n}{M_n}+s+\frac{j-1}{M_n}\right).$$

$$=(2\pi)^{n/2}(2\pi)^{-\left[\frac{M_1+M_2+M_3+....+M_n}{2}\right]}(M_1.M_2\ldots M_n)^{-1/2}\prod_{j=1}^{n}(M_j)^{\frac{1-a_j}{M_j}}(M_j)^s.$$

$$\prod_{j=1}^{M_1}\Gamma\left(\frac{j-a_1}{M_1}+s\right).\prod_{j=1}^{M_2}\Gamma\left(\frac{j-a_2}{M_2}+s\right).\prod_{j=1}^{M3}\Gamma\left(\frac{j-a_3}{M_3}+s\right)\ldots\prod_{j=1}^{M_n}\Gamma\left(\frac{j-a_n}{M_n}+s\right)$$

$$=(2\pi)^{n/2}(2\pi)^{-\frac{[M_1+M_{2+}\ldots M_n]}{2}}.(M_1.M_2\ldots M_n)^{-1/2}\prod_{j=1}^{n}(M_j)^{\frac{1-a_j}{M_j}}(M_j)^s\prod_{i=1}^{M}\Gamma(\alpha_i+s) \tag{2.29}$$

where $\alpha_i=\dfrac{j-a_i}{M_i}+s$

$$M=M_1+M_2+M_3+....+M_n$$

$$\prod_{j=M+1}^{q}\Gamma(N_jY_j)=\Gamma(N_{m+1}Y_{m+1}).\ \Gamma(N_{m+2}Y_{m+2}).\ \Gamma(N_{m+3}Y_{m+3}).......\Gamma(N_qY_q)$$

$$=\left[(2\pi)^{\frac{1-N_{m+1}}{2}}(N_{m+1})^{Y_{m+1}-1/2}\prod_{j=1}^{N_{m+1}}\Gamma\left(Y_{m+1}+\frac{j-1}{N_{m+1}}\right)\right].$$

$$\left[(2\pi)^{\frac{1-N_{m+2}}{2}}(N_{m+2})^{Y_{m+2}-1/2}\prod_{j=1}^{N_{m+2}}\Gamma\left(Y_{m+2}+\frac{j-1}{N_{m+2}}\right)\right].$$

$$\left[(2\pi)^{\frac{1-N_q}{2}}(N_q)^{Y_q-1/2}\prod_{j=1}^{N_q}\Gamma\left(Y_q+\frac{j-1}{N_q}\right)\right]$$

$$=(2\pi)^{\frac{q-m}{2}}.(2\pi)^{-\frac{[N_{m+1}+N_{m+2}+\ldots+N_q]}{2}}.\prod_{j=m+1}^{q}(N_j)^{\frac{1-b_j}{N_j}}(N_j)^s\prod_{j=m+1}^{q}(N_j)^{-1/2}$$

$$\prod_{j=1}^{N_{m+1}}\Gamma\left(\frac{1-b_{m+1}}{N_{m+1}}+s+\frac{j-1}{N_{m+1}}\right).\prod_{j=1}^{N_{m+2}}\Gamma\left(\frac{1-b_{m+2}}{N_{m+2}}+s+\frac{j-1}{N_{m+2}}\right)$$

$$\prod_{j=1}^{N_{m+3}}\Gamma\left(\frac{1-b_{m+3}}{N_{m+3}}+s+\frac{j-1}{N_{m+3}}\right)\cdots\prod_{j=1}^{N_q}\Gamma\left(\frac{1-b_q}{N_q}+s+\frac{j-1}{N_q}\right)$$

$$=(2\pi)^{\frac{q-m}{2}}(2\pi)^{-\frac{[N_{m+1}+N_{m+2}+\ldots+N_q]}{2}}.\prod_{j=m+1}^{q}(N_j)^{\frac{1-b_j}{N_j}}(N_j)^s\prod_{j=m+1}^{q}(N_j)^{-1/2}.$$

$$\prod_{j=1}^{N_{m+1}}\Gamma\left(\frac{j-b_{m+1}}{N_{m+1}}+s\right)\prod_{j=1}^{N_{m+2}}\Gamma\left(\frac{j-b_{m+2}}{N_{m+2}}+s\right)\cdots\prod_{j=1}^{N_q}\Gamma\left(\frac{j-b_q}{N_q}+s\right)$$

$$=(2\pi)^{\frac{q-m}{2}}(2\pi)^{-\frac{[N_{m+1}+N_{m+2}+\ldots+N_q]}{2}}.\prod_{j=m+1}^{q}(N_j)^{\frac{1-b_j}{N_q}}(N_j)^s\prod_{j=m+1}^{q}(N_j)^{-1/2}\prod_{i=m+1}^{Q}\Gamma\left(\frac{j-b_i}{N_i}+s\right)$$

$$=(2\pi)^{\frac{q-m}{2}}(2\pi)^{-\frac{[N_{m+1}+N_{m+2}+\ldots N_q]}{2}}.\prod_{j=m+1}^{q}(N_j)^{\frac{1-b_j}{N_a}}(N_j)^s\prod_{j=m+1}^{q}(N_j)^{-1/2}\prod_{i=m+1}^{Q}\Gamma(\lambda_i+s) \tag{2.30}$$

where

$$Q=N_{m+1}+N_{m+1}+N_{m+3}+\ldots+N_q$$

$$\lambda_i=\Gamma\left(\frac{j-b_i}{N_i}+s\right)$$

$$\prod_{j=n+1}^{P}\Gamma(M_jX_j)=\Gamma(M_{n+1}X_{n+1})\Gamma(M_{n+2}X_{n+2})\Gamma(M_{n+3}X_{n+3})\ldots..\Gamma(M_PX_P)$$

$$=\left[(2\pi)^{\frac{1-M_{n+1}}{2}}(M_{n+1})^{X_{n+1}-1/2}\prod_{j=1}^{M_{n+1}}\Gamma\left(X_{n+1}+\frac{j-1}{M_{n+1}}\right)\right].$$

$$\left[(2\pi)^{\frac{1-M_{n+2}}{2}}(M_{n+2})^{X_{n+2}-1/2}\prod_{j=1}^{M_{n+2}}\Gamma\left(X_{n+2}+\frac{j-1}{M_{n+2}}\right)\right]$$

$$\left[(2\pi)^{\frac{1-M_{n+3}}{2}}(M_{n+3})^{X_{n+3}-1/2}\prod_{j=1}^{M_{n+3}}\Gamma\left(X_{n+3}+\frac{j-1}{M_{n+3}}\right)\right]$$

$$\left[(2\pi)^{\frac{1-M_P}{2}}(M_P)^{X_P-1/2}\prod_{j=1}^{M_P}\Gamma\left(X_P+\frac{j-1}{M_P}\right)\right]$$

$$= (2\pi)^{\frac{p-n}{2}} (2\pi)^{-\frac{[M_{n+1}+M_{n+2}+\ldots+M_p]}{2}} \prod_{j=n+1}^{p} (M_j)^{\frac{a_j}{M_j}} (M_j)^{-s} \prod_{j=n+1}^{R} (M_j)^{-1/2} \prod_{i=n+1}^{p} \Gamma\left(\frac{a_i + j - 1}{M_i} - s\right)$$

where $R = M_{n+1} + M_{n+2} + \ldots + M_p$

$$\delta_i = \frac{a_i + j - 1}{M_i}$$

$$= (2\pi)^{\frac{p-n}{2}} (2\pi)^{-\frac{[M_{n+1}+M_{n+2}+\ldots+M_p]}{2}} \cdot \prod_{j=n+1}^{p} (m_j)^{\frac{a_j}{M_j}} (M_j)^{-s} \prod_{j=m+1}^{p} (M_j)^{-1/2} \prod_{i=n+1}^{R} \Gamma(\delta_i - s) \tag{2.31}$$

Putting this in Equation (2.24) we get

$$H_{p,q}^{m,n}[z] = \int_L (2\pi)^{m/2} (2\pi)^{-\frac{[N_1+N_2+\ldots+N_m]}{2}} \times \prod_{j=1}^{m} (N_j)^{-1/2}.$$

$$\prod_{j=1}^{m} (N_j)^{\frac{b_j}{N_j}} (2\pi)^{n/2} (2\pi)^{-\frac{[M_1+M_2+\ldots+M_n]}{2}} \prod_{j=1}^{n} (M_j)^{-1/2}$$

$$\times \frac{\prod_{j=1}^{n} (M_j)^{\frac{1-a_j}{n_j}} \prod_{i=1}^{N} \Gamma(\beta_i - s) \prod_{i=1}^{M} \Gamma(\alpha_i + s) \prod_{j=1}^{m} (N_j)^{-s} \prod_{j=1}^{n} (M_j)^{s}}{(2\pi)^{\frac{q-m}{2}} (2\pi)^{-\frac{[N_{m+1}+N_{m+2}+\ldots+N_q]}{2}} \prod_{j=m+1}^{q} (N_j)^{\frac{1-b_j}{N_j}} \prod_{j=m+1}^{q} (N_j)^{-1/2} (2\pi)^{\frac{p-n}{2}}}$$

$$\times \frac{1}{(2\pi)^{-\frac{[M_{n+1}+M_{n+2}+..+M_p]}{2}} \prod_{j=m+1}^{p} (M_j)^{\frac{a_j}{M_j}} \prod_{j=m+1}^{q} (M_j)^{-1/2} \prod_{i=m+1}^{Q} \Gamma(\lambda_i + s)}$$

$$\times \frac{1}{\prod_{i=n+1}^{R} \Gamma(\delta_i - s) \prod_{j=m+1}^{q} (N_j)^{s} \prod_{j=n+1}^{p} (M_j)^{-s}} z^s ds$$

$$= (2\pi)^{\frac{m+n-q+m-p+n}{2}} (2\pi)^{-\frac{[(N_1+N_2+\ldots+N_m)+(M_1+M_2+\ldots M_n)+(N_{m+1}+N_{m+2}+\ldots+N_q)+(M_{n+1}+M_{n+2}+\ldots+M_p)]}{2}}$$

$$\times \frac{\prod_{j=1}^{m} (N_j)^{-1/2} \prod_{j=1}^{m} (N_j)^{\frac{b_j}{N_j}} \prod_{j=1}^{n} (M_j)^{-1/2} \prod_{j=1}^{n} (M_j)^{\frac{1-a_j}{n_j}}}{\prod_{j=m+1}^{q} (N_j)^{\frac{1-b_j}{N_j}} \prod_{j=m+1}^{q} (N_j)^{-1/2} \prod_{j=m+1}^{p} (M_j)^{\frac{a_j}{M_j}} \prod_{j=m+1}^{q} (M_j)^{-1/2}}$$

$$\times \int_L \frac{\prod_{i=1}^{N} \Gamma(\beta_i - s) \prod_{i=1}^{M} \Gamma(\alpha_i + s) \prod_{j=1}^{m} (N_j)^{-s} \prod_{j=1}^{n} (M_j)^{s}}{\prod_{i=m+1}^{Q} \Gamma(\lambda_i + s) \prod_{i=n+1}^{R} \Gamma(\delta_i - s) \prod_{j=m+1}^{q} (N_j)^{s} \prod_{j=n+1}^{p} (M_j)^{-s}} z^s ds$$

where $\eta = \sum_{r=1}^{m} N_r,\ \mu = \sum_{r=1}^{n} M_r$

$$\xi = \sum_{r=m+1}^{q} Q_r,\ \psi = \sum_{r=n+1}^{p} R_r$$

$$= \times \frac{1}{2\pi i}\int \frac{\prod_{i=1}^{\eta}\Gamma(\beta_{i-s})\prod_{i=1}^{\mu}\Gamma(\alpha_i+s)}{\prod_{i=m+1}^{\xi}\Gamma(\lambda_i+s)\prod_{i=n+1}^{\psi}\Gamma(\delta_i-s)}\left[\frac{\prod_{j=1}^{m}(N_j)^{-1}\prod_{j=1}^{m}M_j}{\prod_{j=m+1}^{q}(N_j)\prod_{j=m+1}^{p}(M_j)^{-1}}.z\right]^s ds$$

$$=(2\pi)^{\frac{2m+2n-p-q}{2}}$$

$$(2\pi)^{-\left(\frac{\eta+\mu+\xi+\psi}{2}\right)}\frac{\prod_{j=1}^{m}(N_j)^{-1/2}\prod_{j=1}^{m}(N_j)^{\frac{b_j}{N_j}}\prod_{j=1}^{n}(M_j)^{-1/2}\prod_{j=1}^{n}(M_j)^{\frac{1-a_j}{n_j}}}{\prod_{j=m+1}^{q}(N_j)^{\frac{1-b_j}{N_j}}\prod_{j=m+1}^{q}(N_j)^{-1/2}\prod_{j=n+1}^{p}(M_j)^{\frac{a_j}{M_j}}\prod_{j=m+1}^{q}(M_j)^{-1/2}}$$

$$G_{\xi,\psi}^{\eta,\mu}\left[\rho\ \middle|\ \begin{matrix}\alpha_1,\alpha_2,\ldots\alpha_\mu,\delta_{n+1},\ldots\delta_\psi,\\ \beta_1,\beta_2,\ldots\beta_\eta,\lambda_{m+1},\ldots\lambda_\xi\end{matrix}\right] \tag{2.32}$$

Special Cases

If we put r = 1 in I-function, then it reduces to I_H - function

$$I_{p_i,q_i:r}^{m,n}[z] = I_{p_i,q_i:r}^{m,n}\left[z\ \middle|\ \begin{matrix}(q_i,\alpha_i)_{1,n}\cdots & (q_{j1},\alpha_{j1})_{n+1,p}\\ (b_j,\beta_j)_{1,m}\cdots & (b_{j1},\beta_{j1})_{m+1,q}\end{matrix}\right]$$

$$= \frac{1}{2\pi i}\int_L t(s) z^s\, ds \tag{2.33}$$

where

$$t(s) = \frac{\prod_{j=1}^{m}\Gamma(b_j-\beta_j s)\,\Gamma(1-a_j+\alpha_j s)}{\sum_{i=1}^{r}\left\{\prod_{j=m+1}^{q_i}\Gamma(1-b_{j1}+\beta_{j1}s)\prod_{j=n+1}^{p_i}(a_{j1}-\alpha_{j1}s)\right\}} \tag{2.34}$$

By putting $r=1$ in (3.34) we get

$$I^{m,n}_{p_i,q_i:1}:1[z]=I^{m,n}_{p_i,q_i:1}\left[z\left|\begin{matrix}(q_i,\alpha_i)_{1,n}\cdots\cdots & (q_j,\alpha_j)_{n+1,p}\\ (b_j,\beta_j)_{1,m}\cdots\cdots & (b_j,\beta_j)_{m+1,q}\end{matrix}\right.\right]$$

$$=\frac{1}{2\pi i}\int_L\frac{\prod_{j=1}^{m}\Gamma(b_j-\beta_j s)\prod_{j=1}^{n}\Gamma(1-a_j+\alpha_j s)}{\sum_{i=1}^{1}\left\{\prod_{j=m+1}^{q_i}\Gamma(1-b_{j1}+\beta_{j1}s)\prod_{j=n+1}^{p_i}\Gamma(q_{j1}-\alpha_{j1}s)\right\}}z^s\,ds$$

$$=\frac{1}{2\pi i}\int_L\frac{\prod_{j=1}^{m}\Gamma(b_j-\beta_j s)\prod_{j=1}^{n}\Gamma(1-a_j+\alpha_j s)}{\prod_{j=m+1}^{q_i}\Gamma(1-b_j+\beta_j s)\prod_{j=n+1}^{q_i}\Gamma(a_j-\alpha_j)s}z^s \tag{2.35}$$

$$=\frac{1}{2\pi i}\int_L\frac{\prod_{j=1}^{m}\Gamma\left(N_j\left(\frac{b_j}{N_j}-s\right)\right)\prod_{j=1}^{n}\Gamma\left(M_j\left(\frac{1-q_j}{M_j}+s\right)\right)}{\prod_{j=m+1}^{q}\Gamma\left(N_j\left(\frac{1-b_j}{N_j}+s\right)\right)\prod_{j=n+1}^{p}\Gamma\left(M_j\left(\frac{q_j}{M_j}-s\right)\right)}z^s\,ds. \tag{2.36}$$

For $\alpha_j=M_j$, $\beta_j=N_j$

$$=\frac{1}{2\pi i}\int_L\frac{\prod_{j=1}^{m}\Gamma(N_jZ_j)\prod_{j=1}^{n}\Gamma(M_jY_j)}{\prod_{j=m+1}^{q}\Gamma(N_jY_j)\prod_{j=n+1}^{p}\Gamma(M_jX_j)}z^s ds \tag{2.37}$$

For $\frac{b_j}{N_j}-S=Z_j,\ \frac{1-a_j}{M_j}+s=\gamma_j,\ \frac{1-b_j}{N_j}+s=Y_j,\ \frac{a_j}{M_j}-s=X_j$ in (2.25),

By using formula,

$$\Gamma(mZ)=(2\pi)^{\frac{1-m}{2}}m^{z-1/2}\prod_{j=1}^{m}\left(Z+\frac{j-1}{m}\right), \tag{2.38}$$

$$Z\neq 0,\ -1/m,\ 2/m\cdots\cdots m=1,2,3.$$

Then,

$$\prod_{j=1}^{m}\Gamma(N_jZ_j)=\Gamma(N_1Z_1)\Gamma(N_2Z_2)\Gamma(N_3Z_3)\ldots\ldots\ldots\Gamma(N_mZ_m)$$

$$=\left[(2\pi)^{\frac{1-N_1}{2}}(N_1)^{Z_1-\frac{1}{2}}\prod_{j=1}^{N_1}\Gamma\left(Z_1+\frac{j-1}{N_1}\right)\right]$$

$$\left[(2\pi)^{\frac{1-N_2}{2}}(N_2)^{Z_2-\frac{1}{2}}\prod_{j=1}^{N_2}\Gamma\left(Z_2+\frac{j-1}{N_2}\right)\right]$$

$$\left[(2\pi)^{\frac{1-N_3}{2}}(N_3)^{Z_3-\frac{1}{2}}\prod_{j=1}^{N_3}\Gamma\left(Z_3+\frac{j-1}{N_3}\right)\right].$$

$$\left[(2\pi)^{\frac{1-N_m}{2}}(N_m)^{Z_m-\frac{1}{2}}\prod_{j=1}^{N_m}\Gamma\left(Z_m+\frac{j-1}{N_m}\right)\right]$$

$$=(2\pi)^{\left(\frac{1}{2}+\frac{1}{2}+\frac{1}{2}\ldots\ldots m\text{ times}\right)}.(2\pi)^{-\left[\frac{N_1+N_2+N_3+\ldots\ldots+N_m}{2}\right]}.(N_1.N_2.N_3\ldots\ldots N_m)^{-\frac{1}{2}}$$

$$\prod_{j=1}^{m}(N_j)^{Z_j}.\prod_{j=1}^{N_1}\Gamma\left(\frac{b_1}{N_1}-s+\frac{j-1}{N_1}\right).\prod_{j=1}^{N_2}\Gamma\left(\frac{b_2}{N_2}-s+\frac{j-1}{N_2}\right).$$

$$\prod_{j=1}^{N_3}\Gamma\left(\frac{b_3}{N_3}-s+\frac{j-1}{N_3}\right)\ldots\prod_{j=1}^{N_m}\Gamma\left(\frac{b_m}{N_m}-s+\frac{j-1}{N_m}\right)$$

$$=(2\pi)^{\frac{m}{2}}.(2\pi)^{-\left[\frac{N_1+N_2+N_3+\ldots+N_m}{2}\right]}.(N_1.N_2\ldots N_m)^{-\frac{1}{2}}.$$

$$N_1^{\left(\frac{b_1}{N_1}-s\right)}.N_2^{\left(\frac{b_2}{N_2}-s\right)}.N_3^{\left(\frac{b_3}{N_3}-s\right)}\ldots N_m^{\left(\frac{b_m}{N_m}-s\right)}.\prod_{j=1}^{N_1}\Gamma\left(\frac{b_1+j-1}{N_1}-s\right).$$

$$\prod_{j=1}^{N_2}\Gamma\left(\frac{b_2+j-1}{N_2}-s\right).\prod_{j=1}^{N_3}\Gamma\left(\frac{b_3+j-1}{N_3}-s\right)\ldots\prod_{j=1}^{N_m}\left(\frac{b_m+j-1}{N_m}-s\right).$$

$$=(2\pi)^{\frac{m}{2}}.(2\pi)^{\frac{-(N_1+N_2+N_3+\ldots\ldots+N_m)}{2}}.(N_1.N_2.N_3\ldots N_m)^{-\frac{1}{2}}.(N_1.N_2.N_3\ldots\ldots N_m)^{-}$$

$$N_1^{\frac{b_1}{N_1}}.N_2^{\frac{b_2}{N_2}}N_3^{\frac{b_3}{N_3}}\ldots\ldots\ldots\ldots N_m^{\frac{b_m}{N_m}}\prod_{j=1}^{N_1}\Gamma\left(\frac{b_1+j-1}{N_1}-s\right)$$

$$\prod_{j=1}^{N_2}\Gamma\left(\frac{b_2+j-1}{N_2}-s\right).\prod_{j=1}^{N_3}\Gamma\left(\frac{b_3+j-1}{N_3}-s\right)\ldots\prod_{j=1}^{N_m}\left(\frac{b_m+j-1}{N_m}-s\right)$$

$$=(2\pi)^{\frac{m}{2}}.(2\pi)^{\frac{-(N_1+N_2+N_3+\ldots\ldots+N_m)}{2}}.(N_1.N_2.N_3\ldots N_m)^{-\frac{1}{2}}$$

$$\prod_{j=1}^{m}(N_j)^{-s}N_j^{\frac{b_j}{N_j}}\prod_{i=1}^{N}\Gamma(\beta_i-s) \tag{2.39}$$

where $\beta_i = \dfrac{b_i + j - 1}{N_i}$

$N = N_1 + N_2 + N_3 + \ldots\ldots + N_m,\ j = 1$ to N.

$$\prod_{j=1}^{n} \Gamma(M_j \gamma_j) = \Gamma(M_1\gamma_1)\Gamma(M_2\gamma_2).\Gamma(M_3\gamma_3)\ldots\Gamma(M_n\gamma_n)$$

$$= \left[(2\pi)^{\frac{1-M_1}{2}} (M_1)^{\gamma_1 - 1/2} \prod_{j=1}^{M_1} \Gamma\left(\gamma_1 + \frac{j-1}{M_1}\right)\right].$$

$$\left[(2\pi)^{\frac{1-M_2}{2}} M_2^{\gamma_2 - 1/2} \prod_{j=1}^{M_2} \Gamma\left(\gamma_2 + \frac{j-1}{M_2}\right)\right].$$

$$\left[(2\pi)^{\frac{1-M_3}{2}} (M_3)^{\gamma_3 - 1/2} \prod_{j=1}^{M_3} \left(\gamma_3 + \frac{j-1}{M_3}\right)\right]$$

$$\left[(2\pi)^{\frac{1-M_n}{2}} (M_n)^{\gamma_n - 1/2} \prod_{j=1}^{M_n} \left(\gamma_n + \frac{j-1}{M_n}\right)\right]$$

$$= (2\pi)^{\left(1/2 + 1/2 + 1/2 \ldots\ldots n \text{ times}\right)}.$$

$$(2\pi)^{-\left[\frac{M_1+M_2+M_3+\ldots+M_n}{2}\right]} (M_1.M_2.M_3\ldots\ldots M_n)^{-1/2}$$

$$\left[M_1^{\frac{1-a_1}{M_1}} . M_2^{\frac{1-a_2}{M_2}} \ldots . M_n^{\frac{1-a_n}{M_n}}\right] (M_1.M_2.M_3\ldots\ldots M_n)^{+s}$$

$$\prod_{j=1}^{M_1} \Gamma\left(\frac{1-a_1}{M_1} + s + \frac{j-1}{M_1}\right) . \prod_{j=1}^{M_2} \Gamma\left(\frac{1-a_2}{M_2} + s + \frac{j-1}{M_2}\right).$$

$$\prod_{j=1}^{M_3} \Gamma\left(\frac{1-a_3}{M_3} + s + \frac{j-1}{M_3}\right) \ldots . \prod_{j=1}^{M_n} \Gamma\left(\frac{1-a_n}{M_n} + s + \frac{j-1}{M_n}\right).$$

$$= (2\pi)^{n/2} (2\pi)^{-\left[\frac{M_1+M_2+M_3+\ldots+M_n}{2}\right]} (M_1.M_2\ldots M_n)^{-1/2}$$

$$\prod_{j=1}^{n} (M_j)^{\frac{1-a_j}{M_j}} (M_j)^s.$$

$$\prod_{j=1}^{M_1}\Gamma\left(\frac{j-a_1}{M_1}+s\right).\prod_{j=1}^{M_2}\Gamma\left(\frac{j-a_2}{M_2}+s\right).$$

$$\prod_{j=1}^{M3}\Gamma\left(\frac{j-a_3}{M_3}+s\right)\dots\prod_{j=1}^{M_n}\Gamma\left(\frac{j-a_n}{M_n}+s\right)$$

$$=(2\pi)^{n/2}(2\pi)^{-\frac{[M_1+M_{2+\dots}M_n]}{2}}.(M_1.M_2\dots M_n)^{-1/2}$$

$$\prod_{j=1}^{n}(M_j)^{\frac{1-a_j}{M_j}}(M_j)^s$$

$$\prod_{i=1}^{M}\Gamma(\alpha_i+s) \tag{2.40}$$

where $\alpha_i=\dfrac{j-a_i}{M_i}+s$

$$M=M_1+M_2+M_3+\dots.+M_n$$

$$\prod_{j-M+1}^{q}\Gamma(N_jY_j)=\Gamma(N_{m+1}Y_{m+1}).\ \Gamma(N_{m+2}Y_{m+2}).\ \Gamma(N_{m+3}Y_{m+3})\dots\dots.\Gamma(N_qY_q)$$

$$=\left[(2\pi)^{\frac{1-N_{m+1}}{2}}(N_{m+1})^{Y_{m+1}-1/2}\prod_{j=1}^{N_{m+1}}\Gamma\left(Y_{m+1}+\frac{j-1}{N_{m+1}}\right)\right].$$

$$\left[(2\pi)^{\frac{1-N_{m+2}}{2}}(N_{m+2})^{Y_{m+2}-1/2}\prod_{j=1}^{N_{m+2}}\Gamma\left(Y_{m+2}+\frac{j-1}{N_{m+2}}\right)\right].$$

$$\left[(2\pi)^{\frac{1-N_q}{2}}(N_q)^{Y_q-1/2}\prod_{j=1}^{N_q}\Gamma\left(Y_q+\frac{j-1}{N_q}\right)\right]$$

$$=(2\pi)^{\frac{q-m}{2}}.\ (2\pi)^{-\frac{[N_{m+1}+N_{m+2}+\dots+N_q]}{2}}.\prod_{j=m+1}^{q}(N_j)^{\frac{1-b_j}{N_j}}(N_j)^s\prod_{j=m+1}^{q}(N_j)^{-1/2}$$

$$\prod_{j=1}^{N_{m+1}}\Gamma\left(\frac{1-b_{m+1}}{N_{m+1}}+s+\frac{j-1}{N_{m+1}}\right).\prod_{j=1}^{N_{m+2}}\Gamma\left(\frac{1-b_{m+2}}{N_{m+2}}+s+\frac{j-1}{N_{m+2}}\right)$$

$$\prod_{j=1}^{N_{m+3}}\Gamma\left(\frac{1-b_{m+3}}{N_{m+3}}+s+\frac{j-1}{N_{m+3}}\right)\dots..\prod_{j=1}^{N_q}\Gamma\left(\frac{1-b_q}{N_q}+s+\frac{j-1}{N_q}\right)$$

$$= (2\pi)^{\frac{q-m}{2}} (2\pi)^{-\frac{[N_{m+1}+N_{m+2}+\ldots+N_q]}{2}} . \prod_{j=m+1}^{q} (N_j)^{\frac{1-b_j}{N_j}} (N_j)^s \prod_{j=m+1}^{q} (N_j)^{-\frac{1}{2}} .$$

$$\prod_{j=1}^{N_{m+1}} \Gamma\left(\frac{j-b_{m+1}}{N_{m+1}}+s\right) \prod_{j=1}^{N_{m+2}} \Gamma\left(\frac{j-b_{m+2}}{N_{m+2}}+s\right) \ldots\ldots \prod_{j=1}^{N_q} \Gamma\left(\frac{j-b_q}{N_q}+s\right)$$

$$= (2\pi)^{\frac{q-m}{2}} (2\pi)^{-\frac{[N_{m+1}+N_{m+2}+\ldots+N_q]}{2}} . \prod_{j=m+1}^{q} (N_j)^{\frac{1-b_j}{N_q}} (N_j)^s \prod_{j=m+1}^{q} (N_j)^{-\frac{1}{2}} \prod_{i=m+1}^{Q} \Gamma\left(\frac{j-b_i}{N_i}+s\right)$$

$$= (2\pi)^{\frac{q-m}{2}} (2\pi)^{-\frac{[N_{m+1}+N_{m+2}+\ldots N_q]}{2}} . \prod_{j=m+1}^{q} (N_j)^{\frac{1-b_j}{N_a}} (N_j)^s \prod_{j=m+1}^{q} (N_j)^{-\frac{1}{2}} \prod_{i=m+1}^{Q} \Gamma(\lambda_i+s) \tag{2.41}$$

where

$$Q = N_{m+1} + N_{m+1} + N_{m+3} + \ldots + N_q$$

$$\lambda_i = \Gamma\left(\frac{j-b_i}{N_i}+s\right)$$

$$\prod_{j=n+1}^{P} \Gamma(M_j X_j) = \Gamma(M_{n+1}X_{n+1})\Gamma(M_{n+2}X_{n+2})\Gamma(M_{n+3}X_{n+3}) \ldots\ldots \Gamma(M_P X_P)$$

$$= \left[(2\pi)^{\frac{1-M_{n+1}}{2}} (M_{n+1})^{X_{n+1}-1/2} \prod_{j=1}^{M_{n+1}} \Gamma\left(X_{n+1}+\frac{j-1}{M_{n+1}}\right)\right].$$

$$\left[(2\pi)^{\frac{1-M_{n+2}}{2}} (M_{n+2})^{X_{n+2}-1/2} \prod_{j=1}^{M_{n+2}} \Gamma\left(X_{n+2}+\frac{j-1}{M_{n+2}}\right)\right].$$

$$\left[(2\pi)^{\frac{1-M_{n+3}}{2}} (M_{n+3})^{X_{n+3}-1/2} \prod_{j=1}^{M_{n+3}} \Gamma\left(X_{n+3}+\frac{j-1}{M_{n+3}}\right)\right]$$

$$\left[(2\pi)^{\frac{1-M_P}{2}} (M_P)^{X_P-1/2} \prod_{j=1}^{M_P} \Gamma\left(X_P+\frac{j-1}{M_P}\right)\right]$$

$$= (2\pi)^{\frac{p-n}{2}} (2\pi)^{-\frac{[M_{n+1}+M_{n+2}+\ldots+M_p]}{2}} \prod_{j=n+1}^{p} (M_j)^{\frac{a_j}{M_j}} (M_j)^{-s}$$

$$\prod_{j=n+1}^{R} (M_j)^{-\frac{1}{2}} \prod_{i=n+1}^{p} \Gamma\left(\frac{a_i+j-1}{M_i}-s\right)$$

where $R = M_{n+1} + M_{n+2} + \cdots + M_p$

$$\delta_i = \frac{a_i + j - 1}{M_i}$$

$$= (2\pi)^{\frac{p-n}{2}} (2\pi)^{-\frac{[M_{n+1}+M_{n+2}+\ldots+M_p]}{2}} . \prod_{j=n+1}^{p} (m_j)^{\frac{a_j}{M_j}} (M_j)^{-s} \prod_{j=m+1}^{p} (M_j)^{-1/2} \prod_{i=n+1}^{R} \Gamma(\delta_i - s) \tag{2.42}$$

Putting in Equation (2.35), we get

$$I_{p,q;1}^{m,n}[Z] = \int_L (2\pi)^{m/2} (2\pi)^{-\frac{[N_1+N_2+\ldots+N_m]}{2}} \times \prod_{j=1}^{m} (N_j)^{-1/2} . \prod_{j=1}^{m} (N_j)^{\frac{b_j}{N_j}} (2\pi)^{n/2} (2\pi)^{-\frac{[M_1+M_2+\ldots+M_n]}{2}} \prod_{j=1}^{n} (M_j)^{-1/2}$$

$$\times \frac{\prod_{j=1}^{n} (M_j)^{\frac{1-a_j}{n_j}} \prod_{i=1}^{N} \Gamma(\beta_i - s) \prod_{i=1}^{M} \Gamma(\alpha_i + s) \prod_{j=1}^{m} (N_j)^{-s} \prod_{j=1}^{n} (M_j)^{s}}{(2\pi)^{\frac{q-m}{2}} (2\pi)^{-\frac{[N_{m+1}+N_{m+2}+\ldots+N_q]}{2}} \prod_{j=m+1}^{q} (N_j)^{\frac{1-b_j}{N_j}} \prod_{j=m+1}^{q} (N_j)^{-1/2} (2\pi)^{\frac{p-n}{2}}}$$

$$\times \frac{1}{(2\pi)^{-\frac{[M_{n+1}+M_{n+2}+\ldots+M_p]}{2}} \prod_{j=m+1}^{p} (M_j)^{\frac{a_j}{M_j}} \prod_{j=m+1}^{q} (M_j)^{-1/2} \prod_{i=m+1}^{Q} \Gamma(\lambda_i + s)}$$

$$\times \frac{1}{\prod_{i=n+1}^{R} \Gamma(\delta_i - s) \prod_{j=m+1}^{q} (N_j)^{s} \prod_{j=n+1}^{p} (M_j)^{-s}} z^s ds$$

$$= (2\pi)^{\frac{m+n-q+m-p+n}{2}} (2\pi)^{-\frac{[(N_1+N_2+\ldots+N_m)+(M_1+M_2+\ldots M_n)+(N_{m+1}+N_{m+2}+\ldots+N_q)+(M_{n+1}+M_{n+2}+\ldots+M_p)]}{2}}$$

$$\times \frac{\prod_{j=1}^{m} (N_j)^{-1/2} \prod_{j=1}^{m} (N_j)^{\frac{b_j}{N_j}} \prod_{j=1}^{n} (M_j)^{-1/2} \prod_{j=1}^{n} (M_j)^{\frac{1-a_j}{n_j}}}{\prod_{j=m+1}^{q} (N_j)^{\frac{1-b_j}{N_j}} \prod_{j=m+1}^{q} (N_j)^{-1/2} \prod_{j=m+1}^{p} (M_j)^{\frac{a_j}{M_i}} \prod_{j=m+1}^{q} (M_j)^{-1/2}}$$

$$\times \int_L \frac{\prod_{i=1}^{N} \Gamma(\beta_i - s) \prod_{i=1}^{M} \Gamma(\alpha_i + s) \prod_{j=1}^{m} (N_j)^{-s} \prod_{j=1}^{n} (M_j)^{s}}{\prod_{i=m+1}^{Q} \Gamma(\lambda_i + s) \prod_{i=n+1}^{R} \Gamma(\delta_i - s) \prod_{j=m+1}^{q} (N_j)^{s} \prod_{j=n+1}^{p} (M_j)^{-s}} z^s ds$$

where $\eta = \sum_{r=1}^{m} N_r, \ \mu = \sum_{r=1}^{n} M_r$

$$\xi = \sum_{r=m+1}^{q} Q_r, \psi = \sum_{r=n+1}^{p} R_r$$

$$= (2\pi)^{\frac{2m+2n-p-q}{2}} (2\pi)^{-\left(\frac{\eta+\mu+\xi+\psi}{2}\right)} \frac{\prod_{j=1}^{m}(N_j)^{-1/2}\prod_{j=1}^{m}(N_j)^{\frac{b_j}{N_j}}\prod_{j=1}^{n}(M_j)^{-1/2}\prod_{j=1}^{n}(M_j)^{\frac{1-a_j}{n_j}}}{\prod_{j=m+1}^{q}(N_j)^{\frac{1-b_j}{N_j}}\prod_{j=m+1}^{q}(N_j)^{-1/2}\prod_{j=n+1}^{p}(M_j)^{\frac{a_j}{M_j}}\prod_{j=m+1}^{q}(M_j)^{-1/2}}$$

$$\times \frac{1}{2\pi i}\int \frac{\prod_{i=1}^{\eta}\Gamma(\beta_{i-s})\prod_{i=1}^{\mu}\Gamma(\alpha_i+s)}{\prod_{i=m+1}^{\xi}\Gamma(\lambda_i+s)\prod_{i=n+1}^{\psi}\Gamma(\delta_i-s)}\left[\frac{\prod_{j=1}^{m}(N_j)^{-1}\prod_{j=1}^{m}M_j}{\prod_{j=m+1}^{q}(N_j)\prod_{j=m+1}^{p}(M_j)^{-1}}.z\right]^s ds$$

$$= (2\pi)^{\frac{2m+2n-p-q}{2}} (2\pi)^{-\left(\frac{\eta+\mu+\xi+\psi}{2}\right)} \frac{\prod_{j=1}^{m}(N_j)^{-1/2}\prod_{j=1}^{m}(N_j)^{\frac{b_j}{N_j}}\prod_{j=1}^{n}(M_j)^{-1/2}\prod_{j=1}^{n}(M_j)^{\frac{1-a_j}{n_j}}}{\prod_{j=m+1}^{q}(N_j)^{\frac{1-b_j}{N_j}}\prod_{j=m+1}^{q}(N_j)^{-1/2}\prod_{j=n+1}^{p}(M_j)^{\frac{a_j}{M_j}}\prod_{j=m+1}^{q}(M_j)^{-1/2}}$$

$$G_{\xi,\psi}^{\eta,\mu}\left[\rho \middle| \begin{matrix} \alpha_1, \alpha_2, \ldots\ldots \alpha_\mu, \delta_{n+1}, \ldots\ldots \delta_\psi \\ \beta_1, \beta_2, \ldots\ldots \beta_\eta, \lambda_{m+1}, \ldots\ldots \lambda_\xi \end{matrix}\right], \tag{2.43}$$

The transformation relations derived above can generate new properties of the I-function and other higher order generalized hypergeometric functions.

3 Certain Integrals Involving I-Function

In this chapter, integrals involving I-function in the integrand along with other special functions have been evaluated. Derivations are given for few selected cases. Remaining integrals can also be evaluated on the same lines.

3.1 INTEGRALS WITH RESPECT TO THE PARAMETERS

The following four integrals associated with the I-function are presented in which integration is performed with respect to the parameters [35].

i.

$$I_1 \equiv \frac{1}{2\pi i}\int_L \frac{\Gamma(b-x)\Gamma(c-x)}{\Gamma(d-x)} e^{\pm i\pi x} I^{m,n+1}_{p_i+1,q_i+1:r}$$

$$\times\left[y \left| \begin{matrix} (1-a-x,h),(a_j,\alpha_j)_{1,n}; & (a_{ji},\alpha_{ji})_{n+1,p_i} \\ (b_j,\beta_j)_{1,m}; & (b_{ji},\beta_{ji})_{m+1,q_i} \end{matrix} \right. \right] dx \tag{3.1}$$

$$= \frac{e^{\pm i\pi x}}{\Gamma(d-b)\Gamma(d-c)} I^{m,n+1}_{p_i+2,q_i+1:r}$$

$$\times\left[ye^{\pm i\pi x} \left| \begin{matrix} (1-a-b,h),(1-a-c,h),(a_j,\alpha_j)_{1,n}; & (a_{ji},\alpha_{ji})_{n+1,p_i} \\ (d-a-b-c,h), \quad (b_j,\beta_j)_{1,m}; & (b_{ji},\beta_{ji})_{m+1,q_i} \end{matrix} \right. \right]$$

Provided

$$\mathrm{Re}\left[d-a-b-c-h\frac{a_j-1}{a_j} \right] > 0 \text{ for } j=1,\ldots.n.$$

The contour integral form of the I-function defined in (1.23) is put on the left-hand side of (3.1) so that

$$I_1 = (2\pi i)^{-1}\int_L \frac{\Gamma(b-x)\Gamma(c-x)}{\Gamma(d-x)} e^{\pm i\pi x}(2\pi i)^{-1}\int_L y^{\xi}\Gamma(a+x-h\xi)\phi(\xi)d\xi\, dx$$

$$= (2\pi i)^{-1}\int_L y^{\xi}\phi(\xi)d\xi(2\pi i)^{-1}\int_L \frac{\Gamma(b-x)\Gamma(c-x)\Gamma(a+x-h\xi)}{\Gamma(d-x)} e^{\pm i\pi x}\, dx$$

DOI: 10.1201/9781003517757-3

Changing the order of integration we get

$$I_1 = \frac{1}{2\pi i}\int_L \frac{y^{\xi}\phi(\xi)\Gamma(a+b-h\xi)\Gamma(a+c-h\xi)\Gamma(d-a-b-c+h\xi)}{\Gamma(d-b)\Gamma(d-c)} e^{\pm\pi(a-h\xi)d\xi}$$

Expressing in terms of the I-function we arrive at the result (3.1).

ii.

$$I_2 \equiv (2\pi i)^{-1}\int_L \Gamma(a+x)\,\Gamma(b-x)\Gamma(c-x)e^{\pm i\pi x} \times$$

$$I^{m,n}_{p_i,q_i:r}\left[y \left| \begin{matrix} (a_j,\alpha_j)_{1,n}; & (a_{ji},\alpha_{ji})_{n+1,p_i} \\ (b_j,\beta_j)_{1,m}; & (b_{ji},\beta_{ji})_{m+1,q_i} \end{matrix} \right. \right] dx = e^{\pm i\pi\alpha}\Gamma(a+b)\Gamma(b+a) I^{m,n+1}_{p_i+2,q_i+1:r} \quad (3.2)$$

$$\left[q \left| \begin{matrix} (1-d+a+b+c,h),(1-a-c,h),(a_j,\alpha_j)_{1,n}; & (a_{ji},\alpha_{ji})_{n+1,p_i} \\ (b_j,\beta_j)_{1,m};(b_{ji},\beta_{ji})_{m+1,q_i}\ (1-d+b,h) & (1-d+c,h) \end{matrix} \right. \right]$$

provided

$$\operatorname{Re}\left[d-a-b-c+\min_{1\le j\le m}\left(\frac{b_j}{\beta_j}\right)\right] > 0 \ \ for\ \ j=1,....m.$$

iii.

$$I_3 \equiv (2\pi i)^{-1}\int_L \Gamma(a+x)\,\Gamma(b-x)e^{\pm i\pi x}$$

$$\times I^{m,n+1}_{p_i+1,q_i+1:r}\left[y \left| \begin{matrix} (1-c+x,k),\ (a_j,\alpha_j)_{1,n}; & (a_{ji},\alpha_{ji})_{n+1,p_i} \\ (b_j,\beta_j)_{1,m};(b_{ji},\beta_{ji})_{m+1,q_i}, & (1-d+x,h) \end{matrix} \right. \right] dx = e^{\pm i\pi\alpha}\Gamma(a+b)$$

$$I^{m,n+2}_{p_i+2,q_i+2:r}\left[q \left| \begin{matrix} (1-a-c,k),(1-d+a+b+c,h-k),(a_j,\alpha_j)_{1,n}; & (a_{ji},\alpha_{ji})_{n+1,p_i} \\ (b_j,\beta_j)_{1,m};(b_{ji},\beta_{ji})_{m+1,q_i}\ (1-d+b,h) & (1-d+c,h-k) \end{matrix} \right. \right] \quad (3.3)$$

provided

$$\operatorname{Re}\left[d-a-b-c+(h-k)\min_{1\le j\le m}\left(\frac{b_j}{\beta_j}\right)\right] > 0 \text{ for } j=1,\ldots.m;(h,k)<0.$$

iv.

$$I_4 \equiv (2\pi i)^{-1} \int_L \Gamma(a+x)\,\Gamma(b-x)e^{\pm i\pi x}$$

$$\times I^{m,n+1}_{p_i+1,q_i:r}\left[y \left| \begin{matrix} (1-c+x,h),\,(a_j,\alpha_j)_{1,n}; & (a_{ji},\alpha_{ji})_{n+1,p_i},(d-x,k) \\ (b_j,\beta_j)_{1,m}; & (b_{ji},\beta_{ji})_{m+1,q_i} \end{matrix} \right. \right] dx = e^{\pm i\pi a}\Gamma(a+b)$$

$$I^{m+1,n+1}_{p_i+3,q_i+3:r}\left[q \left| \begin{matrix} (1-a-c,h),(a_j,\alpha_j)_{1,n};(a_{ji},\alpha_{ji})_{n+1,p_i},(d-b,k), & (d-c,k+h) \\ (d-a-b-c,h+k)(b_j,\beta_j)_{1,m}; & (b_{ji},\beta_{ji})_{m+1,q_i} \end{matrix} \right. \right] \tag{3.4}$$

provided

$$\mathrm{Re}\left[d-a-b-c+(h+k)\frac{a_j-1}{\alpha_j} \right] > 0 \text{ for } j=1,\ldots.n.$$

3.1.1 Special Cases

The special cases of (3.1) to (3.4) for $r=1$ are given as follows:

$$(2\pi i)^{-1}\int_L \frac{\Gamma(b-x)\Gamma(c-x)}{\Gamma(d-x)} e^{\pm i\pi a} H^{m,n+1}_{p+1,q}\left[y \left| \begin{matrix} (1-a-x,h), & (a_j,\alpha_j)_{1,p} \\ (b_j,\beta_j)_{1,q} & \end{matrix} \right. \right] dx$$

$$= \frac{e^{\pm i\pi a}}{\Gamma(d-b)\Gamma(d-c)} H^{m+1,n+2}_{p+2,q+2}\left[ye^{\pm i\pi h} \left| \begin{matrix} (1-a-b,h),\ (1-a-c,h), & (a_j,\alpha_j)_{1,p} \\ (d-a-b-c,h), & (b_j,\beta_j)_{1,q} \end{matrix} \right. \right] \tag{3.5}$$

provided

$$\psi = \sum_{i=1}^{n}(\alpha_i) - \sum_{i-u+1}^{p}(\alpha_i) + \sum_{j=1}^{m}(\beta_j)\sum_{m+1}^{q}(\beta_j) > 0$$

$$|\arg y| \left\langle \psi\frac{\pi}{2},\ \mathrm{Re}\left[d-a-b-c-h\frac{a_j-1}{a_j} \right] \right\rangle 0 \text{ for } j=1,\ldots.m$$

Also

$$(2\pi i)^{-1}\int_L \Gamma(a+x)\,\Gamma(b+x)\Gamma(c-x)e^{\pm i\pi x}H_{p+1,q}^{m,n+1}\left[y\left|\begin{matrix}(a_j,\alpha_j)_{1,p} & \\ (b_j,\beta_j)_{1,q} & (1-d+x,h)\end{matrix}\right.\right]dx$$

$$=e^{+i\pi a}\Gamma(a+b)\Gamma(c+a)$$

$$H_{p+1,q+1}^{m,n+1}\left[q\left|\begin{matrix}(1-d+a+b+c,h); & (a_j,\alpha_j)_{1,p}\\ (b_j,\beta_j)_{1,q},(1-d+b,h) & (1+d+c,h)\end{matrix}\right.\right] \tag{3.6}$$

provided $\psi<0$,

$$|\arg y|\left\langle \psi\frac{\pi}{2},\ \text{Re}\left[d-a-b-c+h\frac{b_j}{\beta_j}\right]\right\rangle 0,\ \text{for } j=1,....m.$$

The next result is

$$(2\pi i)^{-1}\int_L \Gamma(a+x)\,\Gamma(b+x)e^{\pm i\pi x}H_{p+1,q+1}^{m,n+1}\left[y\left|\begin{matrix}(1-c+x,k), & (a_j,\alpha_j)_{1,p}\\ (b_j,\beta_j)_{1,q}, & (1-d+x,h)\end{matrix}\right.\right]dx$$

$$=e^{\pm i\pi\alpha}\Gamma(a+b)$$

$$H_{p+2,q+2}^{m,n+2}\left[y\left|\begin{matrix}(1-q+c,k),(1-d+a+b+c,h-k), & (a_j,\alpha_j)_{1,p}\\ (b_j,\beta_j)_{1,q},(1-d+b,h) & (1-d+c,h-k)\end{matrix}\right.\right] \tag{3.7}$$

provided $\psi>0$,

$$|\arg y|\left\langle \psi\frac{\pi}{2},\ \text{Re}\left[d-a-b-c+(h-k)\frac{b_j}{\beta_j}\right]\right\rangle 0,\ \text{ as } j=1,....m \text{ with } h>0,\ k>0.$$

$$(2\pi i)^{-1}\int_L \Gamma(a+x)\,\Gamma(b-x)e^{\pm i\pi x}\times H_{p+2,q}^{m,n+1}\left[y\left|\begin{matrix}(1-c+x,h),(a_j,\alpha_j)_{1,p}, & (d-x,k)\\ (b_j,\beta_j)_{1,q}; & \end{matrix}\right.\right]dx$$

$$=e^{\pm i\pi a}\Gamma(a+b)$$

$$H_{p+3,q+1}^{m+1,n+1}\left[y\left|\begin{matrix}(1-a-c,h),(a_j,\alpha_j)_{1,p},(d-b,k), & (d-c,k+h)\\ (d-a-b-c,h+k)\ , & (b_j,\beta_j)_{1,q}\end{matrix}\right.\right] \tag{3.8}$$

provided $\psi > 0$,

$$\left|\arg y\right| \left\langle \psi \frac{\pi}{2}, \operatorname{Re}\left[d-a-b-c+(h+k)\frac{a_j - 1}{\alpha_j} \right] \right\rangle 0, \text{ for } j = 1,\dots.n.$$

In the integral, we put $r = 1, m = 0, n = p_1 = q_1 = h = a_1 = \beta_1 = 1$

And replace y by $\frac{1}{y}$ to obtain the result of Gupta and Jain [2] in the following form and replace a by $2m-a+1$, b_1 *by* $k+m-a+\frac{1}{2}$:

$$(2\pi i)^{-1} \int_L \frac{\Gamma(b-x)\Gamma(c-x)}{\Gamma(d-x)} e^{\pm i\pi x} y^{\frac{x}{2}} W_{k+\frac{1}{2}, m+\frac{1}{2}x}(y)\, dx$$

$$= \frac{y^{m++c+d+\frac{1}{2}} e^{\pm i\pi(b+c-d)+\frac{1}{2}y}}{\Gamma(d-b)\Gamma(d-c)} E\left[\begin{matrix} d-b, d-c, d-b-c-2m : \\ d-b-c-k-m\, {}^{l}\!/_{2}; -y \end{matrix} \right] \tag{3.9}$$

Provided $\operatorname{Re}(d-b-c-2m) > 0$. E denotes MacRobert's E-function.

In a similar manner, the following known result can be obtained

$$(2\pi i)^{-1} \int_L \frac{\Gamma(a+x)\Gamma(b-x)\Gamma(c-x)}{\Gamma(d-x)} e^{\pm i\pi x} {}_pF_{q+1}\left[\begin{matrix} a_1, a, \dots. d_p; \\ d-x-b_1, b_2, \dots. b_q; -y \end{matrix} \right] dx$$

$$dx$$

$$= e^{\pm i\pi a} \frac{\Gamma(a+b)\Gamma(a+c)\Gamma(d-a-b-c)}{\Gamma(d-b)\Gamma(d-c)} \underset{p+1}{F_{q+2}} \left[\begin{matrix} a_1, a, \dots. a_p, d-a-b-c; \\ b_1, b_2, \dots. b_q, d-b, d-c; -y \end{matrix} \right] \tag{3.10}$$

provided $\operatorname{Re}(d-a-b-c) > 0$.

3.2 INTEGRALS WITH ALGEBRAIC FUNCTIONS

Using simple properties of beta and Gamma functions, the following integrals can be obtained

v.

$$I_5 \equiv \int_0^1 y^{-\rho}(1-y)^{\rho-\sigma-1} I^{m,n}_{p_i,q_i:r}\left[xy \middle| \begin{matrix} (a_j,\alpha_j)_{1,n}; & (a_{ji},\alpha_{ji})_{n+1,p_i} \\ (b_j,\beta_j)_{1,m}; & (b_{ji},\beta_{ji})_{m+1,q_i} \end{matrix} \right] dy$$

$$= \Gamma(\rho-\sigma) I^{m,n+1}_{p_i+1,q_i+1:r}\left[\alpha \middle| \begin{matrix} (\rho,1), \ (a_j,\alpha_j)_{1,n}; & (a_{ji},\alpha_{ji})_{n+1,p_i} \\ (b_j,\beta_j)_{1,m}; & (b_{ji},\beta_{ji})_{m+1,q_i}, (\sigma,1) \end{matrix} \right] \tag{3.11}$$

This can be proved by replacing the I-function by its definition (1.23) interchanging the order of integrals and using the definition of beta function.

vi.

$$I_6 \equiv \int_0^1 x^{\rho-1}(1-x)^{\sigma-1} I^{m,n}_{p_i,q_i:r}\left[\alpha x \left| \begin{matrix} (a_j,\alpha_j)_{1,n}; & (a_{ji},\alpha_{ji})_{n+1,p_i} \\ (b_j,\beta_j)_{1,m}; & (b_{ji},\beta_{ji})_{m+1,q_i} \end{matrix} \right.\right] dx$$

$$= \Gamma(\sigma) I^{m,n+1}_{p_i+1,q_i+1:r}\left[\alpha \left| \begin{matrix} (1-\rho,1),\ (a_j,\alpha_j)_{1,n}; & (a_{ji},\alpha_{ji})_{n+1,p_i} \\ (b_j,\beta_j)_{1,m}; & (b_{ji},\beta_{ji})_{m+1,q_i},\ (1-\rho-\sigma,1) \end{matrix} \right.\right] \tag{3.12}$$

vii.

$$I_7 \equiv \int_1^\infty x^{-\rho}(x-1)^{\sigma-1} I^{m,n}_{p_i,q_i:r}\left[\alpha x \left| \begin{matrix} (a_j,\alpha_j)_{1,n}; & (a_{ji},\alpha_{ji})_{n+1,p_i} \\ (b_j,\beta_j)_{1,m}; & (b_{ji},\beta_{ji})_{m+1,q_i} \end{matrix} \right.\right] dx$$

$$= \Gamma(\sigma) I^{m+1,n}_{p_i,q_i+1:r}\left[\alpha \left| \begin{matrix} (a_j,\alpha_j)_{1,n}; & (a_{ji},\alpha_{ji})_{n+1,p_i},\ (\rho,1) \\ (1-\rho-\sigma,1),\ (b_j,\beta_j)_{1,m}; & (b_{ji},\beta_{ji})_{m+1,q_i} \end{matrix} \right.\right] \tag{3.13}$$

viii.

$$I_8 \equiv \int_1^\infty x^{\rho-1}(x+\beta)^{-\sigma} I^{m,n}_{p_i,q_i:r}\left[\alpha x \left| \begin{matrix} (a_j,\alpha_j)_{1,n}; & (a_{ji},\alpha_{ji})_{n+1,p_i} \\ (b_j,\beta_j)_{1,m}; & (b_{ji},\beta_{ji})_{m+1,q_i} \end{matrix} \right.\right] dx \tag{3.14}$$

$$= \frac{\beta^{\rho-\sigma}}{\Gamma(\sigma)} I^{m+1,n+1}_{p_i+1,q_i+1:r}\left[\alpha\beta \left| \begin{matrix} (1-\rho,1),(a_j,\alpha_j)_{1,n}; & (a_{ji},\alpha_{ji})_{n+1,p_i} \\ (\rho-\sigma,1),\ (b_j,\beta_j)_{1,m}; & (b_{ji},\beta_{ji})_{m+1,q_i} \end{matrix} \right.\right]$$

ix.

$$I_9 \equiv \int_{-1}^1 (1-x)^{\rho}(1+x)^{\sigma} I^{m,n}_{p_i,q_i:r}\left[z(1-x)^{\mu} \left| \begin{matrix} (a_j,\alpha_j)_{1,n}; & (a_{ji},\alpha_{ji})_{n+1,p_i} \\ (b_j,\beta_j)_{1,m}; & (b_{ji},\beta_{ji})_{m+1,q_i} \end{matrix} \right.\right] dx$$

$$= 2^{\rho+\sigma+1}\Gamma(\sigma) I^{m,n+1}_{p_i+1,q_i+1:r}\left[2^{\mu} z \left| \begin{matrix} (-\rho,\mu),(a_j,\alpha_j)_{1,n}; & (a_{ji},\alpha_{ji})_{n+1,p_i} \\ (b_j,\beta_j)_{1,m}; & (b_{ji},\beta_{ji})_{m+1,q_i},(-1-\rho-\sigma,\mu) \end{matrix} \right.\right] \tag{3.15}$$

All the above integrals exist under the conditions given in (1.31) and otherwise.

3.3 INTEGRALS WITH ORTHOGONAL POLYNOMIALS AND BESSEL FUNCTIONS

Integrals of the I-function multiplied with Jacobi polynomials, Bessel functions, and generalized associated Legendre functions are given here. With the help of orthogonality of the functions involved, these integrals are further exploited to yield various infinite expansions involving I-functions. Corresponding integrals and expansions for Legendre polynomials are shown to follow as special cases.

The finite integrals involving Jacobi polynomials and the I-function of one variable will be presented in this section.

x.

$$I_{10} \equiv \int_{-1}^{1} x^{-\lambda}(1-x)^{\alpha}(1+x)^{\delta} P_n^{\alpha,\beta}(x)$$

$$I_{p_i,q_i:r}^{m,n}\left[z(1+x)^h \left| \begin{matrix} (a_j,\alpha_j)_{1,n}; & (a_{ji},\alpha_{ji})_{n+1,p_i} \\ (b_j,\beta_j)_{1,m}; & (b_{ji},\beta_{ji})_{m+1,q_i} \end{matrix} \right.\right] dx$$

$$= \frac{(-1)^n 2^{\alpha+\delta+1}\Gamma(\alpha+n+1)}{n!}\sum_{k=0}^{\infty}\frac{(-\lambda)_k 2^k}{k!} I_{p_i+2,q_i+2:r}^{m,n+2}$$

$$\left[2^h z \left| \begin{matrix} (\beta-\delta-k,h),(-\delta-k,h)(a_j,\alpha_j)_{1,n}; & (a_{ji},\alpha_{ji})_{n+1,p_i} \\ (b_j,\beta_j)_{1,m};(b_{ji},\beta_{ji})_{m+1,q_i}, & (\beta-\delta+n-k,h),(-\delta-\alpha-n-k-1,h) \end{matrix} \right.\right]$$

(3.16)

provided $\text{Re}(\lambda) > -1$, $\text{Re}\left[\delta + h\min\left(\frac{b_j}{\beta_j}\right)\right] > -1$, $j = 1,2,\ldots,m$; $|\arg z| < \frac{1}{2}\pi B_i$, $j = 1,2,\ldots,r$

where

$$A_i = \sum_{j=1}^{p_i} a_j - \sum_{j=1}^{q_i} b_j,\ B_i = \sum_{j=1}^{n}\alpha_j - \sum_{j=n+1}^{p_i}\alpha_j + \sum_{j=1}^{m}\beta_j - \sum_{j=m+1}^{q_i}\beta_j$$

To evaluate I_{10} we substitute the value of the I-function from (1.31) on the left-hand side of (3.16) to obtain

$$I_{10} = \int_{-1}^{1} x^{-\lambda}(1-x)^{\alpha}(1+x)^{\delta} P_n^{\alpha,\beta}(x)\left\{\left(\frac{1}{2\pi i}\right)\int_L \phi(\xi) z^{\xi}(1+x)^{h\xi}\, d\xi\right\} dx$$

Interchanging the order of integration, which is justified since the contour integral is absolutely convergent and x-integral is absolutely convergent if $\operatorname{Re}(\lambda) > -1, \operatorname{Re}\left[\delta + h \min\left(\frac{b_j}{\beta_j}\right)\right] > -1,\ j = 1, 2, ..., m,$ the above expression takes the form

$$I_{10} = \frac{1}{2\pi i}\int_L \phi(\xi) z^{\xi} \left\{ \int_{-1}^{1} x^{\lambda}(1-x)^{\alpha}(1+x)^{\delta+h\xi} P_n^{\alpha,\beta}(x)\,dx \right\} d\xi$$

We evaluate the inner integral with the help of the result [32]

$$\int_{-1}^{1} x^{\lambda}(1-x)^{\alpha}(1+x)^{\delta} P_n^{\alpha,\beta}(x)\,dx$$

$$= \frac{(-1)^n 2^{\alpha+\delta+1}\Gamma(\delta+1)\Gamma(\alpha+n+1)\Gamma(\delta+\beta+1)}{n!\Gamma(\delta+\beta+n+1)\Gamma(\delta+\alpha+n+2)} {}_3F_2\left[\begin{matrix} -\lambda, \delta+\beta+1, \delta+1 \\ \delta+\beta+n+1, \delta+\alpha+n+2 \end{matrix}\right]$$

provided $\alpha > -1$ *and* $\beta > -1$.

Further, by simplifying the hypergeometric function in terms of series summation and applying the definition of the I-function, we get the right-hand side of (3.16)

xi.

$$I_{11} \equiv \int_{-1}^{1} (1-x)^{\delta}(1+x)^{\beta} P_n^{\alpha,\beta}(x) P_m^{\rho,\sigma}(x)$$

$$I_{p_i,q_i:r}^{m,n}\left[z(1-x)^h \middle| \begin{matrix} (a_j,\alpha_j)_{1,n};\ (a_{ji},\alpha_{ji})_{n+1,p_i} \\ (b_j,\beta_j)_{1,m};\ (b_{ji},\beta_{ji})_{m+1,q_i} \end{matrix} \right] dx$$

$$= \frac{2^{\beta+\delta+1}\Gamma(\beta+n+1)\Gamma(\rho+m+1)}{n!m!}\sum_{k=0}^{\infty}\frac{(-m)_k\,\Gamma(\rho+\sigma+m+1)}{k!\Gamma(\rho+k+1)} I_{p_i+4,q_i+4:r}^{m+1,n+3}$$

$$\left[2^h z \middle| \begin{matrix} (\alpha-\delta-k,h), (-\delta-k,h), (\alpha+n-\delta,h), \quad (a_j,\alpha_j)_{1,n}; (a_{ji},\alpha_{ji})_{n+1,p_i} (\alpha-\delta,h) \\ (\alpha-\delta+n,h), (b_j,\beta_j)_{1,m}; (b_{ji},\beta_{ji})_{m+1,q_i} \quad (\beta-\delta-k-n-1,h), (\alpha-\delta,h), (\alpha+n-\delta-k,h) \end{matrix} \right] \tag{3.17}$$

provided $\operatorname{Re}(\beta) > -1$, $\operatorname{Re}\left[\delta + h \min\left(\frac{b_j}{\beta_j}\right)\right] > -1, j = 1, 2, ..., m;$ $|\arg z| < \frac{1}{2}\pi B_i$, $j = 1, 2, ..., r.$

$B_i > 0$, $A_i \leq 0$ are defined in (3.16) and h and δ are positive numbers.

xii.

$$I_{12} \equiv \int_{-1}^{1} (1-x)^{\rho}(1-x)^{\sigma} P_n^{\alpha,\beta}(x)$$

$$I_{p_i,q_i:r}^{m,n}\left[z(1-x)^h(1+x)^t \left| \begin{matrix} (a_j,\alpha_j)_{1,n}; & (a_{ji},\alpha_{ji})_{n+1,p_i} \\ (b_j,\beta_j)_{1,m}; & (b_{ji},\beta_{ji})_{m+1,q_i} \end{matrix} \right. \right] dx$$

$$= 2^{\rho+\sigma+1}\sum_{k=0}^{\infty}\frac{(-n)_k(\alpha+\beta+n+1)_k}{(\alpha+1)_k\, k!}\times I_{p_i+2,q_i+1:r}^{m,n+2}$$

$$\left[2^{h+t}z \left| \begin{matrix} (-\sigma,t),\ (-\rho-k,h),(a_j,\alpha_j)_{1,n}; & (a_{ji},\alpha_{ji})_{n+1,p_i} \\ (b_j,\beta_j)_{1,m}; & (b_{ji},\beta_{ji})_{m+1,q_i},(-1-\rho-\sigma-k,h+t) \end{matrix} \right. \right] \tag{3.18}$$

provided

$$\text{Re}\left[\rho+h\min\left(b_j/\beta_j\right)\right]>-1$$

$$\text{Re}\left[\sigma+t\min\left(b_j/\beta_j\right)\right]>-1\ j=1,2,\ldots,m$$

$$|\arg z|<\tfrac{1}{2}\pi B_i\ i=1,2,\ldots,r$$

xiii.

$$I_{13} \equiv \int_{-1}^{1} (1-x)^{\alpha}(1+x)^{\sigma} P_n^{\alpha,\beta}(x)$$

$$I_{p_i,q_i:r}^{m,n}\left[z(1+x)^h \left| \begin{matrix} (a_j,\alpha_j)_{1,n}; & (a_{ji},\alpha_{ji})_{n+1,p_i} \\ (b_j,\beta_j)_{1,m}; & (b_{ji},\beta_{ji})_{m+1,q_i} \end{matrix} \right. \right] dx$$

$$= 2^{\rho+\alpha+1}\Gamma(\alpha+n+1)\Gamma(\alpha-\beta+1)$$

$$I_{p_i+1,q_i+2:r}^{m,n+1}\left[\frac{z}{2^h} \left| \begin{matrix} (-\sigma,h),(a_j,\alpha_j)_{1,n}; & (a_{ji},\alpha_{ji})_{n+1,p_i} \\ (b_j,\beta_j)_{1,m}; & (b_{ji},\beta_{ji})_{m+1,q_i},(-\beta+n-\sigma,h),(\alpha-\sigma-n-1,h) \end{matrix} \right. \right] \tag{3.19}$$

provided

$$\text{Re}\,(\alpha) > -1,\ \text{Re}\left[\rho + h\ \min\left(\frac{b_j}{\beta_j}\right)\right] > -1\ j = 1,2,\ldots,m$$

$$|\arg z| < \tfrac{1}{2}\pi B_i\ i = 1,2,\ldots,r$$

xiv.

$$I_{14} \equiv \int_{-1}^{1} (1-x)^{\rho}(1+x)^{\sigma} P_n^{\alpha,\beta}(x)$$

$$I_{p_i,q_i:r}^{m,n}\left[z(1-x)^h(1+x)^{-t}\,\middle|\,\begin{matrix}(a_j,\alpha_j)_{1,n}; & (a_{ji},\alpha_{ji})_{n+1,p_i}\\ (b_j,\beta_j)_{1,m}; & (b_{ji},\beta_{ji})_{m+1,q_i}\end{matrix}\right]dx$$

$$= 2^{\rho+\sigma+1}\sum_{k=0}^{\infty}\frac{(-n)_k(\alpha+\beta+n+1)_k}{(\alpha+1)_k\,k!}$$

$$\times I_{p_i+2,q_i+1:r}^{m+1,n+1}\left[2^{h-t}z\,\middle|\,\begin{matrix}(-\rho-k,h),(a_j,\alpha_j)_{1,n}; & (a_{ji},\alpha_{ji})_{n+1,p_i},(1-k-\rho-\sigma,h-t)\\ (1+\sigma,t)(b_j,\beta_j)_{1,m}; & (b_{ji},\beta_{ji})_{m+1,q_i}\end{matrix}\right] \tag{3.20}$$

provided

$$\text{Re}\left[\rho + h\ \min\left(\frac{b_j}{\beta_j}\right)\right] > -1$$

$$\text{Re}\left[\sigma - t\ \min\left(\frac{b_j}{\beta_j}\right)\right] > -1\ j = 1,2,\ldots,m$$

$$|\arg z| < \tfrac{1}{2}\pi B_i\ i = 1,2,\ldots,r$$

$B_i > 0,\ A_i \le 0$ are defined in (3.16).

The results (3.17)–(3.20) can be proved by using popular integrals of Jacobi polynomials [3].

3.3.1 Special Cases

i. If we replace δ by $\lambda - 1$ and put $\alpha = \beta = \rho = \sigma = 0$ in the integral (3.17) transforms to the following integral involving product of Legendre polynomials

$$\int_{-1}^{1}(1-x)^{\lambda-1}P_n(x)P_m(x)$$

$$I^{m,n}_{p_i,q_i:r}\left[z(1-x)^h\left|\begin{matrix}(a_j,\alpha_j)_{1,n}; & (a_{ji},\alpha_{ji})_{n+1,p_i}\\ (b_j,\beta_j)_{1,m}; & (b_{ji},\beta_{ji})_{m+1,q_i}\end{matrix}\right.\right]dx=\sum_{k=0}^{\infty}\frac{2^{\lambda}(-m)_k(m+1)_k}{(k!)^2}I^{m+1,n+3}_{p_i+4,q_i+4:r}$$

$$\left[2^h z\left|\begin{matrix}(1-\lambda-k,h),(1-\lambda-k,h),(1-\lambda+n,h), & (a_j,\alpha_j)_{1,n};(a_{ji},\alpha_{ji})_{,n+1,p_i}(1-\lambda,h)\\ (1-\lambda+n,h),(b_j,\beta_j)_{1,m};(b_{ji},\beta_{ji})_{,m+1,q_i} & (-\lambda-k-n,h),(1-\lambda,h),(1-\lambda+n-k,h)\end{matrix}\right.\right] \tag{3.21}$$

$$\text{provided Re}\left[\alpha+h\min\left(b_j/\beta_j\right)\right]\succ 0,\ j=1,2,\ldots,m;$$

$$|\arg z|\langle \tfrac{1}{2}\pi B_i,\quad B_i\rangle 0,\ A_i\le 0$$

where A_i and B_i are defined in (3.16).

If $\alpha=\beta=0$, ρ is replaced by $\rho-1$ and σ simplifying the hypergeometric function by σ - 1 then (3.18) transforms into the following integral involving Legendre polynomials.

$$\int_{-1}^{1}(1-x)^{\rho-1}(1+x)^{\sigma-1}P_n(x)$$

$$I^{m,n}_{p_i,q_i:r}\left[z(1-x)^h(1+x)^t\left|\begin{matrix}(a_j,\alpha_j)_{1,n}; & (a_{ji},\alpha_{ji})_{n+1,p_i}\\ (b_j,\beta_j)_{1,m}; & (b_{ji},\beta_{ji})_{m+1,q_i}\end{matrix}\right.\right]dx \tag{3.22}$$

$$=\sum_{k=0}^{\infty}\frac{2^{\rho+\sigma-1}(-n)_k(n+1)_k}{(k!)^2}I^{m,n+2}_{p_i+2,q_i+2:r}$$

$$\left[2^{h+t}z\left|\begin{matrix}(1-\sigma,t),(1-\rho-k,h), & (a_j,\alpha_j)_{1,n};(a_{ji},\alpha_{ji})_{,n+1,p_i}\\ (b_j,\beta_j)_{1,m};(b_{ji},\beta_{ji})_{,m+1,q_i} & ,(1-\alpha-\beta-k,h+t)\end{matrix}\right.\right]$$

$$\text{provided Re}\left[\rho+h\min\left(b_j/\beta_j\right)\right]\succ 0,$$

$$\text{Re}\left[\sigma+t\min\left(b_j/\beta_j\right)\right]\succ 0,\ j=1,2,\ldots,m$$

$|\arg z| < \frac{1}{2}\pi B_i,$, $\quad i = 1,2,\ldots,r$
and h, t are positive numbers.

Replacing ρ by $\rho-1$, σ by $\sigma-1$ and putting $\alpha=\beta=0$, the integral (3.20) takes the form of the following integral

$$\int_{-1}^{1}(1-x)^{\rho-1}(1+x)^{\sigma-1}P_n(x)$$

$$I^{m,n}_{p_i,q_i:r}\left[z(1-x)^h(1+x)^{-t}\left|\begin{matrix}(a_j,\alpha_j)_{1,n}; & (a_{ji},\alpha_{ji})_{n+1,p_i}\\ (b_j,\beta_j)_{1,m}; & (b_{ji},\beta_{ji})_{m+1,q_i}\end{matrix}\right.\right]dx$$

$$=\sum_{k=0}^{\infty}\frac{2^{\rho+\sigma-1}(-n)_k(n+1)_k}{(k!)^2}I^{m+1,n+1}_{p_i+2,q_i+2:r}$$

$$\left[2^{h-t}z\left|\begin{matrix}(1-\rho-k,h), & (a_j,\alpha_j)_{1,n};(a_{ji},\alpha_{ji}),_{n+1,p_i},(1-\rho-\sigma-k,h-t),\\ (\sigma,t),(b_j,\beta_j)_{1,m};(b_{ji},\beta_{ji}),_{m+1,q_i} & \ldots\ldots\ldots\ldots\ldots\ldots\end{matrix}\right.\right] \tag{3.23}$$

provided $\mathrm{Re}\left[\alpha+h\min\left(b_j/\beta_j\right)\right]\succ 0,$

$\mathrm{Re}\left[\beta+t\min\left(b_j/\beta_j\right)\right]\succ 0,\ \ j=1,2,\ldots,m$

$|\arg z| < \frac{1}{2}\pi B_i,\quad i = 1,2,\ldots,r$

These results can also be proved independently by using well-known integrals[94]. and the definition of the I-function.

xv.

$$I_{15}\equiv\int_0^{\infty}x^tJ_v^{\tau}(x)$$

$$\times I^{m,n}_{p_i,q_i:r}\left[zx^{\gamma}\left|\begin{matrix}(a_j,\alpha_j)_{1,n}; & (a_{ji},\alpha_{ji})_{n+1,p_i}\\ (b_j,\beta_j)_{1,m}; & (b_{ji},\beta_{ji})_{m+1,q_i}\end{matrix}\right.\right]dx \tag{3.24}$$

$$=I^{m,n+1}_{p_i+2,q_i:r}\left[z\left|\begin{matrix}(-l,\gamma), & (a_j,\alpha_j)_{1,n};(a_{ji},\alpha_{ji}),_{n+1,p_i},(1+v-\tau-\tau l,\gamma\tau)\\ (b_j,\beta_j)_{1,m}; & (b_{ji},\beta_{ji}),_{m+1,q_i}\end{matrix}\right.\right]$$

Provided

i. $\lambda > 0, |\arg z| < \frac{1}{2}\pi\lambda$

ii. $\lambda \geq 0, |\arg z| < \frac{1}{2}\pi\lambda$, $\text{Re}(\mu_1 + 1) < 0$, $\text{Re}(\mu + l - v + \tau + \tau l + 2) < 0$

iii. $\gamma - \tau\gamma > 0$, $\text{Re}(l+1) > 0$.

where λ and μ are defined (1.23) and $J_v^\tau(x)$ is Bessel Maitland function defined

$$J_v^\tau(x) = \sum_{r=0}^{\infty} \frac{(-1)^r}{r!\Gamma(1+v+\tau r)}$$

This integral I_{15} is obtained by using the result.

$$\int_0^\infty x^l J_v^\tau(x)\, dx = \frac{\Gamma(l+1)}{\Gamma(1+v-\tau-\tau r)} \tag{3.25}$$

provided $\text{Re}(l) > -1$, $0 < \tau < 1$.

xvi.
$$I_{16} \equiv \int_0^\infty x^{\sigma-1}(1-x^2)^{\frac{\delta}{2}} P_v^\delta(x)$$

$$\times I_{p_i,q_i:r}^{m,n}\left[zx^\gamma \left| \begin{matrix} (a_j,\alpha_j)_{1,n}; & (a_{ji},\alpha_{ji})_{n+1,p_i} \\ (b_j,\beta_j)_{1,m}; & (b_{ji},\beta_{ji})_{m+1,q_i} \end{matrix} \right.\right] dx$$

$$= \frac{2^{-\sigma-\delta}(-1)^\delta \pi^{1/2}\Gamma(1+\delta+v)}{\Gamma(1-\delta+v)} I_{p_i+1,q_i+2:r}^{m,n+1}$$

$$\left[\frac{z}{2^\gamma}\left| \begin{matrix} (1-\sigma,\gamma), & (a_j,\alpha_j)_{1,n};(a_{ji},\alpha_{ji})_{,n+1,p_i} \\ (b_j,\beta_j)_{1,m}; \quad (b_{ji},\beta_{ji})_{,m+1,q_i} & \left(\frac{1}{2}+\frac{v}{2}-\frac{\sigma}{2}-\frac{\delta}{2},\frac{\gamma}{2}\right),\left(-\frac{v}{2}-\frac{\sigma}{2}-\frac{\delta}{2},\frac{\gamma}{2}\right) \end{matrix}\right.\right] \tag{3.26}$$

Provided

i. $\sigma > 0$, δ is non negative integer.

ii. $\lambda > 0, |\arg z| < \frac{1}{2}\pi\lambda$

iii. $\lambda \geq 0, |\arg z| \leq \frac{1}{2}\pi\lambda, \operatorname{Re}(\mu+1)<0, \operatorname{Re}\left(\mu+\delta-\frac{1}{2}\right)<0$

where $P_v^{\delta}(x)$ is Legendre function of the first kind.

This result can be proved by using the result [3].

$$\int_0^1 x^{-1}\left(1-x^2\right)^{\frac{\delta}{2}} P_v^{\delta}(x) = \frac{2^{-\sigma-\delta}(-1)^{\delta}\pi^{1/2}\Gamma(\sigma)\Gamma(1+\delta+v)}{\Gamma(1-\delta+v)\Gamma\left(\frac{1}{2}+\frac{\sigma}{2}+\frac{\delta}{2}-\frac{v}{2}\right)\Gamma\left(\frac{1}{2}+\frac{\sigma}{2}+\frac{\delta}{2}+\frac{v}{2}\right)} \tag{3.27}$$

$\operatorname{Re}(\sigma)>0, \delta=1,2,3....$

xvii.

$$I_{17} \equiv \int_0^1 x^{\sigma-1}\left(1-x^2\right)^{\frac{-\delta}{2}} P_v^{\delta}(x)$$

$$\times I_{p_i,q_i:r}^{m,n}\left[zx^{\gamma}\left| \begin{matrix} (a_j,\alpha_j)_{1,n}; & (a_{ji},\alpha_{ji})_{n+1,p_i} \\ (b_j,\beta_j)_{1,m}; & (b_{ji},\beta_{ji})_{m+1,q_i} \end{matrix} \right.\right]dx$$

$$= \pi^{1/2} 2^{\delta-\sigma} I_{p_i+1,q_i+2:r}^{m,n}$$

$$\left[\frac{z}{2^{\gamma}}\left| \begin{matrix} (1-\sigma,\gamma), & (a_j,\alpha_j)_{1,n};(a_{ji},\alpha_{ji})_{,n+1,p_i},\ldots\ldots\ldots \\ (b_j,\beta_j)_{1,m}; \quad (b_{ji},\beta_{ji})_{,m+1,q_i} & \left(\frac{1}{2}+\frac{v}{2}-\frac{\sigma}{2}+\frac{\delta}{2},\frac{\gamma}{2}\right),\left(-\frac{v}{2}-\frac{\sigma}{2}+\frac{\delta}{2},\frac{\gamma}{2}\right) \end{matrix} \right.\right] \tag{3.28}$$

Provided

i. $\operatorname{Re}(\sigma)>0, \operatorname{Re}(\delta)>1$

ii. $\lambda>0, |\arg z|<\frac{1}{2}\pi\lambda$

iii. $\lambda \geq 0, |\arg z| \leq \frac{1}{2}\pi\lambda, \operatorname{Re}(\mu+1)<0, \operatorname{Re}\left(\mu+\delta-\frac{3}{2}\right)<0$

where

$$\int_0^1 x^{\sigma-1}\left(1-x^2\right)^{\frac{-\delta}{2}} P_v^{\delta}(x)dx$$

$$= \frac{2^{\sigma-\delta}\pi^{1/2}\Gamma(\sigma)}{\Gamma\left(\frac{1}{2}+\frac{\sigma}{2}-\frac{\delta}{2}-\frac{v}{2}\right)\Gamma\left(\frac{1}{2}+\frac{\sigma}{2}-\frac{\delta}{2}-\frac{v}{2}\right)} \tag{3.29}$$

In a similar manner, we can establish the following integrals

xviii.

$$I_{18} \equiv \int_{-1}^{1} \left(1-x^2\right)^{\sigma-1} P_v^{\delta}(x)$$

$$\times I_{p_i,q_i:r}^{m,n}\left[z\left(1-x^2\right)^{\gamma} \middle| \begin{matrix} (a_j,\alpha_j)_{1,n}; & (a_{ji},\alpha_{ji})_{n+1,p_i} \\ (b_j,\beta_j)_{1,m}; & (b_{ji},\beta_{ji})_{m+1,q_i} \end{matrix} \right] dx$$

$$= \frac{2^{\delta}\,\pi}{\Gamma\left(-\frac{\delta}{2}+\frac{v}{2}+1\right)\Gamma\left(-\frac{\delta}{2}-\frac{v}{2}+\frac{1}{2}\right)} I_{p_i+2,q_i+2:r}^{m,n+2}$$

$$\left[z \middle| \begin{matrix} \left(1-\sigma-\frac{\delta}{2},\gamma\right),\left(1-\sigma+\frac{\delta}{2},\gamma\right), & (a_j,\alpha_j)_{1,n};(a_{ji},\alpha_{ji})_{,n+1,p_i} \\ (b_j,\beta_j)_{1,m}; & (b_{ji},\beta_{ji})_{,m+1,q_i}\left(\sigma-\frac{v}{2},\gamma\right),\left(1+\frac{v}{2}-\sigma,\gamma\right) \end{matrix} \right]$$

(3.30)

Provided

i. $\mathrm{Re}(\sigma) > \mathrm{Re}(\delta)$

ii. $\lambda > 0, |\arg z| < \frac{1}{2}\pi\lambda$

iii. $\lambda \geq 0, |\arg z| \leq \frac{1}{2}\pi\lambda, \mathrm{Re}(\mu+1) > 0$

xix.

$$I_{19} \equiv \int_{1}^{\infty} x^{-\rho}\left(x^2-1\right)^{-\delta/2} P_v^{\delta}(x)$$

$$\times I_{p_i,q_i:r}^{m,n}\left[zx^{\gamma} \middle| \begin{matrix} (a_j,\alpha_j)_{1,n}; & (a_{ji},\alpha_{ji})_{n+1,p_i} \\ (b_j,\beta_j)_{1,m}; & (b_{ji},\beta_{ji})_{m+1,q_i} \end{matrix} \right] dx$$

$$= \frac{2^{\rho+\delta-2}}{\pi^{1/2}} I_{p_i+1,q_i+2:r}^{m+2,n}$$

$$\left[\frac{z}{2^{\gamma}} \middle| \begin{matrix} (a_j,\alpha_j)_{1,n}; & (a_{ji},\alpha_{ji})_{,n+1,p_i},(\rho,\gamma),\ldots\ldots\ldots\ldots \\ \left(\frac{\rho}{2}+\frac{\delta}{2}+\frac{v}{2},\frac{\gamma}{2},\frac{\rho}{2}+\frac{\delta}{2}-\frac{v}{2}-\frac{1}{2},\frac{\gamma}{2}\right),\ (b_j,\beta_j)_{1,m}; & (b_{ji},\beta_{ji})_{,m+1,q_i} \end{matrix} \right]$$

(3.31)

Provided

i. $\operatorname{Re}(\delta)\langle 1, \operatorname{Re}(\rho+\delta+v)\rangle 0, \operatorname{Re}(\rho+\delta-v)>1$

ii. $\lambda>0, |\arg z|<\frac{1}{2}\pi\lambda$

iii. $\lambda\geq 0, |\arg z|\leq\frac{1}{2}\pi\lambda, \operatorname{Re}(\mu+1)<0, \operatorname{Re}\left(\mu+\delta-\frac{1}{2}\right)<0$

xx.

$$I_{20}\equiv\int_{1}^{\infty}\left(x^2-1\right)^{\sigma-1}P_v^{\delta}(x)$$

$$\times I_{p_i,q_i:r}^{m,n}\left[z\left(x^2-1\right)^{\gamma}\left|\begin{array}{l}(a_j,\alpha_j)_{1,n};\quad (a_{ji},\alpha_{ji})_{n+1,p_i}\\(b_j,\beta_j)_{1,m};\quad (b_{ji},\beta_{ji})_{m+1,q_i}\end{array}\right.\right]dx$$

$$=\frac{2^{\delta-1}}{\Gamma\left(1-\frac{\delta}{2}+\frac{v}{2}\right)\Gamma\left(1-\frac{\delta}{2}-\frac{v}{2}\right)}I_{p_i+2,q_i+2:r}^{m+2,n+1}$$

$$\left[z\left|\begin{array}{l}\left(1-\sigma+\frac{\delta}{2},\gamma\right),(a_j,\alpha_j)_{1,n};\quad (a_{ji},\alpha_{ji})_{,n+1,p_i},\left(1-\sigma-\frac{\delta}{2},\gamma\right)\\\left(1-\sigma+\frac{v}{2},\gamma\right),\left(\frac{1}{2}-\sigma-\frac{v}{2},\gamma\right),(b_j,\beta_j)_{1,m};\quad (b_{ji},\beta_{ji})_{,m+1,q_i}\end{array}\right.\right] \tag{3.32}$$

Provided

i. $\operatorname{Re}(\sigma)>\operatorname{Re}(\delta), \operatorname{Re}(1+2\sigma-v)>0, \operatorname{Re}(2-2\sigma+v)>0$

ii. $\lambda>0, |\arg z|<\frac{1}{2}\pi\lambda, \gamma\geq 0$

iii. $\lambda\geq 0, |\arg z|\leq\frac{1}{2}\pi\lambda,\ \operatorname{Re}(\mu+1)<0$

xxi.

$$I_{21} \equiv \int_{1}^{\infty} (1-x)^{\rho-1} \left(x^2-1\right)^{\delta/2} P_v^{\delta}(x)$$

$$\times I_{p_i,q_i:r}^{m,n} \left[z(x-1)^{\gamma} \middle| \begin{matrix} \left(a_j,\alpha_j\right)_{1,n}; & \left(a_{ji},\alpha_{ji}\right)_{n+1,p_i} \\ \left(b_j,\beta_j\right)_{1,m}; & \left(b_{ji},\beta_{ji}\right)_{m+1,q_i} \end{matrix} \right] dx$$

$$= \frac{2^{\rho}}{\Gamma(1-\delta+v)\Gamma(-\delta-v)} I_{p_i+2,q_i+2:r}^{m+2,n+1}$$

$$\left[2^{\gamma} z \middle| \begin{matrix} (1-\rho,\gamma),\left(a_j,\alpha_j\right)_{1,n}; & \left(a_{ji},\alpha_{ji}\right)_{,n+1,p_i},(1-\rho-\delta,\gamma) \\ (-\rho-\delta-v,\gamma),(1-\rho-\delta-v,\gamma),\ \left(b_j,\beta_j\right)_{1,m}; & \left(b_{ji},\beta_{ji}\right)_{,m+1,q_i} \end{matrix} \right] \tag{3.33}$$

Provided

i. $\gamma > 0$

ii. $\lambda > 0, |\arg z| < \frac{1}{2}\pi\lambda,\ \gamma \geq 0$

iii. $\lambda \geq 0, |\arg z| \leq \frac{1}{2}\pi\lambda,\ \operatorname{Re}(\mu+1) < 0, \operatorname{Re}(\mu-\delta-2v) < 0$

xxii

$$I_{22} \equiv \int_{-1}^{1} (1+x)^{\rho-1} P_v(x)$$

$$\times I_{p_i,q_i:r}^{m,n} \left[z(1+x)^{\gamma} \middle| \begin{matrix} \left(a_j,\alpha_j\right)_{1,n}; & \left(a_{ji},\alpha_{ji}\right)_{n+1,p_i} \\ \left(b_j,\beta_j\right)_{1,m}; & \left(b_{ji},\beta_{ji}\right)_{m+1,q_i} \end{matrix} \right] dx$$

$$= 2^{\rho}\ I_{p_i+2,q_i+2:r}^{m,n+2} \left[2^{\gamma} z \middle| \begin{matrix} (1-\rho,\gamma),(1-\rho,\gamma),\ \left(a_j,\alpha_j\right)_{1,n}; & \left(a_{ji},\alpha_{ji}\right)_{,n+1,p_i} \\ \left(b_j,\beta_j\right)_{1,m}; & \left(b_{ji},\beta_{ji}\right),\ _{m+1,q_i}(-\rho-v,\gamma),(1-\rho+v,\gamma) \end{matrix} \right] \tag{3.34}$$

Provided

i. $\operatorname{Re}(\rho) > 0$

ii. $\lambda > 0, |\arg z| < \frac{1}{2}\pi\lambda,\ \gamma \geq 0$

iii. $\lambda \geq 0, |\arg z| \leq \frac{1}{2}\pi\lambda, \ \mathrm{Re}(\mu+1) < 0$

xxiii.

$$I_{23} \equiv \int_{-\infty}^{\infty} x^{2\rho} e^{-x^2} H_{2\nu}(x)$$

$$\times I^{m,n}_{p_i,q_i:r}\left[zx^{-2h} \left| \begin{matrix} (a_j,\alpha_j)_{1,n}; & (a_{ji},\alpha_{ji})_{n+1,p_i} \\ (b_j,\beta_j)_{1,m}; & (b_{ji},\beta_{ji})_{m+1,q_i} \end{matrix} \right. \right] dx \tag{3.35}$$

$$= \sqrt{\pi}\ 2^{2(\nu-\rho)} I^{m+1,n}_{p_i+1,q_i+1:r}$$

$$\left[2^{2h} z \left| \begin{matrix} (a_j,\alpha_j)_{1,n}; & (a_{ji},\alpha_{ji})_{,n+1,p_i}, (\rho-\nu+1,h) \\ (2\rho+1,2h),(b_j,\beta_j)_{1,m}; & (b_{ji},\beta_{ji}),\ _{m+1,q_i} \end{matrix} \right. \right]$$

where $h > 0,\ B > 0,\ A \geq 0.$

$|\arg z| < \frac{1}{2}B\pi,\ \rho = 0,1,2,....;\ A = \sum_{j=1}^{p_j} \alpha_{ji} - \sum_{j=1}^{q_j} \beta_{ji}$ and B is given in (3.58).

The alternate form of (3.35) is given as

xxiv.

$$I_{24} \equiv \int_{-\infty}^{\infty} x^{2\rho} e^{-x^2} H_{2\nu}(x)$$

$$\times I^{m,n}_{p_i,q_i:r}\left[zx^{2h} \left| \begin{matrix} (a_j,\alpha_j)_{1,n}; & (a_{ji},\alpha_{ji})_{n+1,p_i} \\ (b_j,\beta_j)_{1,m}; & (b_{ji},\beta_{ji})_{m+1,q_i} \end{matrix} \right. \right] dx \tag{3.36}$$

$$= \sqrt{\pi}\ 2^{2(\nu-\rho)} I^{m,n+1}_{p_i+1,q_i+1:r}$$

$$\left[2^{-2h} z \left| \begin{matrix} (-2\rho,2h),(a_j,\alpha_j)_{1,n}; & (a_{ji},\alpha_{ji})_{,n+1,p_i} \\ (b_j,\beta_j)_{1,m}; & (b_{ji},\beta_{ji}),\ _{m+1,q_i}(\nu-\rho,n) \end{matrix} \right. \right]$$

Provided $h > 0,\ B < 0,\ A \leq 0,\ |\arg z| < \frac{1}{2}B\pi,\ \rho = 0,1,2,.......$

To prove the aforementioned results following orthogonality properties of the Hermite polynomials and Bessel functions have been used [10].

$$\int_{-\infty}^{\infty} e^{-x^2} H_n(x) H_m(x)\,dx = \begin{cases} 0, & m \neq n \\ 2^n n! \sqrt{\pi}, & m = n \end{cases} \tag{3.37}$$

$$\int_{-\infty}^{\infty} x^{-1} J_{v+2n+1}(x) J_{v+2m+1}(x)\,dx = \begin{cases} 0, & m \neq n \\ (4n+2v+2)^{-1}, & m = n \end{cases} \tag{3.38}$$

Provided $\operatorname{Re}(v+m+n) > -1$ along with

$$\int_{-\infty}^{\infty} x^{2\rho} e^{-x^2} H_{2v}(x)\,dx = \frac{\pi^{\frac{1}{2}} 2^{2(v-\rho)} \Gamma(2\rho+1)}{\Gamma(\rho-v+1)} \tag{3.39}$$

xxv.

$$I_{25} \equiv \int_0^1 x^{\sigma-\lambda-\delta} \left(1-x^2\right)^{-\mu/2} P_v^{\mu}(x) J_\delta(xt) J_\lambda(xt)$$

$$I_{p_i,q_i:r}^{m,n}\left[zx^{\rho} \left| \begin{matrix} (a_j,\alpha_j)_{1,n}; & (a_{ji},\alpha_{ji})_{n+1,p_i} \\ (b_j,\beta_j)_{1,m}; & (b_{ji},\beta_{ji})_{m+1,q_i} \end{matrix} \right. \right] dx$$

$$= \frac{2^{\mu-\lambda-\sigma-\delta-1}\, t^{\lambda+\delta}}{\Gamma(\delta+1)\Gamma(\lambda+1)} \pi^{\frac{1}{2}} \sum_{r=0}^{\infty} \frac{2^{-2r} \left(\frac{\lambda+\delta+1}{2}\right)_r \left(\frac{\lambda+\delta+2}{2}\right)_r (-t^2)^r}{(\lambda+1)_r (\delta+1)_r (\lambda+\delta+1)_r\, r!} I_{p_i+1,q_i+2:r}^{m,n+1}$$

$$\left[2^{-\rho} z \left| \begin{matrix} (-\sigma-2r,\rho), (a_j,\alpha_j)_{1,n}; & (a_{ji},\alpha_{ji})_{n+1,p_i} \ldots \\ (b_j,\beta_j)_{1,m}; & (b_{ji},\beta_{ji})_{m+1,q_i} \left(-\frac{\sigma}{2}+\frac{v}{2}+\frac{\mu}{2}-r,\frac{\rho}{2}\right), \left(-\frac{1}{2}-\frac{\sigma}{2}-\frac{v}{2}+\frac{\mu}{2}-r,\frac{\rho}{2}\right) \end{matrix} \right. \right] \tag{3.40}$$

Provided $\operatorname{Re}(\mu)\langle 1,\ \operatorname{Re}(\sigma)\rangle -1,\ \operatorname{Re}(\sigma+1) + \rho \min_{i \le j \le m} \operatorname{Re}\left(b_j / \beta_j \right) > 0.$

xxvi.

$$I_{26} \equiv \int_0^1 x^{\sigma-\lambda-\delta} \left(1-x^2\right)^{-\mu/2} P_v^{\mu}(x) J_\delta(xt) J_\lambda(xt)$$

$$+I_{p_i,q_i:r}^{m,n}\left[zx^{-\rho} \left| \begin{matrix} (a_j,\alpha_j)_{1,n}; & (a_{ji},\alpha_{ji})_{n+1,p_i} \\ (b_j,\beta_j)_{1,m}; & (b_{ji},\beta_{ji})_{m+1,q_i} \end{matrix} \right. \right] dx$$

$$= \frac{2^{\mu-\lambda-\delta-\sigma-1}\, t^{\lambda+\delta}}{\Gamma(\delta+1)\Gamma(\lambda+1)} \pi^{\frac{1}{2}} \sum_{r=0}^{\infty} \frac{2^{-2r} \left(\frac{\lambda+\delta+1}{2}\right)_r \left(\frac{\lambda+\delta+2}{2}\right)_r (-t^2)^r}{(\lambda+1)_r (\delta+1)_r (\lambda+\delta+1)_r\, r!} I_{p_i+2,q_i+1:r}^{m,n+1}$$

$$\left[2^{\rho} z \left| \begin{matrix} (-\sigma-2r,\rho), (a_j,\alpha_j)_{1,n}; & (a_{ji},\alpha_{ji})_{n+1,p_i} \left(1-\frac{v}{2}-\frac{\mu}{2}+r+\frac{\sigma}{2}+\frac{\rho}{2}\right), \left(\frac{3}{2}+\frac{v}{2}+\frac{\mu}{2}+\frac{\sigma}{2}+\frac{\rho}{2}\right) \\ (1+2r+\sigma,\rho), (b_j,\beta_j)_{1,m}; & (b_{ji},\beta_{ji})_{m+1,q_i} \end{matrix} \right. \right] \tag{3.41}$$

Provided $\text{Re}(\mu)\langle 1,\ \text{Re}(\sigma)\rangle - 1.$

xxvii.

$$I_{27} \equiv \int_0^\infty x^{\nu+1}\left(\alpha^2+x^2\right)^{-\frac{\lambda}{2}-\delta/4} \exp\left(\frac{-p^2\alpha}{\alpha^2+x^2}\right) J_\nu\left(\frac{p^2 x}{\alpha^2+x^2}\right)$$

$$I_{p_i,q_i:r}^{m,n}\left[z\left(\alpha^2+x^2\right)^{\mu} \middle| \begin{matrix} (a_j,\alpha_j)_{1,n}; & (a_{ji},\alpha_{ji})_{n+1,p_i} \\ (b_j,\beta_j)_{1,m}; & (b_{ji},\beta_{ji})_{m+1,q_i} \end{matrix} \right] dx$$

$$= 2^{-\lambda-\nu-\frac{1}{2}} p^{2\nu} \pi^{\frac{1}{2}} \alpha^{-\lambda-\frac{1}{2}} \sum_{r=0}^{\infty} \frac{\left(-p^2/2\alpha\right)^r}{r!} I_{p_i+2,q_i+1:r}^{m+1,n}$$

$$\left[(2\alpha)^{2\mu} z \middle| \begin{matrix} (a_j,\alpha_j)_{1,n}; & (a_{ji},\alpha_{ji})_{n+1,p_i}\left(\frac{\lambda}{2}+\frac{3}{4},\mu\right),\left(\nu+\frac{\lambda}{2}+\frac{\delta}{4}+r,\mu\right) \\ \left(\lambda+\frac{1}{2}+r,\ 2\mu\right),(b_j,\beta_j)_{1,m}; & (b_{ji},\beta_{ji}),_{m+1,q_i} \end{matrix} \right] \tag{3.42}$$

Provided $\text{Re}(\alpha)>0, p>0, \text{Re}\left(\lambda+\frac{1}{2}\right)>0,\ \text{Re}(\nu+1)>0.$

xxviii.

$$I_{28} \equiv \int_0^\infty x^{\nu+1}\left(\alpha^2+x^2\right)^{-\frac{\lambda}{2}-\delta/4} \exp\left(\frac{-p^2\alpha}{\alpha^2+x^2}\right) J_\nu\left(\frac{p^2 x}{\alpha^2+x^2}\right)$$

$$I_{p_i,q_i:r}^{m,n}\left[z\left(\alpha^2+x^2\right)^{-\mu} \middle| \begin{matrix} (a_j,\alpha_j)_{1,n}; & (a_{ji},\alpha_{ji})_{n+1,p_i} \\ (b_j,\beta_j)_{1,m}; & (b_{ji},\beta_{ji})_{m+1,q_i} \end{matrix} \right] dx$$

$$= 2^{-\lambda-\nu-\frac{1}{2}} p^{2\nu} \pi^{\frac{1}{2}} \alpha^{-\lambda-\frac{1}{2}} \sum_{r=0}^{\infty} \frac{\left(-p^2/2\alpha\right)^r}{r!} I_{p_i+1,q_i+2:r}^{m,n+1}$$

$$\left[(2\alpha)^{-2\mu} z \middle| \begin{matrix} \left(1-\frac{\lambda}{2}-r,2\mu\right),(a_j,\alpha_j)_{1,n}; & (a_{ji},\alpha_{ji}),_{n+1,p_i} \\ (b_j,\beta_j)_{1,m}; & (b_{ji},\beta_{ji}),_{m+1,q_i}\left(\frac{1}{4}-\frac{\lambda}{2},\ \mu\right),\left(\frac{1}{4}-\nu-\frac{\lambda}{2}-r,\ \mu\right) \end{matrix} \right] \tag{3.43}$$

Provided $\text{Re}(\alpha)>0, p>0, \text{Re}\left(\lambda+\frac{1}{2}\right)>0,\ \text{Re}(\nu+1)>0.$

xxix.

$$I_{29} \equiv \int_0^\infty x^{-\rho} J_\delta(x) J_v(x)$$

$$I_{p_i,q_i:r}^{m,n}\left[z(x)^\gamma \middle| \begin{matrix} (a_j,\alpha_j)_{1,n}; & (a_{ji},\alpha_{ji})_{r+1,p_i} \\ (b_j,\beta_j)_{1,m}; & (b_{ji},\beta_{ji})_{r1+1,q_i} \end{matrix} \right] dx$$

$$= 2^{-\rho} I_{p_i+4,q_i+1:r}^{m+1,n+1} \tag{3.44}$$

$$\left[2^\gamma z \middle| \begin{matrix} \left(\frac{1}{2}+\frac{\rho}{2}-\frac{\delta}{2}-\frac{v}{2},\frac{\gamma}{2}\right),(a_j,\alpha_j)_{1,n}; & (a_{ji},\alpha_{ji})_{,n+1,p_i} \\ (\rho,\gamma),(b_j,\beta_j)_{1,m}; & (b_{ji},\beta_{ji}),\ {}_{m+1,q_i},\left(\frac{1}{2}+\frac{\rho}{2}+\frac{\delta}{2}+\frac{v}{2},\frac{\gamma}{2}\right)\left(\frac{1}{2}+\frac{\rho}{2}-\frac{\delta}{2}+\frac{v}{2},\frac{\gamma}{2}\right),\left(\frac{1}{2}+\frac{\rho}{2}+\frac{\delta}{2}-\frac{v}{2},\frac{\gamma}{2}\right) \end{matrix} \right]$$

Provided

i. $0 < \mathrm{Re}(\rho) < \mathrm{Re}(\delta + v + 1)$

ii. $\lambda > 0, |\arg z| < \frac{1}{2}\pi\lambda$

iii. $\lambda \geq 0, |\arg z| \leq \frac{1}{2}\pi\lambda,\ \mathrm{Re}(\mu+1) < 0.\ \mathrm{Re}(\mu-\rho+2) < 0$

3.4 INTEGRALS INVOLVING HYPERGEOMETRIC FUNCTIONS

This section presents certain integrals involving the product of the I-function with Gauss hypergeometric functions and its generalized forms.

xxx.

$$I_{30} \equiv \int_0^\infty x^{-\rho}(1-x)^{\sigma-1}{}_2F_1\left[\begin{matrix} k+\sigma-\rho, \lambda+\sigma-\rho; \\ \sigma; \end{matrix} (1-x)\right]$$

$$I_{p_i,q_i:r}^{m,n}\left[\alpha x \left| \begin{matrix} (a_j,\alpha_j)_{1,n}; & (a_{ji},\alpha_{ji})_{n+1,p_i} \\ (b_j,\beta_j)_{1,m}; & (b_{ji},\beta_{ji})_{m+1,q_i} \end{matrix}\right.\right] dx$$

$$= \Gamma(\sigma) I_{p_i+3,q_i+3:r}^{m+1,n+2}$$

$$\times\left[\alpha \left| \begin{matrix} (\rho-\sigma,1),(k+\sigma+\lambda-\rho,1)(a_j,\alpha_j)_{1,n}; & (a_{ji},\alpha_{ji})_{n+1,p_i},(\rho,1) \\ (\rho-\sigma,1),(b_j,\beta_j)_{1,m}; & (b_{ji},\beta_{ji}),_{m+1,q_i},(k,1),(\lambda,1) \end{matrix}\right.\right] \tag{3.45}$$

We have

$$I_{30} \equiv \int_0^\infty x^{-\beta}(1-x)^{\sigma-1}{}_2F_1\left[\begin{matrix} k+\sigma-\rho, \lambda+\sigma-\rho; \\ \sigma; \end{matrix} (1-x)\right]$$

$$\times\frac{1}{2\pi i}\int_L \frac{\prod_{j=1}^{m}\Gamma(b_j-\beta_j\xi)\prod_{j=1}^{n}\Gamma(1-a_j-\alpha_j\xi)\alpha^{\xi}x^{\xi}}{\sum_{i=1}^{r}\left\{\prod_{j=m+1}^{q_i}\Gamma(1-b_{ji}-\beta_{ji}\xi)\prod_{j=n+1}^{p_i}\Gamma(a_{ji}-\alpha_{ji}\xi)\right\}} d\xi dx$$

Putting $x = t+1$ and $dx = dt$, we get

$$I_{30} = \int_0^\infty t^{\sigma-1}(t+1)^{-\rho} {}_2F_1\left[\begin{matrix} k+\sigma-\rho, \lambda+\sigma-\rho; \\ \sigma; \end{matrix} \; -t\right]$$

$$\times \frac{1}{2\pi i}\int_L \frac{\prod_{j=1}^{m}\Gamma(b_j-\beta_j\xi)\prod_{j=1}^{n}\Gamma(1-a_j-\alpha_j\xi)\alpha^{\xi}(t+1)^{\xi}}{\sum_{i=1}^{r}\left\{\prod_{j=m+1}^{q_i}\Gamma(1-b_{ji}-\beta_{ji}\xi)\prod_{j=n+1}^{p_i}\Gamma(a_{ji}-\alpha_{ji}\xi)\right\}}d\xi dt \quad = \sum_{v=0}^{\infty}\frac{(k+\rho+\sigma)_k(\lambda+\sigma-\rho)_v(-1)^v}{(\sigma)_v\, v!}$$

$$\times \frac{1}{2\pi i}\int_L \frac{\prod_{j=1}^{m}\Gamma(b_j-\beta_j\xi)\prod_{j=1}^{n}\Gamma(1-a_j-\alpha_j\xi)\alpha^{\xi}}{\sum_{i=1}^{r}\left\{\prod_{j=m+1}^{q_i}\Gamma(1-b_{ji}-\beta_{ji}\xi)\prod_{j=n+1}^{p_i}\Gamma(a_{ji}-\alpha_{ji}\xi)\right\}} \times \int_0^\infty t^{v+\sigma-1}(t+1)^{\xi-\rho}\, d\xi dt$$

$$= \sum_{v=0}^{\infty}\frac{(k+\rho+\sigma)_v(\lambda+\sigma-\rho)_v(-1)^v}{(\sigma)_v\, v!} \times \frac{1}{2\pi i}\int_L \frac{\prod_{j=1}^{m}\Gamma(b_j-\beta_j\xi)\prod_{j=1}^{n}\Gamma(1-a_j-\alpha_j\xi)\alpha^{\xi}}{\sum_{i=1}^{r}\left\{\prod_{j=m+1}^{q_i}\Gamma(1-b_{ji}-\beta_{ji}\xi)\prod_{j=n+1}^{p_i}\Gamma(a_{ji}-\alpha_{ji}\xi)\right\}} \times B(v+\sigma, \rho-s-v-\sigma)d\xi$$

$$= \frac{\Gamma(\sigma)}{2\pi i}\int_L \frac{\prod_{j=1}^{m}\Gamma(b_j-\beta_j\xi)\prod_{j=1}^{n}\Gamma(1-a_j-\alpha_j\xi)\Gamma(\rho-\sigma-\xi)\;\alpha^{\xi}}{\sum_{i=1}^{r}\left\{\prod_{j=m+1}^{q_i}\Gamma(1-b_{ji}-\beta_{ji}\xi)\prod_{j=n+1}^{p_i}\Gamma(a_{ji}-\alpha_{ji}\xi)\right\}\Gamma(\rho-\xi)} \times {}_2F_1\left[\begin{matrix} k+\sigma-\rho, \lambda+\sigma-\rho; \\ 1-\rho+\xi+\sigma; \end{matrix} \; 1\right]d\xi$$

$$= \frac{\Gamma(\sigma)}{2\pi i}\int_L \frac{\prod_{j=1}^{m}\Gamma(b_j-\beta_j\xi)\prod_{j=1}^{n}\Gamma(1-a_j-\alpha_j\xi)\Gamma(\rho-\sigma-\xi)}{\sum_{i=1}^{r}\left\{\prod_{j=m+1}^{q_i}\Gamma(1-b_{ji}-\beta_{ji}\xi)\prod_{j=n+1}^{p_i}\Gamma(a_{ji}-\alpha_{ji}\xi)\right\}\Gamma(\rho-\xi)} \times \frac{\Gamma(1-\rho+\xi+\sigma)\Gamma(1-k-\sigma-\lambda+\rho+\xi)\alpha^{\xi}}{\Gamma(1-k-\xi)\Gamma(1-\lambda+\xi)}d\xi$$

Provided $\mu>0, v>0, \operatorname{Re}(\rho)-\mu \max_{i\le j\le n}\left[\operatorname{Re}\left(1-a_j/\alpha_j\right)\right]>0, \operatorname{Re}(\sigma)-v \max_{i\le j\le n}\left[\operatorname{Re}\left(1-a_j/\alpha_j\right)\right]>0$

xxxiii.

$$I_{33} \equiv \int_0^t x^{\rho-1}(t-x)^{\sigma-1}{}_pF_q\left[(g_p);(h_q);ax^{\mu}(t-x)^{\eta}\right]$$

$$I^{m,n}_{p_i,q_i:r}\left[zx^{\mu}(t-x)^{-v}\left|\begin{matrix}(a_j,\alpha_j)_{1,n}; & (a_{ji},\alpha_{ji})_{n+1,p_i}\\(b_j,\beta_j)_{1,m}; & (b_{ji},\beta_{ji})_{m+1,q_i}\end{matrix}\right.\right]dx$$

$$= t^{\rho+\sigma-1}\sum_{k=0}^{\infty} f(k)t^{(\mu+\eta)k}\times I^{m+1,n+1}_{p_i+1,q_i+2:r}$$

$$\left[zt^{\mu-v}\left|\begin{matrix}(1-\rho-\mu k,\mu),(a_j,\alpha_j)_{1,n}; & (a_{ji},\alpha_{ji})_{n+1,p_i}\\(\sigma+\eta k,v),(b_j,\beta_j)_{1,m}; & (b_{ji},\beta_{ji})_{m+1,q_i}, (1-\rho-\sigma-\mu k-\eta k,\mu-v)\end{matrix}\right.\right] \tag{3.50}$$

Provided $\mu>0,\ v\ge 0$, such that $\mu-v\ge 0, \operatorname{Re}(\rho)+\mu \min_{i\le j\le m}\left[\operatorname{Re}\left(b_j/\beta_j\right)\right]>0$,

$\operatorname{Re}(\sigma)-v \max_{i\le j\le n}\left[\operatorname{Re}\left(1-a_j/\alpha_j\right)\right]>0$

Along with the set of conditions (b), (c) and (d) given by I_{30} and $f(k)$ is given by (3.48).

xxxiv.

$$I_{34} \equiv \int_0^t x^{\rho-1}(t-x)^{\sigma-1}{}_pF_q\left[(g_p);(h_q);ax^{\mu}(t-x)^{\eta}\right]$$

$$I^{m,n}_{p_i,q_i:r}\left[zx^{\mu}(t-x)^{-v}\left|\begin{matrix}(a_j,\alpha_j)_{1,n}; & (a_{ji},\alpha_{ji})_{n+1,p_i}\\(b_j,\beta_j)_{1,m}; & (b_{ji},\beta_{ji})_{m+1,q_i}\end{matrix}\right.\right]dx \tag{3.51}$$

$$= t^{\rho+\sigma-1}\sum_{k=0}^{\infty} f(k)t^{(\mu+\eta)k}\times I^{m+1,n+1}_{p_i+2,q_i+1:r}$$

$$\left[zt^{v-\mu}\left|\begin{matrix}(1-\rho-\mu k,\mu),(a_j,\alpha_j)_{1,n}; & (a_{ji},\alpha_{ji})_{n+1,p_i},(\rho+\sigma+\eta k+\mu k, v-\mu)\\(\sigma+\eta k,v),(b_j,\beta_j)_{1,m}; & (b_{ji},\beta_{ji})_{m+1,q_i}\end{matrix}\right.\right]$$

Provided $\mu \geq 0,\ v > 0$, such that $v - \mu \geq 0$, $\operatorname{Re}(\rho) - \mu \max_{i \leq j \leq n}\left[\operatorname{Re}\left(1 - {a_j}/{\alpha_j}\right)\right] > 0$,

$\operatorname{Re}(\sigma) + v \min_{i \leq j \leq m}\left[\operatorname{Re}\left({b_j}/{\beta_j}\right)\right] > 0$ along with the set of conditions (b), (c), and (d) given by (3.47) and $f(k)$ is given by (3.48).

The integrals (3.49)–(3.51) can be proved on lines similar to those of integrals (3.46).

xxxv.

$$I_{35} \equiv \int_0^{\pi} (\sin x)^{\rho-1} e^{imx} {}_pF_q\left[(\alpha_p);(\beta_q);a(\sin x)^{2h}\right] \times {}_rF_s\left[(\gamma_r);(\delta_s);b(\sin x)^{2k}\right]$$

$$I^{m,n}_{p_i,q_i:r}\left[z(\sin x)^{-2\lambda}\left|\begin{matrix}(a_j,\alpha_j)_{1,n}; & (a_{ji},\alpha_{ji})_{n+1,p_i}\\ (b_j,\beta_j)_{1,m}; & (b_{ji},\beta_{ji})_{m+1,q_i}\end{matrix}\right.\right]dx$$

$$= \sqrt{\pi} e^{im\pi/2} \sum_{l,t=0}^{\infty} f(l) g(t) \times I^{m+2,n}_{p_i+2,q_i+2:r}$$

$$\left[z\left|\begin{matrix}(a_j,\alpha_j)_{1,n}; & (a_{ji},\alpha_{ji})_{n+1,p_i}, \left(\dfrac{1+2hl+2kt \pm m+1}{2},\lambda\right)\\ \left(\dfrac{\sigma+2hl+2kt}{2},\lambda\right),\left(\dfrac{\sigma+2hl+2kt+1}{2},\lambda\right)(b_j,\beta_j)_{1,m}; & (b_{ji},\beta_{ji}),\ _{m+1,q_i}\end{matrix}\right.\right] \tag{3.52}$$

Provided

$$A_i > 0 \text{ and } B_i \leq 0; |\arg z| < \frac{1}{2}\pi A_i,\ \forall i \in 1,2,....,r$$

where

$$f(l) = \frac{\prod_{j=1}^{p}(a_p)_l}{\prod_{j=1}^{q}(\beta_q)_l}\frac{a^l}{l!}\ g(t) = \frac{\prod_{j=1}^{r}(\gamma_r)_t}{\prod_{j=1}^{q}(\delta_s)_t}\frac{b^t}{t!}$$

And $\Gamma(x \pm y)$ represents $\Gamma(x+y), \Gamma(x-y)$; h and k are non-negative integers; $p<q$(or $p=q+1$ and $|z|<1$); β_q and δ_s are neither zero nor negative integers; and $\operatorname{Re}(\sigma)>0$.

Here A_i and B_i are given by (3.47).

These results can be proved if we substitute the value of the hypergeometric function and the I-function, and use the following integral

$$\int_0^{\pi} (\sin x)^{\omega-1} e^{imx} {}_pF_q\left[(\alpha_p);(\beta_q);a(\sin x)^{2h}\right] \times {}_uF_v\left[(\gamma_u);(\delta_v);b(\sin x)^{2k}\right]dx$$

$$= \frac{\pi e^{im\pi/2}}{2^{\omega-1}} \times \sum_{r,t=0}^{\infty} \frac{(\alpha_p)_r a^r (\gamma_u)_t \ b^t \sqrt{(\omega+2hr+2kt)}}{(\beta_q)_r r!(\delta_v)_t t! 2^{2(hr+kt)} \sqrt{\frac{(\omega+2hr+2kt \pm m+1)}{2}}}$$

Along with the multiplication formula for the Gamma function

$$\Gamma(mz) = (2\pi)\frac{1-m}{2} m^{z-1/2} \prod_{j=1}^{m}\left(z+\frac{\xi-1}{m}\right), \ z \neq 0, -\frac{1}{m}, \frac{2}{m}, \ldots; m = 1,2,3\ldots$$

3.5 MISCELLANEOUS RESULTS

In this section, we write integrals involving products of H-function and I-function and other mixed forms.

xxxvi.

$$I_{36} \equiv \int_0^t x^{\eta+1} I_{p_i,q_i:r}^{m_1,n_1}\left[zx^{\sigma}\left|\begin{matrix}(a_j,\alpha_j)_{1,n}; & (a_{ji},\alpha_{ji})_{n+1,p_i}\\(b_j,\beta_j)_{1,m}; & (b_{ji},\beta_{ji})_{m+1,q_i}\end{matrix}\right.\right] \times H_{p,q}^{m,n}\left[wx\left|\begin{matrix}(c_j,\gamma_j)_{1,n}; & (a_j,\alpha_j)_{n+1,p}\\(d_j,\delta_j)_{1,m}; & (b_j,\beta_j)_{m+1,q}\end{matrix}\right.\right]dx$$

$$= w^{-\eta} I_{p_i+q,q_i+p:r}^{m_1+n,n_1+m}$$

$$\times\left[zw^{-\sigma}\left|\begin{matrix}(a_j,\alpha_j)_{1,n_1},(1-d_j-\eta\delta_j,\delta_j\sigma)_{1,m}; & (a_{ji},\alpha_{ji})_{n+1,p_i},(1-d_j-\eta\delta_j,\delta_j\sigma)_{m+1,q}\\(b_j,\beta_j)_{1,m_1},(1-c_j-\eta\gamma_j,\gamma_j\sigma)_{1,n}; & (b_{ji},\beta_{ji}),_{m+1,q_i},(1-c_j-\eta\gamma_j,\gamma_j\sigma)_{n+1,p}\end{matrix}\right.\right] \quad (3.53)$$

Provided

a. $\lambda > 0, |\arg z| < \frac{1}{2}\pi\lambda$

b. $\lambda \geq 0, |\arg z| \leq \frac{1}{2}\pi\lambda, \ \mathrm{Re}(\mu+1) < 0$

c. $\lambda_1 > 0, |\arg w| < \frac{1}{2}\pi\lambda_1$

d. $\lambda_1 \geq 0, |\arg w| \leq \frac{1}{2}\pi\lambda_1, \ \mathrm{Re}(\mu_1+1) < 0$

e. $\sigma > 0, -\sigma \min_{1\leq i\leq m_1} \mathrm{Re}\left(b_j/\beta_j\right) - \min_{1\leq i\leq m} \mathrm{Re}\left(d_j/\delta_j\right) < \mathrm{Re}(\eta) < \sigma < \min_{1\leq i\leq n_1}\left[\mathrm{Re}\left(1-a_j/\alpha_j\right)\right]$

$+\min_{1\leq i\leq n}\left[\mathrm{Re}\left(1-c_j/\gamma_j\right)\right]$

$$\lambda_1 = \sum_{j=1}^{m}\delta_j + \sum_{j=1}^{n}\gamma_j - \sum_{j=m+1}^{q}\delta_j - \sum_{j=n+1}^{p}\gamma_j$$

$$\mu_1 = \frac{1}{2}(p-q) + \sum_{j=1}^{q}d_j - \sum_{j=1}^{p}c_j$$

We have

$$I_{36} = \frac{1}{2\pi i}\int_L \phi(\xi) z^{\xi} \int_0^t x^{\eta+\sigma\xi-1} H_{p,q}^{m,n}\left[wx \left| \begin{matrix} (c_j,\gamma_j)_{1,n}; & (a_j,\alpha_j)_{n+1,p} \\ (d_j,\delta_j)_{1,m}; & (b_j,\beta_j)_{m+1,q} \end{matrix} \right. \right] dx d\xi \quad (3.54)$$

We use the Mellin transform of H-function, so that

$$I_{36} = \frac{1}{2\pi i}\int_L \phi(\xi) z^{\xi} w^{-(\eta+\sigma\xi)} \frac{\prod_{j=1}^{m}\Gamma(d_j+\delta_j)(\eta+\sigma\xi)\prod_{j=1}^{n}\Gamma(1-c_j-\gamma_j)(\eta+\sigma\xi)}{\prod_{j=m+1}^{m}\Gamma(1-d_j-\delta_j)(\eta+\sigma\xi)\prod_{j=n+1}^{p}\Gamma(c_j+\gamma_j)(\eta+\sigma\xi)} d\xi$$

$$= w^{\eta}\frac{1}{2\pi i}\int_L \phi(\xi) \frac{\prod_{j=1}^{m}\Gamma(d_j+\delta_j\eta+\delta_j\sigma\xi)\prod_{j=1}^{n}\Gamma(1-c_j-\eta c_j-\sigma\xi\gamma_j) w^{-(\sigma\xi)}}{\prod_{j=m+1}^{m}\Gamma(1-d_j-\delta_j\eta-\delta_j\sigma\xi)\prod_{j=n+1}^{p}\Gamma(c_j+\gamma_j\eta+\gamma_j\sigma\xi)} z^{\xi} d\xi$$

$$= w^{\eta} I_{p_i+q,q_i+p,r}^{m_1+n,n_1+m}$$

$$\times\left[zx^{-\sigma} \left| \begin{matrix} (a_j,\alpha_j)_{1,n_1}, (1-d_j-\eta\delta_j,\delta_j\sigma)_{1,m}; & (a_{ji},\alpha_{ji})_{n+1,p_i}, (1-d_j-\eta\delta_j,\delta_j\sigma)_{m+1,q} \\ (b_j,\beta_j)_{1,m_1}, (1-c_j-\eta\gamma_j,\gamma_j\sigma)_{1,n}; & (b_{ji},\beta_{ji}), {}_{m+1,q_i}, (1-c_j-\eta\gamma_j,\gamma_j\sigma)_{n+1,p} \end{matrix} \right. \right]$$

Hence the integral (3.53).

We know the following well-known relation

$$H_{p,q}^{m,n}\left[x \left| \begin{matrix} (a_j,\alpha_j)_{1,p} \\ (b_j,\beta_j)_{1,q} \end{matrix} \right. \right] = H_{q,p}^{n,m}\left[x^{-1} \left| \begin{matrix} (1-b_j,\beta_j)_{1,q} \\ (1-a_j,\alpha_j)_{1,p} \end{matrix} \right. \right]$$

If we apply this result in (3.53), then we get the following important formula.

xxxvii.

$$I_{37} \equiv \int_0^\infty x^{\eta-1} I_{p_i,q_i:r}^{m_1,n_1}\left[zx^{-\sigma} \left| \begin{matrix} (a_j,\alpha_j)_{1,n}; & (a_{ji},\alpha_{ji})_{n+1,p_i} \\ (b_j,\beta_j)_{1,m}; & (b_{ji},\beta_{ji})_{m+1,q_i} \end{matrix} \right.\right]$$

$$\times H_{p,q}^{m,n}\left[wx \left| \begin{matrix} (c_j,\gamma_j)_{1,n}; & (a_j,\alpha_j)_{n+1,p} \\ (d_j,\delta_j)_{1,m}; & (b_j,\beta_j)_{m+1,q} \end{matrix} \right.\right] dx$$

$$= w^{-\eta} I_{p_i+p,q_i+q:r}^{m_1+n,n_1+m}$$

$$\times\left[zw^{\sigma} \left| \begin{matrix} (a_j,\alpha_j)_{1,n_1},(c_j+\eta\gamma_j,\gamma_j\sigma)_{1,n}; & (a_{ji},\alpha_{ji})_{n+1,p_i},(c_j+\eta\gamma_j,\gamma_j\sigma)_{n+1,p} \\ (b_j,\beta_j)_{1,m_1},(d_j+\eta\delta_j,\delta_j\sigma)_{1,m}; & (b_{ji},\beta_{ji}),_{m+1,q_i},(d_j+\eta\delta_j,\delta_j\sigma)_{m+1,q} \end{matrix} \right.\right] \tag{3.55}$$

Provided

a. $\lambda > 0, |\arg z| < \frac{1}{2}\pi\lambda$

b. $\lambda \geq 0, |\arg z| \leq \frac{1}{2}\pi\lambda, \ \mathrm{Re}(\mu+1) < 0$

c. $\lambda_1 > 0, |\arg w| < \frac{1}{2}\pi\lambda_1$

d. $\lambda_1 \geq 0, |\arg w| \leq \frac{1}{2}\pi\lambda_1, \ \mathrm{Re}(\mu_1+1) < 0$

e. $\sigma \min_{1\leq j\leq n_1}\left[\mathrm{Re}^{(a_j-1)}\big/_{\alpha_j}\right] - \min_{1\leq j\leq m}\left[\mathrm{Re}\left({}^{d_j}\big/_{\delta_j}\right)\right] < \mathrm{Re}(\eta) < \sigma < \min_{1\leq j\leq m_1}\left[\mathrm{Re}\left({}^{b_j}\big/_{\beta_j}\right)\right] -$
$\min_{1\leq j\leq n}\left[\mathrm{Re}^{(c_j-1)}\big/_{\gamma_j}\right]$

$$\lambda_1 = \sum_{j=1}^{m}\delta_j + \sum_{j=1}^{n}\gamma_j - \sum_{j=m+1}^{q}\delta_j - \sum_{j=n+1}^{p}\gamma_j$$

$$\mu_1 = \frac{1}{2}(p-q) + \sum_{j=1}^{q} d_j - \sum_{j=1}^{p} c_j$$

xxxviii.

$$I_{38} \equiv \int_0^\infty x^{\eta-1} I_{p_{l_1},q_{l_1}i:r}^{m_1,n_1}\left[zx \left| \begin{matrix} (a_j,\alpha_j)_{1,n_1}; & (a_{ji},\alpha_{ji})_{n_1+1,p_{l_1}} \\ (b_j,\beta_j)_{1,m_1}; & (b_{ji},\beta_{ji})_{m_1+1,q_{l_1}} \end{matrix} \right.\right]$$

$$\times I_{p_{l_2},q_{l_2}i:r_2}^{m_1,n_1}\left[wx \left| \begin{matrix} (c_j,\gamma_j)_{1,n_2}; & (c_{ji},\gamma_{ji})_{n_2+1,p_{l_2}} \\ (d_j,\delta_j)_{1,m_2}; & (d_{ji},\delta_{ji})_{m_2+1,q_{l_2}} \end{matrix} \right.\right] dx \tag{3.56}$$

$$= w^{-\eta} I_{P_l,Q_l:r}^{M,N}\left[\frac{z}{w} \left| \begin{matrix} (A_j,\alpha_j)_{1,N}; & (A_{jl},\alpha_{jl})_{N+l,P_l} \\ (B_j,\beta_j)_{1,M}; & (B_{jl},\beta_{jl})_{M+l,Q_l} \end{matrix} \right.\right]$$

where $M = m_1 + n_2$, $N = n_1 + m_2$, $Q_i = q_{l_1} + p_{l_2} p_i + p_i$; $p_i = p_{l_1} + q_{l_2} : r = r_1 r_2$

$$A_j = a_j;\ \alpha_j = \alpha_j;\quad j = 1,2,3,.....n_1$$

$$A_{m_1+j} = 1 - d_j - \eta\delta_j;\ \alpha_{m_1+j} = \delta_j;\ j = 1,2,3,....\ m_2$$

$$B_j = b_j; \beta_j = \beta_j;\quad j = 1,2,3,.....m_1$$

$$B_{m_1+j} = 1 - c_j - \eta\gamma_j;\ \beta_{m_1+j} = \gamma_j;\ j = 1,2,3,.....n_2$$

And

$$A_{N+j,l} = a_{m_1+j,l_1},\ \alpha_{N+j,l} = \alpha_{n_1+j,l_1};\ j = 1,2,3,.....p_i - n_1$$

$$A_{p_i+m_2+j,l} = 1 - d_{m_2+j,l_2} - \eta\delta_{m_2+j,l_2}; \alpha_{p_l+m_2+j,l} = \delta_{m_2+j,l_2}; \forall j = 1,2,....q_{l_2} - m_2$$

$$B_{m+j} = b_{m_1+j,l_1}; \beta_{m+j,l} = \beta_{m_1+j,l};\quad j = 1,2,3,.....q_{l_1} - m_1$$

$$B_{q_1+n_2+j,l} = 1 - c_{n_2+j,l_2} - \eta\gamma_{n_2+j,l_2}; \beta_{q_{l_1}+n_2+j,l} = \gamma_{n_2+j,l_2}; \forall j = 1,2,....p_{l_2} - n_2$$

Here

$$I = (I_1 - 1)r_2 + I_2;\ I_1 = 1,2,\ldots,\ r_1 \text{ and } I_2 = 1,2,\ldots,\ r_2$$

i. $\lambda_1 > 0, |\arg z| < \frac{1}{2}\pi\lambda_1$

ii. $\lambda \geq 0, |\arg z| \leq \frac{1}{2}\pi\lambda,\ \operatorname{Re}(\mu_1 + 1) < 0$

iii. $\lambda_2 > 0, |\arg w| < \frac{1}{2}\pi\lambda_2$

iv. $\lambda_2 \geq 0, |\arg w| \leq \frac{1}{2}\pi\lambda_2, \mathrm{Re}(\mu_2 + 1) < 0$

v. $\min_{1 \leq j \leq m_1}\left[\mathrm{Re}\left(b_j / \beta_j\right)\right] - \min_{1 \leq j \leq m_2}\left[\mathrm{Re}\left(d_j / \delta_j\right)\right] < \mathrm{Re}(\eta) < \min_{1 \leq j \leq n_1}\left[\mathrm{Re}^{(1-a_j)}/\alpha_j\right] + \min_{1 \leq j \leq n_2}\left[\mathrm{Re}^{(1-c_j)}/\gamma_j\right]$

where

$$\lambda_1 = \sum_{j=1}^{n_1}\alpha_j + \sum_{j=1}^{m_1}\beta_j + \max_{1 \leq l_1 \leq r_1}\left[\sum_{j=n_1+1}^{p_{l_1}}\alpha_{jl_1} + \sum_{j=m_1+1}^{q_{l_1}}\beta_{jl_1}\right].$$

$$\mu_1 = \sum_{j=1}^{m_1}b_j - \sum_{j=1}^{n_1}a_j - \min_{1 \leq l_1 \leq r_1}\left[\sum_{j=n_1+1}^{p_{l_1}}a_{jl_1} + \sum_{j=m_1+1}^{q_{l_1}}b_{jl_1} - \left(\frac{p_{l_1} - q_{l_1}}{2}\right)\right]$$

$$\lambda_2 = \sum_{j=1}^{n_2}\gamma_j + \sum_{j=1}^{m_2}\delta_j - \max_{1 \leq l_2 \leq r_2}\left[\sum_{j=n_2+1}^{p_{l_2}}\gamma_{jl_2} + \sum_{j=m_2+1}^{q_{l_2}}\beta_{jl_2}\right].$$

$$\mu_2 = \sum_{j=1}^{m_2}d_j - \sum_{j=1}^{n_2}c_j - \min_{1 \leq l_2 \leq r_2}\left[\sum_{j=n_2+1}^{p_{l_2}}c_{jl_2} - \sum_{j=m_2+1}^{q_{l_2}}d_{jl_2} - \left(\frac{p_{l_2} - q_{l_2}}{2}\right)\right]$$

This above formula can be extended to the following integral of product of K I-functions.

xxxix.

$$I_{39} \equiv \int_0^\infty \dots \int_0^\infty X^{\eta-1} I_{p_{l_1},q_{l_1}i:r}^{m_1,n_1}\left[XZ \middle| \begin{matrix} \left(a^1{}_j,\alpha^1{}_j\right)_{1,n_1}; & \left(a^1{}_{jl_1},\alpha^1{}_{jl_1}\right)_{n_1+1,p_{l_1}} \\ \left(b^1{}_j,\beta^1{}_j\right)_{1,m_1}; & \left(b^1{}_{jl_1},\beta^1{}_{jl_1}\right)_{m_1+1,q_{l_1}} \end{matrix}\right]$$

$$\times \prod_{t=2}^{K} I_{p_{l_t},q_{l_t}:r_t}^{m_t,n_t}\left[w_t x_t \middle| \begin{matrix} \left(a^t{}_j,\alpha^t{}_j\right)_{1,n_t}; & \left(a^t{}_{jl_t},\alpha^t{}_{jl_t}\right)_{n_t+1,p_{l_t}} \\ \left(b^t{}_j,\beta^t{}_j\right)_{1,m_t}; & \left(b^t{}_{jl_t},\beta^t{}_{jl_t}\right)_{m_t+1,q_{l_t}} \end{matrix}\right] dX \tag{3.57}$$

$$= W^{-\eta} I_{P_l,Q_l:r}^{M,N}\left[\frac{Z}{W} \middle| \begin{matrix} \left(A_j,\alpha_j\right)_{1,N}; & \left(A_{jl},\alpha_{jl}\right)_{N+l,P_l} \\ \left(B_j,\beta_j\right)_{1,M}; & \left(B_{jl},\beta_{jl}\right)_{M+l,Q_l} \end{matrix}\right]$$

$$X = x_2 x_3 x_4 \ldots x_K, \quad W = w_1 w_2 w_3 \ldots w_K, \quad dX = dx_2 dx_3 dx_4 \ldots dx_K$$

$$M = m_1 + n_2 + n_3 + \ldots. + n_K, \, N = n_1 + m_2 + m_3 + \ldots. + m_K$$

$$Q_i = q_{li} + p_{2i} + p_{3i} + \ldots. p_{Ki}, \, P_i = p_{li} + q_{2i} + q_{3i} + \ldots. q_{Ki}$$

$$r = r_1 r_2 r_3 \ldots r_k$$

$$A_j = a^1{}_j; \, \alpha_j = \alpha^1{}_j; \quad j = 1,2,3,\ldots\ldots n_1$$

$$A_{n_1+j} = 1 - b^2{}_j - \eta\beta^2{}_j; \, \alpha_{n_1+j} = \beta^2{}_j; \, j = 1,2,3,\ldots. \, m_2$$

$$A_{n_1+m_2+j} = 1 - b^3{}_j - \eta\beta^3{}_j; \, \alpha_{n_1+m_2+j} = \beta^3{}_j; \, j = 1,2,3,\ldots. \, m_3$$

$$A_{n_1+m_2+m_3+\ldots\ldots m_k-1+j} = 1 - b^k{}_j - \eta\beta^k{}_j; \, \alpha_{n_1+m_2+m_3+\ldots\ldots m_k-1+j} = \beta^k{}_j; \, j = 1,2,3,\ldots. \, m_k$$

$$B_j = b^1{}_j; \beta_j = \beta^1{}_j; \quad j = 1,2,3,\ldots\ldots m_1$$

$$B_{m_1+j} = 1 - a^2{}_j - \eta\alpha^2{}_j; \, \beta_{m_1+j} = \alpha^2{}_j; \, j = 1,2,3,\ldots\ldots n_2$$

$$B_{m_1+n_2+j} = 1 - a^3{}_j - \eta\alpha^3{}_j; \, \beta_{m_1+n_2+j} = \alpha^3{}_j; \, j = 1,2,3,\ldots\ldots n_3$$

$$B_{m_1+n_2+n_3+\ldots\ldots n_{k-1}+j} = 1 - a^k{}_j - \eta\alpha^k{}_j; \, \beta_{m_1+n_2+n_3+\ldots\ldots n_{k-1}+j} = \alpha^k{}_j; \, j = 1,2,3,\ldots\ldots n_k$$

And

$$A_{N+j,l} = a^1{}_{n_1+j,}, \, \alpha_{N+j,l} = \alpha^1{}_{n_1+j,l,1}; \quad j = 1,2,3,\ldots\ldots p_{l_1} - n_1$$

$$A_{p_{l_1} + \sum_{h=2}^{t-1} q l_h \sum_{s=t}^{k} m_s + j, l} = 1 - b^t{}_{m_t + j l_t} - \eta\beta^t{}_{m_t+j,l_t}; \quad \alpha_{p_{l_1} + \sum_{h=2}^{t-1} q l_h \sum_{s=t}^{k} m_s + j, l}$$

$$= \beta^t{}_{m_t+j,l_t}; \, j = 1,2,\ldots. q_{l_t} - m_t, \quad \forall t = 2,3,4,\ldots\ldots,k$$

$$B_{M+j,l} = b^1{}_{M_1+j,l_1}; \beta_{M+j,l} = \beta^1{}_{m_1+j,l}; \quad j = 1,2,3,\ldots\ldots q_{l_1} - m_1$$

$$B_{q_{l_1}+\sum_{h=2}^{t-1} p_{l_h} \sum_{s=t}^{k} n_s+j,l} = 1 - a^t{}_{n_t+j,l_t} - \eta \alpha^t{}_{n_t+j,l_t};\ \beta_{q_{l_1}+\sum_{h=2}^{t-1} p_{l_h} \sum_{s=t}^{k} n_s+j,l} = \alpha_{n_t+j,l_t};\ j$$

$$= 1,2,....p_{l_t} - n_t,\ ,\ \forall t = 2,3,4,.....,k$$

Here

$$I = \left(...\left(\left(\left((I_1-1)r_2 + I_2 - 1\right)r_3 + I_3 - 1\right)r_4 + I_4 - 1\right)r_{5.......}\right)r_k + I_k$$

Also

$$I_1 = 1,2,3,....,\ r_1$$

$$I_2 = 1,2,3,....,\ r_2$$

$$I_k = 1,2,3,....,\ r_k$$

i. $\lambda_1 > 0,\ |\arg z| < \frac{1}{2}\pi\lambda_1$

ii. $\lambda \geq 0,\ |\arg z| \leq \frac{1}{2}\pi\lambda,\ \operatorname{Re}(\mu_1 + 1) < 0$

iii. $\lambda_t > 0,\ |\arg w_t| < \frac{1}{2}\pi\lambda_t, \forall t = 2,3,....k$

iv. $\lambda_t \geq 0,\ |\arg w_t| \leq \frac{1}{2}\pi\lambda_t,\ \operatorname{Re}(\mu_t + 1) < 0 \forall t = 2,3,....k$

v. $\sum_{i=1}^{k}\left\{-\min_{1\leq j\leq m_i}\left[\operatorname{Re}\left(b^i{}_j / \beta^i{}_j\right)\right]\right\}\operatorname{Re}(\eta) < \sum_{i=1}^{k}\left\{-\min_{1\leq j\leq n_i}\left[\operatorname{Re}^{(1-a^i{}_j)}/\alpha^i{}_j\right]\right\}$

where

$$\lambda_i = \sum_{j=1}^{n_i}\alpha^i{}_j + \sum_{j=1}^{m_i}\beta^i{}_j - \max_{1\leq l_i\leq r_i}\left[\sum_{j=n_i+1}^{p_{l_i}}\alpha^i{}_{jl_i} + \sum_{j=m_i+1}^{q_{l_i}}\beta^i{}_{jl_i}\right].$$

$$\mu_i = \sum_{j=1}^{m_i} b^i{}_j - \sum_{j=1}^{n_i} a^i{}_j - \min_{1\leq l_i\leq r_i}\left[\sum_{j=n_i+1}^{p_{l_i}} a^i{}_{jl_i} + \sum_{j=m_i+1}^{q_{l_i}} b^i{}_{jl_i} - \left(\frac{p_{l_i} - q_{l_i}}{2}\right)\right]\forall i = 1,2,3...,l$$

XXXX.

$$I_{40} \equiv \int_0^1\int_0^1 x^{\lambda-1}(1-x)^{\mu-1} y^{\rho-1}(1-y)^{a-2\rho}(1+ty)^{\rho-a-1} {}_2F_1\left[\begin{matrix} a,b; & \frac{(1+t)y}{(1+ty)} \\ 1+a-b; & \end{matrix}\right]$$

$$\times J_s^{(\alpha,\beta)}(1-2x;k)S_n^m\left[cy^{\gamma}(1-y)^{-2\gamma}(1+ty)^{r}\right]$$

$$I_{p_i,q_i:r}^{m,n}\left[zx^{\sigma}(1-x)^{v}\left\{\frac{(1+t)y}{(1-y)^2}\right\}^T \middle| \begin{matrix} (a_j,\alpha_j)_{1,n}; & (a_{ji},\alpha_{ji})_{n+1,p_i} \\ (b_j,\beta_j)_{1,m}; & (b_{ji},\beta_{ji})_{m+1,q_i} \end{matrix}\right]dxdy \tag{3.58}$$

$$= \frac{2^{\alpha}\left\{4(1+t)^{-\rho}\right\}\Gamma\left(1+\frac{a}{2}\right)\Gamma(1+a-b)(a+1)_{ks}}{\sqrt{\pi}\Gamma(1+a)\Gamma\left(1+\frac{a}{2}-b\right)S!}\sum_{u=0}^{[n/m]}\sum_{t=0}^{s}\frac{(-n)_{mu}}{u!}\times\frac{(\alpha+\beta+S+1)_{kl}(-S)_l C^u}{(\alpha+1)_{kl}\, l!\{4(1+t)\}^{\gamma u}}A_{n,u}I_{p_i+4,q_i+3:r}^{m+2,n+3}$$

$$\left[z\{4(1+t)\}^{-T} \middle| \begin{matrix} (1-\mu,v),(1-\lambda-kl,\sigma),(1-\rho-\gamma u,T)(a_j,\alpha_j)_{1,n}; & (a_{ji},\alpha_{ji})_{n+1,p_i},(1+a-b-\rho-\gamma u,T) \\ \left(\frac{1}{2}+\frac{a}{2}-\rho-\sigma u,T\right),\left(1+\frac{a}{2}-b-\rho-\sigma u,T\right),(b_j,\beta_j)_{1,m}; & (b_{ji},\beta_{ji})_{m+1,q_i},(1-\lambda-\mu-kl,\sigma+v) \end{matrix}\right]$$

The integral (3.58) is valid if the following sets of (sufficient) conditions are satisfied:

i. m is arbitrary positive integer and the coefficient $A_{n,u}(n,u\geq 0)$ are arbitrary constants, real or complex.

ii. $B=\sum_{j=1}^{n}\alpha_j-\sum_{j=n+1}^{p_i}\alpha_{ji}+\sum_{j=1}^{m}\beta_j-\sum_{j=m+1}^{q_i}\beta_{ji}>0$ and $|\arg z|<\frac{1}{2}\pi B$.

iii. $\mathrm{Re}(1+a-b)>0$, $\mathrm{Re}(\alpha)\succ -1$, $\mathrm{Re}(\beta)\succ -1$, $t\succ -1$, $\gamma\geq 0$, $\mathrm{Re}(\lambda+\sigma\xi)\succ 0$,
$\mathrm{Re}(\mu+v\xi)\succ 0$, $\mathrm{Re}(\rho+\sigma u+T\xi)\succ 0$, $\mathrm{Re}(1+a-2\rho-2\gamma u-2T\xi)\succ 0$

where $\xi=\min\limits_{1\leq j\leq m}\left[\mathrm{Re}\left(b_j/\beta_j\right)\right]$ and $u=0,1,2,....,\left(\frac{n}{m}\right)$, $\mathrm{Re}(1-2b)>0$.

xxxxi.

$$I_{41}\equiv\int_0^1\int_0^1 x^{\lambda-1}(1-x)^{\mu-1}y^{\rho-1}(1-y)^{a-2\rho}(1+ty)^{\rho-a-1}{}_3F_2\left[\begin{matrix}a,\frac{a}{2}+1,b; \\ \frac{a}{2},1+a-b;\end{matrix}\ \frac{(1+t)y}{(1+ty)}\right]\times J_s^{(\alpha,\beta)}(1-2x;k)S_n^m\left[cy^{\gamma}(1-y)^{-2\gamma}(1+ty)^{r}\right]$$

$$I_{p_i,q_i:r}^{m,n}\left[zx^{\sigma}(1-x)^{v}\left\{\frac{(1+t)y}{(1-y)^2}\right\}^{T}\left|\begin{matrix}(a_j,\alpha_j)_{1,n}; & (a_{ji},\alpha_{ji})_{n+1,p_i} \\ (b_j,\beta_j)_{1,m}; & (b_{ji},\beta_{ji})_{m+1,q_i}\end{matrix}\right.\right]dxdy \tag{3.59}$$

$$=\frac{2^{\alpha}\left\{4(1+t)^{-\rho}\right\}\Gamma(1+a-b)(a+1)_{ks}}{\sqrt{\pi}\Gamma(1+a)S!}\sum_{u=0}^{[n/m]}\sum_{t=0}^{s}\frac{(-n)_{mu}}{u!}\times\frac{(\alpha+\beta+S+1)_{kl}(-S)_l C^u}{(\alpha+1)_{kl}\,l!\{4(1+t)\}^{\gamma U}}A_{n,u}I_{p_i+4,q_i+3:r}^{m+2,n+3}$$

$$\left[z\{4(1+t)\}^{-T}\left|\begin{matrix}(1-\mu,v),(1-\lambda-kl,\sigma),(1-\rho-\gamma u,T)(a_j,\alpha_j)_{1,n}; & (a_{ji},\alpha_{ji})_{n+1,p_i},(1+a-b-\rho-\gamma u,T) \\ \left(\frac{1}{2}+\frac{a}{2}-\rho-\gamma u,T\right),\left(1+\frac{a}{2}-b-\rho-\gamma u,T\right),(b_j,\beta_j)_{1,m}; & (b_{ji},\beta_{ji}),_{m+1,q_i},(1-\lambda-\mu-kl,\sigma+v)\end{matrix}\right.\right]$$

The validity of (3.59) is the same as those mentioned with (3.58) except that here we have $0\leq b<1$ instead of $Re(1-2b)>0$.

xxxxii.

$$I_{42} \equiv \int_0^1\int_0^1 x^{\lambda-1}(1-x)^{\mu-1}\, y^{\rho-1}(1-y)^{a-2\rho}(1+ty)^{\rho-a-1}\, {}_cF_3\left[\begin{matrix} a,\frac{a}{2}+1,c,b; \\ \frac{a}{2},1+a-c,1+a-d; \end{matrix}\ \frac{(1+t)y}{(1+ty)}\right] \times J_s^{(\alpha,\beta)}(1-2x;k)\,S_n^m\left[cy^{\gamma}(1-y)^{-2\gamma}(1+ty)^{\gamma}\right]$$

$$I_{p_i,q_i:r}^{m,n}\left[zx^{\sigma}(1-x)^{v}\left\{\frac{(1+t)y}{(1-y)^2}\right\}^{T}\left|\begin{matrix}(a_j,\alpha_j)_{1,n};\ (a_{ji},\alpha_{ji})_{n+1,p_i}\\ (b_j,\beta_j)_{1,m};\ (b_{ji},\beta_{ji})_{m+1,q_i}\end{matrix}\right.\right]dxdy$$

$$=\frac{2^{\alpha}\left\{4(1+t)^{-\rho}\right\}\Gamma(1+a-c)\Gamma(1+a-d)(a+1)_{ks}}{\sqrt{\pi}\,\Gamma(1+a)\Gamma(1+a-c-d)S!}\sum_{u=0}^{[n/m]}\sum_{t=0}^{s}\frac{(-n)_{mu}}{u!}\times\frac{(\alpha+\beta+S+1)_{kl}(-S)_l\,C^u}{(\alpha+1)_{kl}\,l!\{4(1+t)\}^{\gamma u}}A_{n,u}\,I_{p_i+5,q_i+4:r}^{m+3,n+3}$$

$$\left[z\{4(1+t)\}^{-T}\left|\begin{matrix}(1-\mu,v),(1-\lambda-kl,\sigma),(1-\rho-\gamma u,T)(a_j,\alpha_j)_{1,n};\quad (a_{ji},\alpha_{ji})_{n+1,p_i},(1+a-c-\rho-\gamma u,T),(1+a-d-\rho-\gamma u,T)\\ \left(\frac{1}{2}+\frac{a}{2}-\rho-\gamma u,T\right),\left(1+\frac{a}{2}-b-\rho-\gamma u,T\right),(1+a-c-d-\rho-\gamma u,T),(b_j,\beta_j)_{1,m};\quad (b_{ji},\beta_{ji})_{m+1,q_i},(1-\lambda-\mu-kl,\sigma+v)\end{matrix}\right.\right] \tag{3.60}$$

The integral (3.60) is valid if the following set of (sufficient) conditions are satisfied

i. $\mathrm{Re}(1+a-c)>0$, $\mathrm{Re}(1+a-2c-2d)>0$, $\mathrm{Re}(1+a-d)>0$

ii. The sets of conditions (i) to (ii) mentioned in (3.56) also hold

To establish (3.58), we use series representation of $S_n^m(x)$ as given below and for the I-function we use Mellin-Barnes types of contour integral on the left-hand side of (3.58) and change the order of integration and summation

$$S_n^m(x)=\sum_{u=0}^{[n/m]}\frac{(-n)_{mu}}{u!}A_{n,u}x^4,\ n=0,1,2,\ldots.$$

where m is arbitrary positive integer and the coefficient $A_{n,u}$, $(n,u \geq 0)$ are arbitrary constants, real or complex.

$$I_{40} = \sum_{u=0}^{[n/m]} \frac{(-n)_{mu}}{u!} C^u A_{n,u} \int_L \left[\phi(\xi) z^{\xi} \times \int_0^1 x^{\lambda+\sigma\xi-1} (1-x)^{\mu+\nu\xi-1} J_s^{(\alpha,\beta)}(1-2x;k)dx \times \int_0^1 y^{\rho+\gamma u+T\zeta-1} (1-y)^{a-T\xi-2\gamma U-2\rho} (1+ty)^{\rho-a+\gamma u+T\xi-1} \times {}_2F_1 dy \right] d\xi$$

Now, using the known results of Erdelyi[22] and interpreting the resulting contour integral as the I-function, at once, we get the right-hand side of (3.58).

The proofs of (3.59) and (3.60) integrals are similar to that of (3.58).

xxxxiii.

$$I_{43} \equiv \int_0^{\infty} z^{\rho+\lambda-1} \sin 2z F \left[\begin{matrix} (a),(a');(c);(c'); \\ (b),(b');(d);(d'); \end{matrix} \; -x^2z^2, -y^2z^2 \right] \times I_{pi,qi:r}^{m,n} \left[\beta z^{2l} \middle| \begin{matrix} (a_j,\alpha_j)_{1,n}; & (a_{ji},\alpha_{ji})_{n+1,p_i} \\ (b_j,\beta_j)_{1,m}; & (b_{ji},\beta_{ji})_{m+1,q_i} \end{matrix} \right] dz = \sum_{k=0}^{\infty} \frac{(\lambda+2k)\Gamma(\lambda+k)}{2^{l+\rho-\lambda-2k}} F \left[\begin{matrix} -k,\lambda+k,(a),(a');(c);(c'); \\ (b),(b');(d);(d'); \end{matrix} \; -x^2, y^2 \right] \tag{3.61}$$

$$\times I_{p_i+3,q_i+1:r}^{m+1,n+1} \left[\beta z^{-2l} \middle| \begin{matrix} \left(\frac{1}{2}-k-\frac{\lambda}{2}-\frac{\rho}{2},l\right),(a_j,\alpha_j)_{1,n}; & (a_{ji},\alpha_{ji})_{n+1,p_i},(1+2k+\lambda-\rho,2l),\left(1-k-\frac{\lambda}{2}-\frac{\rho}{2},l\right) \\ \left(\frac{1}{2}-\rho,2l\right),(b_j,\beta_j)_{1,m}; & (b_{ji},\beta_{ji})_{m+1,q_i} \end{matrix} \right]$$

where $A + A' + C \leq B + B' + D$, $A + A' + C' \leq B + B' + D'$

$$R\left(\rho+\lambda+\frac{2lb_j}{\beta_j}\right) > -1 \quad j = 1,2,3.....,m$$

$$R\left(\rho+\lambda+\frac{2l(a_j-1)}{\alpha_j}\right)<1 \quad j=1,2,3.....,n$$

$$M_2 \equiv \sum_{j=1}^{p_i}\alpha_{ji}-\sum_{j=1}^{q_i}\beta_{ji} \leq 0$$

$$\sum_{j=1}^{n}\alpha_j-\sum_{j=l+1}^{p_i}\alpha_{ji}+\sum_{j=1}^{m}\beta_j-\sum_{j=k+1}^{q_i}\beta_{ji} \equiv M_1>0 \text{ and } |\arg(\beta)|<\frac{1}{2}\pi M$$

xxxxiv.

$$I_{44} \equiv \int_0^\infty z^{\rho+\lambda-1}\sin 2zF\left[\begin{matrix}(a),(a');(c);(c');\\(b),(b');(d);(d');\end{matrix}\quad -x^2z^2,-y^2z^2\right]\times I^{m,n}_{p_i,q_i:r}$$

$$\left[\beta z^{2l}\left|\begin{matrix}(a_j,\alpha_j)_{1,n}; & (a_{ji},\alpha_{ji})_{n+1,p_i}\\(b_j,\beta_j)_{1,m}; & (b_{ji},\beta_{ji})_{m+1,q_i}\end{matrix}\right.\right]dz=\sum_{k=0}^{\infty}\frac{(\lambda+2k)\Gamma(\lambda+k)}{2^{l+\rho-\lambda-2k}k!}F\left[\begin{matrix}-k,\lambda+k,(a),(a');(c);(c');\\(b),(b');(d);(d');\end{matrix}\quad -x^2,y^2\right] \tag{3.62}$$

$$\times I^{m+1,n+1}_{p_i+3,q_i+1:r}\left[\beta(xy)^{-2l}\left|\begin{matrix}\left(\frac{1}{2}-k-\frac{\lambda}{2}-\frac{\rho}{2},l\right),(a_j,\alpha_j)_{1,n}; & (a_{ji},\alpha_{ji})_{n+1,p_i},(1+2k+\lambda-\rho,2l),\left(\frac{1}{2}-k-\frac{\lambda}{2}-\frac{\rho}{2},l\right)\\\left(\frac{1}{2}-\rho,2l\right),(b_j,\beta_j)_{1,m}; & (b_{ji},\beta_{ji})_{m+1,q_i}\end{matrix}\right.\right]$$

Which is valid under the following conditions

$$A+A'+C \leq B+B'+D,\ A+A'+C' \leq B+B'+D'$$

$$R\left(\rho+\lambda+\frac{2lb_j}{\beta_j}\right)>-1 \quad j=1,2,3.....,m$$

$$R\left(\rho+\lambda+\frac{2l(a_j-1)}{\alpha_j}\right)<1 \quad j=1,2,3.....,n$$

$M_1>0,\ M_2\le 0$ and $|\arg(\beta)|<\frac{1}{2}\pi M_1$

XXXXV.

$$I_{45}\equiv\int_0^\infty z^{\rho+\lambda-1}\sin 2zF\left[\begin{matrix}(a),(a');(c);(c');\\(b),(b');(d);(d');\end{matrix}\ -x^2z^2,-y^2z^2\right]\times I^{m,n}_{p_i,q_i:r}\left[\beta z^{2l}\left|\begin{matrix}(a_j,\alpha_j)_{1,n}; & (a_{ji},\alpha_{ji})_{n+1,p_i}\\(b_j,\beta_j)_{1,m}; & (b_{ji},\beta_{ji})_{m+1,q_i}\end{matrix}\right.\right]$$

$$dz=\frac{\Gamma(\lambda+1)}{2^{l+\rho-\lambda}}\sum_{k=0}^{\infty}\frac{1}{2^k k!}F\left[\begin{matrix}-k,\lambda+1,(a),(a');(c);(c');\\(b),(b');(d);(d');\end{matrix}\ x^2,y^2\right]\times$$

$$I^{m+1,n+1}_{p_i+3,q_i+1:r}\left[\beta(xy)^{-2l}\left|\begin{matrix}\left(\frac{1}{2}-k-\frac{\lambda}{2}-\frac{\rho}{2},l\right),(a_j,\alpha_j)_{1,n}; & (a_{ji},\alpha_{ji})_{n+1,p_i},(1+\lambda-\rho,2l),\left(1-k-\frac{\lambda}{2}-\frac{\rho}{2},l\right)\\\left(\frac{1}{2}-\rho-k,2l\right),(b_j,\beta_j)_{1,m}; & (b_{ji},\beta_{ji})_{m+1,q_i}\end{matrix}\right.\right] \tag{3.63}$$

which is valid under the following conditions as in (3.58)

xxxxvi.

$$I_{46} \equiv \int_0^\infty z^{\rho+\lambda-1} \cos 2zF\left[\begin{matrix}(a),(a');(c);(c'); \\ (b),(b');(d);(d');\end{matrix} \; -x^2z^2,-y^2z^2\right] \times I^{m,n}_{p_i,q_i:r}\left[\beta z^{2l} \left| \begin{matrix}(a_j,\alpha_j)_{1,n}; & (a_{ji},\alpha_{ji})_{n+1,p_i} \\ (b_j,\beta_j)_{1,m}; & (b_{ji},\beta_{ji})_{m+1,q_i}\end{matrix}\right.\right]$$

$$dz = \frac{\Gamma(\lambda+1)}{2^{l+\rho-\lambda}} \sum_{k=0}^{\infty} \frac{1}{2^k k!} F\left[\begin{matrix}-k,\lambda+1,(a),(a');(c);(c'); \\ (b),(b');(d);(d');\end{matrix} \; x^2,y^2\right]$$

$$\times I^{m+1,n+1}_{p_i+3,q_i+1:r}\left[\beta(xy)^{-2l} \left| \begin{matrix}\left(\frac{1}{2}-k-\frac{\lambda}{2}-\frac{\rho}{2},l\right),(a_j,\alpha_j)_{1,n}; & (a_{ji},\alpha_{ji})_{n+1,p_i},(1+\lambda-\rho,2l),\left(1-k-\frac{\lambda}{2}-\frac{\rho}{2},l\right) \\ \left(\frac{1}{2}-\rho-k,2l\right),(b_j,\beta_j)_{1,m}; & (b_{ji},\beta_{ji})_{m+1,q_i}\end{matrix}\right.\right] \tag{3.64}$$

The conditions of (3.64) are same as for (3.59).

For $a=b$ and $a'=b'$, the double hypergeometric function in the left breaks up into the product of two generalized hypergeometric functions and from (3.61) we get

xxxxvii.

$$I_{47} \equiv \int_0^\infty z^{\rho+\lambda-1} \sin 2z_C F_D\left[\begin{matrix}(c); \\ (d);\end{matrix} \; -x^2z^2\right]_{C'} F_{D'}\left[\begin{matrix}(c'); \\ (d');\end{matrix} \; -y^2z^2\right] \times I^{m,n}_{p_i,q_i:r}\left[\beta z^{2l} \left| \begin{matrix}(a_j,\alpha_j)_{1,n}; & (a_{ji},\alpha_{ji})_{n+1,p_i} \\ (b_j,\beta_j)_{1,m}; & (b_{ji},\beta_{ji})_{m+1,q_i}\end{matrix}\right.\right]$$

$$dz = \sum_{k=0}^{\infty} \frac{(\lambda+2k)\Gamma(\lambda+k)}{2^{l+\rho-\lambda-2k} k!} F\left[\begin{matrix}-k,\lambda+k;(c);(c'); \\ (d);(d');\end{matrix} \; -x^2,y^2\right]$$

$$\times I^{m+1,n+1}_{p_i+3,q_i+1:r}\left[\beta(xy)^{-2l} \left| \begin{matrix}\left(\frac{1}{2}-k-\frac{\lambda}{2}-\frac{\rho}{2},l\right),(a_j,\alpha_j)_{1,n}; & (a_{ji},\alpha_{ji})_{n+1,p_i},(1+2k+\lambda-\rho,2l),\left(1-k-\frac{\lambda}{2}-\frac{\rho}{2},l\right) \\ \left(\frac{1}{2}-\rho,2l\right),(b_j,\beta_j)_{1,m}; & (b_{ji},\beta_{ji})_{m+1,q_i}\end{matrix}\right.\right] \tag{3.65}$$

The conditions of (3.65) are same (with $A=B,\ A'=B'$) as given in (3.61) and

$$R\left(\rho+\frac{2l\left(a_j-1\right)}{\alpha_j}\right)<\frac{1}{2}\quad j=1,2,3.....,n$$

xxxxviii.

$$I_{48}\equiv\int_0^{\frac{\pi}{2}}e^{i(2\rho-1)}(\sin\theta)^{\rho-1}(\cos\theta)^{\rho-2}{}_2F_1\left[\alpha,\beta;\frac{1}{2}(\alpha+\beta);e^{i\left(\theta-\frac{\pi}{2}\right)}\sin\theta\right]$$

$$\times I_{p_i,q_i:r}^{m,n}\left[ze^{2i\lambda\theta}(\sin\theta)^{\lambda}(\cos\theta)^{\lambda}\left|\begin{matrix}(a_j,\alpha_j)_{1,n};&(a_{ji},\alpha_{ji})_{n+1,p_i}\\(b_j,\beta_j)_{1,m};&(b_{ji},\beta_{ji})_{m+1,q_i}\end{matrix}\right.\right]d\theta=\frac{e^{i\pi\frac{\rho}{2}}\Gamma\left(\frac{\alpha}{2}+\frac{\beta}{2}\right)}{2^{\rho-\alpha-\beta+1}\Gamma(\alpha)\Gamma(\beta)}$$

$$\left\{\Gamma\left(\frac{\alpha}{2}+\frac{1}{2}\right)\Gamma\left(\frac{\beta}{2}\right)I_{p_i+3,q_i+3:r}^{m,n+3}\left[\frac{ze^{i\pi\frac{\lambda}{2}}}{2^{\lambda}}\left|\begin{matrix}(2-\rho,\lambda),\left(2-\rho+\frac{\alpha}{2}+\frac{\beta}{2},\lambda\right),(2+\alpha-\beta-2\rho,2\lambda),(a_j,\alpha_j)_{1,n};&(a_{ji},\alpha_{ji})_{n+1,p_i}\\(b_j,\beta_j)_{1,m};&(b_{ji},\beta_{ji})_{m+1,q_i},\left(1-\rho+\frac{\alpha}{2},\lambda\right),\left(\frac{3}{2}+\frac{\beta}{2}-\rho,\lambda\right),(3-2\rho+\alpha-\beta,2\lambda)\end{matrix}\right.\right]\right.$$

$$\left.+\Gamma\left(\frac{\alpha}{2}\right)\Gamma\left(\frac{\beta}{2}+\frac{1}{2}\right)I_{p_i+3,q_i+3:r}^{m,n+3}\left[\frac{ze^{i\pi\frac{\lambda}{2}}}{2^{\lambda}}\left|\begin{matrix}(2-\rho,\lambda),\left(2-\rho+\frac{\alpha}{2}+\frac{\beta}{2},\lambda\right),(2-\alpha+\beta-2\rho,2\lambda),(a_j,\alpha_j)_{1,n};&(a_{ji},\alpha_{ji})_{n+1,p_i}\\(b_j,\beta_j)_{1,m};&(b_{ji},\beta_{ji})_{m+1,q_i},\left(1-\rho+\frac{\beta}{2},\lambda\right),\left(\frac{3}{2}-\rho+\frac{\alpha}{2},\lambda\right),(3-2\rho-\alpha+\beta,2\lambda)\end{matrix}\right.\right]\right\}\quad(3.66)$$

Provided $\mathrm{Re}(\rho)>1$, $\mathrm{Re}(2\rho-\alpha-\beta)>2$, $\lambda\geq 0$, $|\arg z|<\frac{1}{2}\pi B$, where B is given in (3.58).

xxxxix.

$$I_{48} \equiv \int_0^{\frac{\pi}{2}} x^{\rho-1}(1-x)^{\rho-2}\left[1+ax+b(1-x)\right]^{-2\rho+1} {}_2F_1\left[\alpha,\beta;\frac{1}{2}(\alpha+\beta);\frac{x(1+a)}{1+ax+b(1-x)}\right] \times I^{m,n}_{p_i,q_i:r}\left[\frac{x^{\lambda}(1-x)^{\lambda}z}{\{1+ax+b(1-x)\}^{2\lambda}}\middle|\begin{matrix}(a_j,\alpha_j)_{1,n}; & (a_{ji},\alpha_{ji})_{n+1,p_i}\\ (b_j,\beta_j)_{1,m}; & (b_{ji},\beta_{ji})_{m+1,q_i}\end{matrix}\right]$$

$$dx = \frac{2^{-2\rho+\alpha+\beta-1}\Gamma\left(\frac{\alpha}{2}+\frac{\beta}{2}\right)}{(1+a)^{\rho}(1-b)^{\rho-1}\Gamma(\alpha)\Gamma(\beta)}$$

$$\Bigg\{\Gamma\left(\frac{\alpha}{2}+\frac{1}{2}\right)\Gamma\left(\frac{\beta}{2}\right) I^{m,n+3}_{p_i+3,q_i+3:r}\left[\frac{z2^{-2\lambda}}{(1+a)^{\lambda}(1+b)^{\lambda}}\middle|\begin{matrix}(2-\rho,\lambda),\left(2-\rho+\frac{\alpha}{2}+\frac{\beta}{2},\lambda\right),(2+\alpha-\beta-2\rho,2\lambda),(a_j,\alpha_j)_{1,n}; & (a_{ji},\alpha_{ji})_{n+1,p_i}\\ (b_j,\beta_j)_{1,m}; & (b_{ji},\beta_{ji})_{m+1,q_i},\left(1+\rho+\frac{\alpha}{2},\lambda\right),\left(\frac{3}{2}+\frac{\beta}{2}-\rho,\lambda\right),(3-2\rho+\alpha-\beta,2\lambda)\end{matrix}\right]$$

$$+\Gamma\left(\frac{\alpha}{2}\right)\Gamma\left(\frac{\beta}{2}+\frac{1}{2}\right) I^{m,n+3}_{p_i+3,q_i+3:r}\left[\frac{z2^{-2\lambda}}{(1+a)^{\lambda}(1+b)^{\lambda}}\middle|\begin{matrix}(2-\rho,\lambda),\left(2-\rho+\frac{\alpha}{2}+\frac{\beta}{2},\lambda\right),(2+\alpha+\beta-2\rho,2\lambda),(a_j,\alpha_j)_{1,n}; & (a_{ji},\alpha_{ji})_{n+1,p_i}\\ (b_j,\beta_j)_{1,m}; & (b_{ji},\beta_{ji})_{m+1,q_i},\left(1-\rho+\frac{\beta}{2},\lambda\right),\left(\frac{3}{2}-\rho+\frac{\alpha}{2},\lambda\right),(3-2\rho-\alpha+\beta,2\lambda)\end{matrix}\right]\Bigg\} \tag{3.67}$$

where

$\mathrm{Re}(\rho)>1$, $\mathrm{Re}(2\rho-\alpha-\beta)>2$, a and b are such that none of the exprssions $1+a$, $1+b$, $1+ax+b(1-x)$; $0\le x\le 1$, is zero and $\lambda\ge 0$, $|\arg z|<\frac{1}{2}\pi B$, where B is given in (3.58).

XXXXX.

$$I_{50} \equiv \int_0^{\frac{\pi}{2}} e^{i(2\rho-1)\theta} (\sin\theta)^{\rho-1} (\cos\theta)^{\rho-2} {}_2F_1\left[\alpha,\beta;\frac{1}{2}(\alpha+\beta+2); e^{i\left(\theta-\frac{\pi}{2}\right)}\sin\theta\right]$$

$$\times I^{m,n}_{p_i,q_i:r}\left[ze^{2i\lambda\theta}(\sin\theta)^{\lambda}(\cos\theta)^{\lambda}\left|\begin{matrix}(a_j,\alpha_j)_{1,n}; & (a_{ji},\alpha_{ji})_{n+1,p_i}\\ (b_j,\beta_j)_{1,m}; & (b_{ji},\beta_{ji})_{m+1,q_i}\end{matrix}\right.\right]$$

$$d\theta = \frac{e^{\frac{i\pi\rho}{2}}\Gamma\left(\frac{\alpha}{2}+\frac{\beta}{2}+1\right)}{2^{2\rho-\alpha-\beta+2}(\alpha-\beta)\Gamma(\alpha)\Gamma(\beta)}$$

$$\left\{\Gamma\left(\frac{\alpha}{2}+\frac{1}{2}\right)\Gamma\left(\frac{\beta}{2}\right)I^{m,n+3}_{p_i+3,q_i+3:r}\left[\frac{ze^{\frac{i\pi\lambda}{2}}}{4^{\lambda}}\left|\begin{matrix}(1-\rho,\lambda),\left(1-\rho+\frac{\alpha}{2}+\frac{\beta}{2},\lambda\right),(\alpha-\beta-2\rho,2\lambda),(a_j,\alpha_j)_{1,n}; & (a_{ji},\alpha_{ji})_{n+1,p_i}\\ (b_j,\beta_j)_{1,m}; & (b_{ji},\beta_{ji})_{m+1,q_i},\left(-\rho+\frac{\alpha}{2},\lambda\right),\left(\frac{1}{2}+\frac{\beta}{2}-\rho,\lambda\right),(1-2\rho+\alpha-\beta,2\lambda)\end{matrix}\right.\right]\right.$$

$$\left.-\Gamma\left(\frac{\alpha}{2}\right)\Gamma\left(\frac{\beta}{2}+\frac{1}{2}\right)I^{m,n+3}_{p_i+3,q_i+3:r}\left[\frac{ze^{\frac{i\pi\lambda}{2}}}{4^{\lambda}}\left|\begin{matrix}(1-\rho,\lambda),\left(1-\rho+\frac{\alpha}{2}+\frac{\beta}{2},\lambda\right),(-\alpha+\beta-2\rho,2\lambda),(a_j,\alpha_j)_{1,n}; & (a_{ji},\alpha_{ji})_{n+1,p_i}\\ (b_j,\beta_j)_{1,m}; & (b_{ji},\beta_{ji})_{m+1,q_i},\left(-\rho+\frac{\beta}{2},\lambda\right),\left(\frac{1}{2}-\rho+\frac{\alpha}{2},\lambda\right),(1-2\rho+\alpha+\beta,2\lambda)\end{matrix}\right.\right]\right\} \tag{3.68}$$

Provided $\text{Re}(\rho)>0$, $\text{Re}(2\rho-\alpha-\beta)>2$, $\lambda\geq 0$, $|\arg z|<\frac{1}{2}\pi B$, where B is given in (3.58).

xxxxxi.

$$I_{51} \equiv \int_0^{\frac{\pi}{2}} x^{\rho-1}(1-x)^{\rho}\, {}_2F_1\left[\alpha,\beta;\frac{1}{2}(\alpha+\beta+2);x\right] \times I^{m,n}_{p_i,q_i:r}\left[x^{\lambda}(1-x)^{\lambda} z \,\middle|\, \begin{matrix} (a_j,\alpha_j)_{1,n}; & (a_{ji},\alpha_{ji})_{n+1,p_i} \\ (b_j,\beta_j)_{1,m}; & (b_{ji},\beta_{ji})_{m+1,q_i} \end{matrix}\right]$$

$$dx = \frac{2^{-2\rho+\alpha+\beta-2}\Gamma\left(\frac{\alpha}{2}+\frac{\beta}{2}+1\right)}{(\alpha-\beta)\Gamma(\alpha)\Gamma(\beta)}$$

$$\left\{\Gamma\left(\frac{\alpha}{2}+\frac{1}{2}\right)\Gamma\left(\frac{\beta}{2}\right) I^{m,n+3}_{p_i+3,q_i+3:r}\left[\frac{z}{4^{\lambda}} \,\middle|\, \begin{matrix} (1-\rho,\lambda),\left(1-\rho+\frac{\alpha}{2}+\frac{\beta}{2},\lambda\right),(\alpha-\beta-2\rho,2\lambda),(a_j,\alpha_j)_{1,n}; & (a_{ji},\alpha_{ji})_{n+1,p_i} \\ (b_j,\beta_j)_{1,m}; & (b_{ji},\beta_{ji})_{m+1,q_i},\left(-\rho+\frac{\alpha}{2},\lambda\right),\left(\frac{1}{2}-\rho+\frac{\beta}{2},\lambda\right),(1-2\rho+\alpha-\beta,2\lambda) \end{matrix}\right]\right. \tag{3.69}$$

$$\left. -\Gamma\left(\frac{\alpha}{2}\right)\Gamma\left(\frac{\beta}{2}+\frac{1}{2}\right) I^{m,n+3}_{p_i+3,q_i+3:r}\left[\frac{z}{4^{\lambda}} \,\middle|\, \begin{matrix} (1-\rho,\lambda),\left(1-\rho+\frac{\alpha}{2}+\frac{\beta}{2},\lambda\right),(-\alpha+\beta-2\rho,2\lambda),(a_j,\alpha_j)_{1,n}; & (a_{ji},\alpha_{ji})_{n+1,p_i} \\ (b_j,\beta_j)_{1,m}; & (b_{ji},\beta_{ji})_{m+1,q_i},\left(-\rho+\frac{\beta}{2},\lambda\right),\left(\frac{1}{2}-\rho+\frac{\alpha}{2},\lambda\right),(1-2\rho-\alpha+\beta,2\lambda) \end{matrix}\right]\right\}$$

Provided $\mathrm{Re}(\rho)>0$, $\mathrm{Re}(2\rho-\alpha-\beta)>2$, $\lambda \geq 0$, $|\arg z| < \frac{1}{2}\pi B$, where B is given in (3.58).

xxxxxii.

$$I_{52} \equiv \int_0^{\frac{\pi}{2}} e^{i\pi(2\rho-1)\theta} (\sin\theta)^{\rho-2} (\cos\theta)^{\rho-1} {}_2F_1\left[\alpha, \beta; \frac{1}{2}(\alpha+\beta); e^{i\theta}\cos\theta\right]$$

$$\times I^{m,n}_{p_i,q_i:r}\left[ze^{2i\pi\lambda\theta}(\sin\theta)^{\lambda}(\cos\theta)^{\lambda}\left| \begin{matrix} (a_j,\alpha_j)_{1,n}; & (a_{ji},\alpha_{ji})_{n+1,p_i} \\ (b_j,\beta_j)_{1,m}; & (b_{ji},\beta_{ji})_{m+1,q_i} \end{matrix} \right.\right]$$

$$d\theta = \frac{e^{i\pi\frac{(\rho-1)}{2}}\Gamma\left(\frac{\alpha}{2}+\frac{\beta}{2}\right)}{2^{2\rho-\alpha-\beta+1}\Gamma(\alpha)\Gamma(\beta)}$$

$$\left\{\Gamma\left(\frac{\alpha}{2}+\frac{1}{2}\right)\Gamma\left(\frac{\beta}{2}\right) I^{m,n+3}_{p_i+3,q_i+3:r}\left[\frac{ze^{i\pi\frac{\lambda}{2}}}{4^{\lambda}}\left| \begin{matrix} (2-\rho,\lambda),\left(2-\rho+\frac{\alpha}{2}+\frac{\beta}{2},\lambda\right),(\alpha-\beta-2\rho+2,2\lambda),(a_j,\alpha_j)_{1,n}; & (a_{ji},\alpha_{ji})_{n+1,p_i} \\ (b_j,\beta_j)_{1,m}; & (b_{ji},\beta_{ji})_{m+1,q_i},\left(1-\rho+\frac{\alpha}{2},\lambda\right),\left(\frac{3}{2}+\frac{\beta}{2}-\rho,\lambda\right),(3-2\rho+\alpha-\beta,2\lambda) \end{matrix} \right.\right]+\right.$$

$$\left.\Gamma\left(\frac{\alpha}{2}\right)\Gamma\left(\frac{\beta}{2}+\frac{1}{2}\right) I^{m,n+3}_{p_i+3,q_i+3:r}\left[\frac{ze^{i\pi\frac{\lambda}{2}}}{4^{\lambda}}\left| \begin{matrix} (2-\rho,\lambda),\left(2-\rho+\frac{\alpha}{2}+\frac{\beta}{2},\lambda\right),(-\alpha+\beta-2\rho+2,2\lambda),(a_j,\alpha_j)_{1,n}; & (a_{ji},\alpha_{ji})_{n+1,p_i} \\ (b_j,\beta_j)_{1,m}; & (b_{ji},\beta_{ji})_{m+1,q_i},\left(1-\rho+\frac{\beta}{2},\lambda\right),\left(\frac{3}{2}-\rho+\frac{\alpha}{2},\lambda\right),(3-2\rho-\alpha+\beta,2\lambda) \end{matrix} \right.\right]\right\} \tag{3.70}$$

For $\mathrm{Re}(\rho)>1$, $\mathrm{Re}(2\rho-\alpha-\beta)>2$, $\lambda\geq 0$, $|\arg z|<\frac{1}{2}\pi B$, where B is given in (3.58).

xxxxxiii.

$$I_{53} \equiv \int_0^{\frac{\pi}{2}} e^{i\pi(2\rho+1)\theta} (\sin\theta)^{\rho} (\cos\theta)^{\rho-1} {}_2F_1\left[\alpha, \beta; \frac{1}{2}(\alpha+\beta); e^{i\theta}\cos\theta\right]$$

$$\times I^{m,n}_{p_i,q_i:r}\left[ze^{2i\pi\lambda\theta}(\sin\theta)^{\lambda}(\cos\theta)^{\lambda} \left| \begin{matrix} (a_j,\alpha_j)_{1,n}; & (a_{ji},\alpha_{ji})_{n+1,p_i} \\ (b_j,\beta_j)_{1,m}; & (b_{ji},\beta_{ji})_{m+1,q_i} \end{matrix}\right.\right]$$

$$d\theta = \frac{e^{i\pi\frac{(\rho-1)}{2}}\Gamma\left(\frac{\alpha}{2}+\frac{\beta}{2}+1\right)}{2^{2\rho-\alpha-\beta+2}\Gamma(\alpha-\beta)\Gamma(\alpha)\Gamma(\beta)}$$

$$\left\{\Gamma\left(\frac{\alpha}{2}+\frac{1}{2}\right)\Gamma\left(\frac{\beta}{2}\right) I^{m,n+3}_{p_i+3,q_i+3:r}\left[\frac{ze^{i\pi\frac{\lambda}{2}}}{4^{\lambda}} \left| \begin{matrix} (1-\rho,\lambda),\left(1-\rho+\frac{\alpha}{2}+\frac{\beta}{2},\lambda\right),(\alpha-\beta-2\rho+2,2\lambda),(a_j,\alpha_j)_{1,n}; & (a_{ji},\alpha_{ji})_{n+1,p_i} \\ (b_j,\beta_j)_{1,m}; & (b_{ji},\beta_{ji})_{m+1,q_i},\left(1-\rho+\frac{\alpha}{2},\lambda\right),\left(\frac{3}{2}+\frac{\beta}{2}-\rho,\lambda\right),(3-2\rho+\alpha-\beta,2\lambda) \end{matrix}\right.\right]\right.$$

$$\left.+\Gamma\left(\frac{\alpha}{2}\right)\Gamma\left(\frac{\beta}{2}+\frac{1}{2}\right) I^{m,n+3}_{p_i+3,q_i+3:r}\left[\frac{ze^{i\pi\frac{\lambda}{2}}}{4^{\lambda}} \left| \begin{matrix} (2-\rho,\lambda),\left(2-\rho+\frac{\alpha}{2}+\frac{\beta}{2},\lambda\right),(-\alpha+\beta-2\rho+2,2\lambda),(a_j,\alpha_j)_{1,n}; & (a_{ji},\alpha_{ji})_{n+1,p_i} \\ (b_j,\beta_j)_{1,m}; & (b_{ji},\beta_{ji})_{m+1,q_i},\left(-\rho+\frac{\alpha}{2},\lambda\right),\left(\frac{1}{2}-\rho+\frac{\beta}{2},\lambda\right),(1-2\rho-\alpha+\beta,2\lambda) \end{matrix}\right.\right]\right\}$$

(3.71)

For $\text{Re}(\rho)>1$, $\text{Re}(2\rho-\alpha-\beta)>2$, $\lambda\geq 0$, $|\arg z|<\frac{1}{2}\pi B$, where B is given in (3.58).

4 Some Expansion Formulas of I-Function

In this chapter, we have given recurrence relations, generating series, multiplication theorems and finite and infinite series of I-function. Some special cases have also been discussed.

4.1 RECURRENCE RELATIONS

We established ten recurrence relations of the I-function. No special condition, other than the trivial restrictions on the parameters, is needed for the validity of the results.

The following results [94] will be required in deriving some of these recurrence relations.

$$\tau x J_{v+\tau}^{\tau}(x) = -\, J_{v-j}^{\tau}(x) + v J_{v}^{\tau}(x) \tag{4.1}$$

$$(2v+1)xP_{v}^{\delta}(x) = -\,(v-\delta+1)\, P_{v+1}^{\delta}(x) + (v+\delta)\, P_{v-1}^{\delta}(x) \tag{4.2}$$

$$P_{v-1}^{\delta}(x) - P_{v+1}^{\delta}(x) = (2v+1)\left(x^2-1\right)^{1/2} P_{v}^{\delta-1}(x) \tag{4.3}$$

$$(v-\delta)(v-\delta+1)P_{v+1}^{\delta}(x) - (v+\delta)(v+\delta+1)P_{v-1}^{\delta}(x) = (2v+1)\left(x^2-1\right)^{1/2} P_{v}^{\delta+1}(x) \tag{4.4}$$

$$2vJ_{v}(x) = xJ_{v-1}(x) + J_{v+1}(x) \tag{4.5}$$

The recurrence relations are

DOI: 10.1201/9781003517757-4

i.

$$R_1 \equiv (a_{p_i}\alpha_1 + \alpha_{p_i} - \alpha_{p_i}a_1 - \alpha_1) I^{m,n}_{p_i,q_i:r}\left[z \middle| \begin{matrix} (a_j,\alpha_j)_{1,n}; & (a_{ji},\alpha_{ji})_{n+1,p_i} \\ (b_j,\beta_j)_{1,m}; & (b_{ji},\beta_{ji})_{m+1,q_i} \end{matrix} \right]$$

$$= \alpha_{p_i} I^{m,n}_{p_i,q_i:r}\left[z \middle| \begin{matrix} (a_1-1,\alpha_j),(a_j,\alpha_j)_{2,n}; & (a_{ji},\alpha_{ji})_{n+1,p_i} \\ (b_j,\beta_j)_{1,m}; & (b_{ji},\beta_{ji})_{m+1,q_i} \end{matrix} \right] \tag{4.6}$$

$$+\alpha_1 I^{m,n}_{p_i,q_i:r}\left[z \middle| \begin{matrix} (a_j,\alpha_j)_{1,n}; & (a_{ji},\alpha_{ji})_{n+1,p_i-1}(\alpha_{p_i}-1,\alpha_{p_i}) \\ (b_j,\beta_j)_{1,m}; & (b_{ji},\beta_{ji})_{m+1,q_i} \end{matrix} \right]$$

This can be proved by multiplying

$$x^l I^{m,n}_{p_i,q_i:r}\left[zx^{\gamma} \middle| \begin{matrix} (a_j,\alpha_j)_{1,n}; & (a_{ji},\alpha_{ji})_{n+1,p_i} \\ (b_j,\beta_j)_{1,m}; & (b_{ji},\beta_{ji})_{m+1,q_i} \end{matrix} \right]$$

On both the side of (4.1), integrating with respect to x from 0 to ∞ on both sides and using the integral (3.24). Accordingly, we get

$$\tau I^{m,n+1}_{p_i+2,q_i:r}\left[z \middle| \begin{matrix} (-l.-1,\gamma),(a_j,\alpha_j)_{1,n}; & (a_{ji},\alpha_{ji})_{n+1,p_i}(1+v-\tau-\tau l,\ \tau\gamma) \\ (b_j,\beta_j)_{1,m}; & (b_{ji},\beta_{ji})_{m+1,q_i} \end{matrix} \right]$$

$$= I^{m,n+1}_{p_i+2,q_i:r}\left[z \middle| \begin{matrix} (-l.\gamma),(a_j,\alpha_j)_{1,n}; & (a_{ji},\alpha_{ji})_{n+1,p_i}(v-\tau-\tau l,\ \tau\gamma) \\ (b_j,\beta_j)_{1,m}; & (b_{ji},\beta_{ji})_{m+1,q_i} \end{matrix} \right] \tag{4.7}$$

$$+v\ I^{m,n+1}_{p_i+2,q_i:r}\left[z \middle| \begin{matrix} (-l.-\gamma),(a_j,\alpha_j)_{1,n}; & (a_{ji},\alpha_{ji})_{n+1,p_i}(1+v-\tau-\tau l,\ \tau\gamma) \\ (b_j,\beta_j)_{1,m}; & (b_{ji},\beta_{ji})_{m+1,q_i} \end{matrix} \right]$$

Putting

i. $N = n+1,\ p_i + 2 = p_i$

ii. $-l = a_i,\ a_1 = a_2,\ a_2 = a_3,\ a_3 = a_4 \ldots,\ a_n = a_N,\ a_{n+1} = a_{N+1}, \ldots, a_{p_1} = a_{p_1-1},$
$1+\left(1+v-\tau-\dfrac{t}{a_{p_i}}\right)$

iii. $\gamma = \alpha_i,\ \alpha_1 = \alpha_2,\ \alpha_2 = \alpha_N,\ a_{n+1} = a_{N+1}, \ldots, a_{p_1} = a_{p_1-1},\ \tau\gamma = \alpha_{p_i}$

$$a_2 = a_3$$

Meaning

$$\tau = \frac{\alpha_{p_i}}{\alpha_1},\ v = \left(a_{p_i}\alpha_1 + \alpha_{p_i} - \alpha_{p_i}a_1 - \alpha_1 \right) / \alpha_1$$

We arrive at the result. On the same lines, we can derive the recurrence relations R_2 to R_1 using the formulas (4.2) to (4.5) and the integrals (3.26), (3.28), (3.30), (3.31), (3.32) and (3.44), respectively.

ii.

$$R_2 \equiv \left(b_{q_{i-1}} - b_{q_i}\right) I^{m,n}_{p_i,q_i:r}\left[x \,\middle|\, \begin{matrix} (a_j,\alpha_j)_{1,n}; & (a_{ji},\alpha_{ji})_{n+1,p_i} \\ (b_j,\beta_j)_{1,m}; & (b_{ji},\beta_{ji})_{m+1,q_i-2}\left(b_{q_i-1},\frac{\alpha_1}{2}\right)\left(b_{q_i},\frac{\alpha_1}{2}\right) \end{matrix} \right]$$

$$= \left(a_1 - 2b_{q_i}\right) I^{m,n}_{p_i,q_i:r}\left[x \,\middle|\, \begin{matrix} (1+a_1,\alpha_1)(a_j,\alpha_j)_{2,n}; & (a_{ji},\alpha_{ji})_{n+1,p_i} \\ (b_j,\beta_j)_{1,m}; & (b_{ji},\beta_{ji})_{m+1,q_i-2}\left(1+b_{q_i-1},\frac{\alpha_1}{2}\right)\left(b_{q_i},\frac{\alpha_1}{2}\right) \end{matrix} \right]$$

$$+ \left(2b_{q_i} - a_1\right) I^{m,n}_{p_i,q_i:r}\left[x \,\middle|\, \begin{matrix} (1+a_1,\alpha_1)(a_j,\alpha_j)_{2,n}; & (a_{ji},\alpha_{ji})_{n+1,p_i} \\ (b_j,\beta_j)_{1,m}; & (b_{ji},\beta_{ji})_{m+1,q_i-2}\left(b_{q_i-1},\frac{\alpha_1}{2}\right)\left(1+b_{q_i},\frac{\alpha_1}{2}\right) \end{matrix} \right] \tag{4.8}$$

iii.

$$R_3 \equiv \left(b_{q_{i-1}} - b_{q_i}\right) I^{m,n}_{p_i,q_i:r}\left[x \,\middle|\, \begin{matrix} (a_j,\alpha_j)_{1,n}; & (a_{ji},\alpha_{ji})_{n+1,p_i} \\ (b_j,\beta_j)_{1,m}; & (b_{ji},\beta_{ji})_{m+1,q_i-2}\left(b_{q_i-1},\frac{\alpha_1}{2}\right)\left(b_{q_i},\frac{\alpha_1}{2}\right) \end{matrix} \right]$$

$$= \left(a_1 - 2b_{q_i}\right) I^{m,n}_{p_i,q_i:r}\left[x \,\middle|\, \begin{matrix} (1+a_1,\alpha_1),(a_j,\alpha_j)_{2,n}; & (a_{ji},\alpha_{ji})_{n+1,p_i} \\ (b_j,\beta_j)_{1,m}; & (b_{ji},\beta_{ji})_{m+1,q_i-2}\left(1+b_{q_i-1},\frac{\alpha_1}{2}\right)\left(b_{q_i},\frac{\alpha_1}{2}\right) \end{matrix} \right]$$

$$+ \left(2b_{q_i} - a_1\right) I^{m,n}_{p_i,q_i:r}\left[x \,\middle|\, \begin{matrix} (1+a_1,\alpha_1),(a_j,\alpha_j)_{2,n}; & (a_{ji},\alpha_{ji})_{n+1,p_i} \\ (b_j,\beta_j)_{1,m}; & (b_{ji},\beta_{ji})_{m+1,q_i-2}\left(b_{q_i-1},\frac{\alpha_1}{2}\right)\left(1+b_{q_i},\frac{\alpha_1}{2}\right) \end{matrix} \right] \tag{4.9}$$

iv.

$$R_4 \equiv (b_1 - b_2) I^{m,n}_{p_i,q_i:r}$$

$$\times\left[x\left|\begin{array}{cc}(a_j,\alpha_j)_{1,n}; & (a_{ji},\alpha_{ji})_{n+1,p_i-1}(a_p-\alpha_p)\\ (b_1,\alpha_p/2),(b_2,\alpha_p/2)\,(b_j,\beta_j)_{3,m}; & (b_{ji},\beta_{ji})_{m+1,q_i-2}\end{array}\right.\right]$$

$$=(a_p-2b_2) I^{m,n}_{p_i,q_i:r}$$

$$\times\left[x\left|\begin{array}{cc}(a_j,\alpha_j)_{1,n}; & (a_{ji},\alpha_{ji})_{n+1,p_i-1}(1+a_p,\alpha_p)\\ \left(1+b_1:\dfrac{\alpha_p}{2}\right),\left(b_2,\dfrac{\alpha_p}{2}\right),(b_j,\beta_j)_{1,m}; & (b_{ji},\beta_{ji})_{m+1,q_i}\end{array}\right.\right]$$

$$+(2b_2-a_p) I^{m,n}_{p_i,q_i:r}$$

$$\times\left[x\left|\begin{array}{cc}(a_j,\alpha_j)_{1,n}; & (a_{ji},\alpha_{ji})_{n+1,p_i-1}(1+a_p,\alpha_p)\\ (b_1,\alpha_p/2),(1+b_2,\alpha_p/2)\,(b_j,\beta_j)_{1,m}; & (b_{ji},\beta_{ji})_{m+1,q_i}\end{array}\right.\right] \tag{4.10}$$

v.

$$R_5 \equiv (a-b_{q_i-1}-1) I^{m,n}_{p_i,q_i:r}$$

$$\times\left[x\left|\begin{array}{cc}(a_1,\alpha_1),(a_2,\alpha_1)\,(a_j,\alpha_j)_{3,n}; & (a_{ji},\alpha_{ji})_{n+1,p_i-1}\\ (b_j,\beta_j)_{1,m}; & (b_{ji},\beta_{ji})_{m+1,q_i-2}(1+b_{q_i-1},\alpha_1)(a_1+a_2-b_{q_i-1}-2,\alpha_1)\end{array}\right.\right]$$

$$-(b_{q_i-1}-a_2+3/2) I^{m,n}_{p_i,q_i:r}$$

$$\times\left[x\left|\begin{array}{cc}(a_1,\alpha_1),(a_2,\alpha_1)\,(a_j,\alpha_j)_{1,n}; & (a_{ji},\alpha_{ji})_{n+1,p_i-1}\\ (b_j,\beta_j)_{1,m}; & (b_{ji},\beta_{ji})_{m+1,q_i-2}(b_{q_i-1},\alpha_1);(a_1+a_2-b_{q_i-1}-1,\alpha_1)\end{array}\right.\right]$$

$$=(-1)^{\frac{3}{2}}(a_1+a_2-2b_{q_i-1}-5/2) I^{m,n}_{p_i,q_i:r}$$

$$\times\left[x\left|\begin{array}{cc}(a_1,\alpha_1),(a_2-1,\alpha_1)\,(a_j,\alpha_j)_{3,n}; & (a_{ji},\alpha_{ji})_{n+1,p_i-1}\,\ldots\ldots\ldots\\ (b_j,\beta_j)_{1,m}; & (b_{ji},\beta_{ji})_{m+1,q_i}(b_{q_i-1},\alpha_1);(a_1+a_2-b_{q_i-1}-2,\alpha_1)\end{array}\right.\right] \tag{4.11}$$

vi.

$$R_6 \equiv (1+b_1-a_1) I_{p_i,q_i:r}^{m,n} \times \left[x \left| \begin{matrix} (a_j,\alpha_j)_{1,n}; & (a_{ji},\alpha_{ji})_{n+1,p_i-1}, \left(b_1+b_2-a_1+\frac{3}{2},\alpha_1\right) \\ (b_1,\alpha_1),(1+b_2,\alpha_1),(b_j,\beta_j)_{3,m}; & (b_{ji},\beta_{ji})_{m+1,q_i} \end{matrix} \right. \right]$$

$$+(a_1-b_2-1) I_{p_i,q_i:r}^{m,n} \times \left[x \left| \begin{matrix} (a_j,\alpha_j)_{1,n};; & (a_{ji},\alpha_{ji})_{n+1,p_i-1}, \left(b_1+b_2-a_1+\frac{3}{2},\alpha_1\right) \\ (1+b_2,\alpha_1),(b_2,\alpha_1),(b_j,\beta_j)_{3,m}; & (b_{ji},\beta_{ji})_{m+1,q_i} \end{matrix} \right. \right]$$

$$=(b_2-b_1) I_{p_i,q_i:r}^{m,n} \times \left[x \left| \begin{matrix} (a_1-1,\alpha_1),(a_j,\alpha_j)_{2,n}; & (a_{ji},\alpha_{ji})_{n+1,p_i-1}, \left(b_1-b_2-a_1+\frac{3}{2},\alpha_1\right) \\ (b_1,\alpha_1),(b_2,\alpha_1),(b_j,\beta_j)_{3,m}; & (b_{ji},\beta_{ji})_{m+1,q_i} \end{matrix} \right. \right] \tag{4.12}$$

vii.

$$R_7 \equiv (a_1-b_{q_i-1}-3/2) I_{p_i,q_i:r}^{m,n} \times \left[x \left| \begin{matrix} (a_1,\alpha_1),(a_2,\alpha_1)\ (a_j,\alpha_j)_{3,n}; & (a_{ji},\alpha_{ji})_{n+1,p_i} \\ (b_j,\beta_j)_{1,m}; & (b_{ji},\beta_{ji})_{m+1,q_i-2} (b_{q_i-1},\alpha_1)(a_1-a_2-b_{q_i-1}-1,\alpha_1) \end{matrix} \right. \right]$$

$$+(a_2-b_{q_i-1}-1) I_{p_i,q_i:r}^{m,n} \times \left[x \left| \begin{matrix} (a_1,\alpha_1),(a_2,\alpha_1)\ (a_j,\alpha_j)_{3,n}; & (a_{ji},\alpha_{ji})_{n+1,p_i} \\ (b_j,\beta_j)_{1,m}; & (b_{ji},\beta_{ji})_{m+1,q_i-2} (1+b_{q_i-1},\alpha_1),(a_1-a_2-b_{q_i-1}-1,\alpha_1) \end{matrix} \right. \right]$$

$$(-1)^{\frac{1}{2}} (a_1+a_2-2b_{q_i-1}-5/2) I_{p_i,q_i:r}^{m,n} \times \left[x \left| \begin{matrix} (a_1-1,\alpha_1),(a_2,\alpha_1),(a_j,\alpha_j)_{3,n}; & (a_{ji},\alpha_{ji})_{n+1,p_i} \ldots\ldots\ldots\ldots \\ (b_j,\beta_j)_{1,m}; & (b_{ji},\beta_{ji})_{m+1,q_i-2} (b_{q_i-1},\alpha_1)(a_1+a_2-2-b_{q_i-1},\alpha_1) \end{matrix} \right. \right] \tag{4.13}$$

viii.

$$R_8 \equiv (b_1 - a_1 + 1/2) I^{m,n}_{p_i,q_i:r} \times \left[x \left| \begin{matrix} (a_j,\alpha_j)_{1,n}; & (a_{ji},\alpha_{ji})_{n+1,p_i-1}, \left(b_1 - b_2 - a_1 + \frac{3}{2}, \alpha_1\right) \\ (1+b_1,\alpha_1),(b_2,\alpha_1),(b_j,\beta_j)_{3,m}; & (b_{ji},\beta_{ji})_{m+1,q_i} \end{matrix} \right. \right]$$

$$+(a_1 - b_2 - 1/2) I^{m,n}_{p_i,q_i:r} \times \left[x \left| \begin{matrix} (a_j,\alpha_j)_{1,n}; & (a_{ji},\alpha_{ji})_{n+1,p_i-1}, \left(b_1 - b_2 - a_1 + \frac{3}{2}, \alpha_1\right) \\ (b_1,\alpha_1),(1+b_2,\alpha_1),(b_j,\beta_j)_{3,m}; & (b_{ji},\beta_{ji})_{m+1,q_i} \end{matrix} \right. \right]$$

$$= (b_1 - b_2) I^{m,n}_{p_i,q_i:r} \times \left[x \left| \begin{matrix} (a_j,\alpha_j)_{1,n}; & (a_{ji},\alpha_{ji})_{n+1,p_i-1}, \left(b_1 - b_2 - a_1 + \frac{1}{2}, \alpha_1\right) \\ (b_1,\alpha_1),(b_2,\alpha_1),(b_j,\beta_j)_{3,m}; & (b_{ji},\beta_{ji})_{m+1,q_i} \end{matrix} \right. \right] \tag{4.14}$$

ix.

$$R_9 \equiv 2(a_{p_{i-1}} - a_1) I^{m,n}_{p_i,q_i:r} \times \left[x \left| \begin{matrix} (a_j,\alpha_j)_{1,n}; (a_{ji},\alpha_{ji})_{n+1,p_i-1}, (a_{p_{i-2}},\alpha_1); & (a_{p_{-1}},\alpha_1)(a_1 + a_{p_{i-2}} - a_{p_{-1}},\alpha_1) \\ (a_1 + a_{p_{i-2}} - I, 2\alpha_1),(b_j,\beta_j)_{2,m}; & (b_{ji},\beta_{ji})_{m+1,q_i} \end{matrix} \right. \right]$$

$$= I^{m,n}_{p_i,q_i:r} \times \left[x \left| \begin{matrix} (a_j,\alpha_j)_{1,n}; (a_{ji},\alpha_{ji})_{n+1,p_i-3}, (a_{p_{i-2}} - 1,\alpha_1) & (a_{p_{i-2}},\alpha_1)(a_1 + a_{p_{i-2}} - a_{p_{-1}},\alpha_1) \\ (a_1 + a_{p_{i-2}} - 2, 2\alpha_1),(b_j,\beta_j)_{2,m}; (b_{ji},\beta_{ji})_{m+1,q_i} & \ldots\ldots\ldots\ldots \end{matrix} \right. \right]$$

$$+ I^{m,n}_{p_i,q_i:r} \times \left[x \left| \begin{matrix} (a_1 - 1,\alpha_j), (a_j,\alpha_j)_{2,n}; (a_{ji},\alpha_{ji})_{n+1,p_i-3}, (a_{p_{i-2}},\alpha_1) & (a_{p_{i-1}},\alpha_1)(a_1 + a_{p_{i-2}} - a_{p_{-1}},\alpha_1) \\ (a_1 + a_{p_{i-2}} - I, 2\alpha_1),(b_j,\beta_j)_{2,m}; (b_{ji},\beta_{ji})_{m+1,q_i} & \ldots\ldots\ldots\ldots \end{matrix} \right. \right] \tag{4.15}$$

x.

$$R_{10} \equiv \left(b_1\beta_{q_i} - p_{q_i}\beta_1\right) I^{m,n}_{p_i,q_i:r}\left[z \left| \begin{matrix} \left(a_j,\alpha_j\right)_{1,n}; & \left(a_{ji},\alpha_{ji}\right)_{n+1,p_i} \\ \left(b_j,\beta_j\right)_{1,m}; & \left(b_{ji},\beta_{ji}\right)_{m+1,q_i} \end{matrix} \right.\right]$$

$$= \beta_{q_i} I^{m,n}_{p_i,q_i:r}\left[z \left| \begin{matrix} \left(a_j,\alpha_j\right)_{1,n}; & \left(a_{ji},\alpha_{ji}\right)_{n+1,p_i} \\ \left(b_1+1,\beta_1\right),\left(b_j,\beta_j\right)_{2,m}; & \left(b_{ji},\beta_{ji}\right) \end{matrix} \right.\right] \quad (4.16)$$

$$+\beta_1 I^{m,n}_{p_i,q_i:r}\left[z \left| \begin{matrix} \left(a_j,\alpha_j\right)_{1,n}; & \left(a_{ji},\alpha_{ji}\right) \\ \left(b_j,\beta_j\right)_{1,m};\left(b_{ji},\beta_{ji}\right)_{m+1,q_i}, & \left(b_{q1}+1,\ \beta_{q_i}\right) \end{matrix} \right.\right]$$

Where, $m \geq 1$, $q_i \geq 2$.

The proof is

$$\left(b_1 - b_{q_i}\beta_1\right) I^{m,n}_{p_i,q_i:r}\left[z \left| \begin{matrix} \left(a_j,\alpha_j\right)_{1,n}; & \left(a_{ji},\alpha_{ji}\right)_{n+1,p_i} \\ \left(b_j,\beta_j\right)_{1,m}; & \left(b_{ji},\beta_{ji}\right)_{m+1,q_i} \end{matrix} \right.\right]$$

$$= \frac{1}{2\pi i}\int_L \frac{\left(b_1 - b_{q_i}\beta_1\right)\prod_{j=1}^{m}\Gamma\left(b_j-\beta_j\xi\right)\prod_{j=1}^{n}\Gamma\left(1-a_j-\alpha_j\xi\right)z^{\xi}d\xi}{\sum_{i=1}^{r}\left\{\prod_{j=m+1}^{q_i}\Gamma\left(1-b_{ij}-\beta_{ji}\xi\right)\prod_{j=n+1}^{p_i}\Gamma\left(a_{ji}-\alpha_{ji}\xi\right)\right\}}$$

$$= \frac{1}{2\pi i}\int_L \frac{\left(b_1\beta_{q_i} - b_{q_i}\beta_1 + \beta_1\beta_{q_i}\xi\right)\prod_{j=1}^{m}\Gamma\left(b_j-\beta_j\xi\right)\prod_{j=1}^{n}\Gamma\left(1-a_j-\alpha_j\xi\right)}{\sum_{i=1}^{r}\left\{\prod_{j=m+1}^{q_i}\Gamma\left(1-b_{ij}-\beta_{ji}\xi\right)\prod_{j=n+1}^{p_i}\Gamma\left(a_{ji}-\alpha_{ji}\xi\right)\right\}}z^{\xi}d\xi$$

$$= \frac{\beta_{q_i}}{2\pi i}\int_L \frac{\left(b_1 - \beta_1\xi\right)\Gamma(b_1-\beta_1\xi)\prod_{j=2}^{m}\Gamma\left(b_j-\beta_j\xi\right)\prod_{j=1}^{n}\Gamma\left(1-a_j-\alpha_j\xi\right)}{\sum_{i=1}^{r}\left\{\prod_{j=m+1}^{q_i}\Gamma\left(1-b_{ij}-\beta_{ji}\xi\right)\prod_{j=n+1}^{p_i}\Gamma\left(a_{ji}-\alpha_{ji}\xi\right)\right\}}z^{\xi}d\xi$$

$$+ \frac{\beta_1}{2\pi i}\int_L \frac{\left(-b_{q_i} + \beta_{q_i}\xi\right)\prod_{j=2}^{m}\Gamma\left(b_j-\beta_j\xi\right)\prod_{j=1}^{n}\Gamma\left(1-a_j-\alpha_j\xi\right)}{\sum_{i=1}^{r}\left\{\prod_{j=m+1}^{q_i-1}\Gamma\left(1-b_{ij}+\beta_{ji}\xi\right)\Gamma\left(1-b_{q_i i}+\beta_{q_i i}\xi\right)\prod_{j=n+1}^{p_i}\Gamma\left(a_{ji}-\alpha_{ji}\xi\right)\right\}}z^{\xi}d\xi$$

$$= \frac{\beta_{q_i}}{2\pi i}\int_L \frac{\left(b_1 + 1 - \beta_1\xi\right)\prod_{j=2}^{m}\Gamma\left(b_j-\beta_j\xi\right)\prod_{j=1}^{n}\Gamma\left(1-a_j-\alpha_j\xi\right)}{\sum_{i=1}^{r}\left\{\prod_{j=m+1}^{q_i}\Gamma\left(1-b_{ij}+\beta_{ji}\xi\right)\prod_{j=n+1}^{p_i}\Gamma\left(a_{ji}-\alpha_{ji}\xi\right)\right\}}z^{\xi}d\xi$$

$$+ \frac{\beta_1}{2\pi i}\int_L \frac{\prod_{j=2}^{m}\Gamma\left(b_j-\beta_j\xi\right)\prod_{j=1}^{n}\Gamma\left(1-a_j-\alpha_j\xi\right)}{\sum_{i=1}^{r}\left\{\prod_{j=m+1}^{q_i-1}\Gamma\left(1-b_{ij}+\beta_{ji}\xi\right)\Gamma\left(-b_{q_i i}+\beta_{q_i i}\xi\right)\Gamma\left(-b_{q_i i}+\beta_{q_i i}\xi\right)\prod_{j=n+1}^{p_i}\Gamma\left(a_{ji}-\alpha_{ji}\xi\right)\right\}}z^{\xi}d\xi$$

Thus result is obtained using the definition.

4.1.1 Special Cases

a. If a_{p_i} the recurrence relation R_1 reduces to the following recurrence relation

$$\left(a_{p_i}-a_1\right)I^{m,n}_{p_i,q_i:r}\left[z\left|\begin{matrix}\left(a_j,\alpha_j\right)_{1,n}; & \left(a_{ji},\alpha_{ji}\right)_{n+1,p_i-3}\left(a_{p_i}-\alpha_1\right)\\ \left(b_j,\beta_j\right)_{1,m}; & \left(b_{ji},\beta_{ji}\right)_{m+1,q_i}\end{matrix}\right.\right]$$

$$=I^{m,n}_{p_i,q_i:r}\left[z\left|\begin{matrix}\left(a_1-1,\alpha_j\right)\left(a_j,\alpha_j\right)_{2,n}; & \left(a_{ji},\alpha_{ji}\right)_{n+1,p_i-1}\left(a_{p_i},\alpha_1\right)\\ \left(b_j,\beta_j\right)_{1,m}; & \left(b_{ji},\beta_{ji}\right)_{m+1,q_i}\end{matrix}\right.\right] \tag{4.17}$$

$$+I^{m,n}_{p_i,q_i:r}\left[z\left|\begin{matrix}\left(a_j,\alpha_j\right)_{2,n}; & \left(a_{ji},\alpha_{ji}\right)_{n+1,p_i-1},\left(a_{p_i}-1,\alpha_1\right)\\ \left(b_j,\beta_j\right)_{1,m}; & \left(b_{ji},\beta_{ji}\right)_{m+1,q_i}\end{matrix}\right.\right]$$

b. If $r=1$ then R_1 reduces to recurrence relation of H-function given by Saxena [31]

$$\left(a_{p_i}\alpha_1-\alpha_{p_i}a_1+\alpha_p-\alpha_1\right)H^{m,n}_{p,q}\left[z\left|\begin{matrix}\left(a_j,\alpha_j\right)_{1,n}; & \left(a_{ji},\alpha_{ji}\right)_{n+1,p_i}\\ \left(b_j,\beta_j\right)_{1,m}; & \left(b_{ji},\beta_{ji}\right)_{m+1,q_i}\end{matrix}\right.\right]$$

$$=\alpha_p H^{m,n}_{p,q}\left[z\left|\begin{matrix}\left(a_1-1,\alpha_j\right),\left(a_j,\alpha_j\right)_{2,n}; & \left(a_{ji},\alpha_{ji}\right)_{n+1,p}\\ \left(b_j,\beta_j\right)_{1,m}; & \left(b_j,\beta_j\right)_{m+1,q}\end{matrix}\right.\right] \tag{4.18}$$

$$+\alpha_1 H^{m,n}_{p.q}\left[z\left|\begin{matrix}\left(a_j,\alpha_j\right)_{1,n}; & \left(a_j,\alpha_j\right)_{n+1,p-1}\left(\alpha_p-1,\alpha_p\right)\\ \left(b_j,\beta_j\right)_{1,m}; & \left(b_j,\beta_j\right)_{m+1,q}\end{matrix}\right.\right]$$

c. If $r=1$ then the recurrence relation R_2 reduces to the following recurrence relation of H-function

$$\left(b_{q-1}-b_q\right)H^{m,n}_{p,q}\left[x\left|\begin{matrix}\left(a_j,\alpha_j\right)_{1,n}; & \left(a_j,\alpha_j\right)_{n+1,p}\\ \left(b_j,\beta_j\right)_{1,m}; & \left(b_j,\beta_j\right)_{m+1,q-2}\left(b_{q-1},\frac{\alpha_1}{2}\right)\left(b_q,\frac{\alpha_1}{2}\right)\end{matrix}\right.\right]$$

$$= \left(a_1 - 2b_q\right) H_{p,q}^{m,n}\left[x \,\middle|\, \begin{matrix} (1+a_1,\alpha_1),\left(a_j,\alpha_j\right)_{2,n}; & \left(a_j,\alpha_j\right)_{n+1,p} \\ \left(b_j,\beta_j\right)_{1,m}; & \left(b_j,\beta_j\right)_{m+1,q-2}\left(1+b_{q-1},\dfrac{\alpha_1}{2}\right)\left(b_q,\dfrac{\alpha_1}{2}\right) \end{matrix}\right]$$

$$+\left(2b_q - a_1\right) H_{p,q}^{m,n}\left[x \,\middle|\, \begin{matrix} (1+a_1,\alpha_1),\left(a_j,\alpha_j\right)_{2,n}; & \left(a_j,\alpha_j\right)_{n+1,p} \\ \left(b_j,\beta_j\right)_{1,m}; & \left(b_j,\beta_j\right)_{m+1,q-2}\left(b_{q-1},\dfrac{\alpha_1}{2}\right)\left(1+b_q,\dfrac{\alpha_1}{2}\right) \end{matrix}\right] \tag{4.19}$$

Similarly, the recurrence relations R_3 to R_9 can be reduced to the corresponding recurrence relation of H-function by putting $r=1$.

d. Substituting $r=1$, relation (4.19) takes the form

$$\left(b_{q-1} - b_q\beta_1\right) H_{p,q}^{m,n}\left[z \,\middle|\, \begin{matrix} \left(a_j,\alpha_j\right)_{1,p} \\ \left(b_j,\beta_j\right)_{1,q} \end{matrix}\right] = \beta_q H_{p,q}^{m,n}\left[z \,\middle|\, \begin{matrix} \left(a_j,\alpha_j\right)_{1,p} \\ \left(b_j+1,\beta_1\right),\left(b_j,\beta_j\right)_{2,q} \end{matrix}\right]$$

$$= \beta_q H_{p,q}^{m,n}\left[z \,\middle|\, \begin{matrix} \left(a_j,\alpha_j\right)_{1,p} \\ \left(b_j,\beta_j\right)_{1,q-1},\left(b_q+1,\beta_q\right) \end{matrix}\right] \tag{4.20}$$

Provided $m \geq 1,\ q \geq 2$.

4.2 MULTIPLICATION THEOREMS

In this section, we describe six multiplication theorems of the I-function. These Theorem are valid under the normal existence conditions enunciated in (1.23). Only extra required conditions are stated with the results.

i.

$$M_1 \equiv I_{p_i,q_i:r}^{m,n}\left[\lambda z \,\middle|\, \begin{matrix} \left(a_j,\alpha_j\right)_{1,n}; & \left(a_{ji},\alpha_{ji}\right)_{n+1,p_i} \\ \left(b_j,\beta_j\right)_{1,m}; & \left(b_{ji},\beta_{ji}\right)_{m+1,q_i} \end{matrix}\right]$$

$$= \lambda^{(a_1-1)/\alpha_1} \sum_{k=0}^{\infty}\left(1-\lambda^{-\frac{1}{\alpha_1}}\right)^k \alpha_{p_i} I_{p_i,q_i:r}^{m,n}\left[z \,\middle|\, \begin{matrix} \left(a_1-k,\alpha_1\right),\left(a_j,\alpha_j\right)_{2,n}; & \left(a_{ji},\alpha_{ji}\right)_{n+1,p_i} \\ \left(b_j,\beta_j\right)_{1,m}; & \left(b_{ji},\beta_{ji}\right)_{m+1,q_i} \end{matrix}\right] \tag{4.21}$$

Provided $\left(1-\lambda^{-\frac{1}{\alpha_1}}\right)^k < 1, n \geq 1.$

We have

$$M_1 = \frac{1}{2\pi i}\int_L \frac{\prod_{j=1}^{m}\Gamma(\mathrm{b_j}-\beta_j\xi)\prod_{j=1}^{n}\Gamma(1-\mathrm{a_j}-\alpha_j\xi)\lambda^{\xi}z^{\xi}d\xi}{\sum_{i=1}^{r}\left\{\prod_{j=m+1}^{q_i}\Gamma(1-\mathrm{b_{ij}}-\beta_{ji}\xi)\prod_{j=n+1}^{p_i}\Gamma(\mathrm{a_{ji}}-\alpha_{ji}\xi)\right\}}$$

For $\lambda=(1-h)^{-\alpha_1}=h=1-\lambda^{-\frac{1}{\alpha_1}}$ so that the above expression is

$$= \frac{1}{2\pi i}\int_L \frac{\prod_{j=1}^{m}\Gamma(\mathrm{b_j}-\beta_j\xi)\prod_{j=1}^{n}\Gamma(1-\mathrm{a_j}-\alpha_j\xi)(1-h)^{-\alpha_1\xi}z^{\xi}d\xi}{\sum_{i=1}^{r}\left\{\prod_{j=m+1}^{q_i}\Gamma(1-\mathrm{b_{ij}}-\beta_{ji}\xi)\prod_{j=n+1}^{p_i}\Gamma(\mathrm{a_{ji}}-\alpha_{ji}\xi)\right\}}$$

Or

$$=(1-h)^{1-\alpha_1}\frac{1}{2\pi i}\int_L \frac{\prod_{j=1}^{m}\Gamma(\mathrm{b_j}-\beta_j\xi)\prod_{j=1}^{n}\Gamma(1-\mathrm{a_j}-\alpha_j\xi)(1-h)^{(1-\alpha_1\xi+\alpha_1\xi)}z^{\xi}d\xi}{\sum_{i=1}^{r}\left\{\prod_{j=m+1}^{q_i}\Gamma(1-\mathrm{b_{ij}}-\beta_{ji}\xi)\prod_{j=n+1}^{p_i}\Gamma(\mathrm{a_{ji}}-\alpha_{ji}\xi)\right\}}$$

Using the binomial theorem we get M_1.

For $r=1$, relation (4.21) takes the form given by Saxena and Singh [35]

$$H_{p,q}^{m,n}\left[\lambda z\left|\begin{matrix}(a_j,\alpha_j)_{1,p}\\(b_j,\beta_j)_{1,q}\end{matrix}\right.\right]=\lambda^{(a_1-1)/a_1}\sum_{k=0}^{\infty}\frac{\left(1-\lambda^{-\frac{1}{\alpha_1}}\right)^k}{k!}=H_{p,q}^{m,n}\left[\lambda z\left|\begin{matrix}(a_1-k,\alpha_1),(a_j,\alpha_j)_{1,p}\\(b_j,\beta_j)_{1,q}\end{matrix}\right.\right] \tag{4.22}$$

Provided $|1-\lambda|^{-1/\alpha_1}<1, n\geq 1.$

On substituting $\alpha_1=\alpha_2=\ldots=\alpha_p=\beta_1=\beta_2.....=\beta_q=1$, this result reduces to the well-known multiplication theorem for G function.

ii.

$$M_2 \equiv I_{p_i,q_i:r}^{m,n}\left[\lambda z\left|\begin{matrix}(a_j,\alpha_j)_{1,n};&(a_{ji},\alpha_{ji})_{n+1,p_i}\\(b_j,\beta_j)_{1,m};&(b_{ji},\beta_{ji})_{m+1,q_i}\end{matrix}\right.\right]=\lambda^{b_1/\beta_1}\sum_{k=0}^{\infty}\frac{\left(1-\lambda^{-\frac{1}{\beta_1}}\right)^k}{k!} \tag{4.23}$$

$$I_{p_i,q_i:r}^{m,n}\left[z\left|\begin{matrix}(a_j,\alpha_j)_{1,n};&(a_{ji},\alpha_{ji})_{n+1,p_i}\\(b_1+k,\beta_1),(b_j,\beta_j)_{2,m};&(b_{ji},\beta_{ji})_{m+1,q_i}\end{matrix}\right.\right]$$

Provided $|1-\lambda|^{-1/\beta_1} < 1, \mathrm{m} \geq 1.$

$$M_2 = \frac{1}{2\pi i}\int_L \frac{\prod_{j=1}^{m}\Gamma(\mathrm{b_j}-\beta_j\xi)\prod_{j=1}^{n}\Gamma(1-\mathrm{a_j}-\alpha_j\xi)(\lambda)^{\xi} z^{\xi} d\xi}{\sum_{i=1}^{r}\left\{\prod_{j=m+1}^{q_i}\Gamma(1-\mathrm{b_{ij}}+\beta_{ji}\xi)\prod_{j=n+1}^{p_i}\Gamma(\mathrm{a_{ji}}-\alpha_{ji}\xi)\right\}} \tag{4.24}$$

Putting $\lambda = (1-h)^{\beta_1}$ for $h = (1-\lambda)^{1/\beta_1}$ the above integral can be transformed to

$$= \frac{1}{2\pi i}\int_L \frac{\prod_{j=1}^{m}\Gamma(\mathrm{b_j}-\beta_j\xi)\prod_{j=1}^{n}\Gamma(1-\mathrm{a_j}-\alpha_j\xi)(1-h)^{\beta_1\xi} z^{\xi}}{\sum_{i=1}^{r}\left\{\prod_{j=m+1}^{q_i}\Gamma(1-\mathrm{b_{ij}}+\beta_{ji}\xi)\prod_{j=n+1}^{p_i}\Gamma(\mathrm{a_{ji}}-\alpha_{ji}\xi)\right\}} d\xi$$

On using the binomial theorem, it will lead to the theorem.

Substituting $r = 1$, we get the result for the H-function [31]

$$H_{p,q}^{m,n}\left[\lambda z \left| \begin{matrix} (a_1,\alpha_1),(a_2,\alpha_2),\ldots,(a_p,\alpha_p) \\ (b_1,\beta_1),(b_1,\beta_1)\ldots,(b_q,\beta_q) \end{matrix} \right.\right] = \lambda^{(b_1)/\beta_1}\sum_{k=0}^{\infty}\frac{\left(1-\lambda^{-\frac{1}{\beta_1}}\right)^k}{k!}$$

$$= H_{p,q}^{m,n}\left[\lambda z \left| \begin{matrix} (a_1,\alpha_1),(a_2,\alpha_2),\ldots,(a_p,\alpha_p) \\ (b_1,\beta_1),(b_1,\beta_1)\ldots,(b_q,\beta_q) \end{matrix} \right.\right] \tag{4.25}$$

Provided $|1-\lambda|^{-1/\beta_1} < 1, m \geq 1.$

This multiplication theorem reduces to the well-known multiplication theorem of the G function taking all α's and β's as unity.

iii.

$$M_3 \equiv x^{2\rho} y^{\sigma} I_{p_i,q_i:r}^{m,n}\left[zx^{-2h}y^{2k} \left| \begin{matrix} (a_j,\alpha_j)_{1,n}; & (a_{ji},\alpha_{ji})_{n+1,p_i} \\ (b_j,\beta_j)_{1,m}; & (b_{ji},\beta_{ji})_{m+1,q_i} \end{matrix} \right.\right] = x^{\sigma-2\rho}\sum_{\mu,t=0}^{\infty}\frac{(2t+v+1)}{2u!} H_{2u}(x) J_{v+2t+1}(y)$$

$$I_{p_i+3,q_i+1:r}^{m+1,n+1}\left[2^{2(h+k)}z \left| \begin{matrix} \left(\frac{1-\sigma-v-2t}{2},k\right),(a_j,\alpha_j)_{1,n}; & (a_{ji},\alpha_{ji})_{n+1,p_i}\,(\rho-u+1,h),\left(\frac{3-\sigma+v+2t}{2},k\right) \\ (2\rho+1,2h),\,(b_j,\beta_j)_{1,m}; & (b_{ji},\beta_{ji})_{m+1,q_i}\ \ldots\ldots\ldots\ldots\ldots\ldots\ldots\ldots \end{matrix} \right.\right] \tag{4.26}$$

iv.

$$M_4 \equiv x^{2\rho} y^{\sigma} I^{m,n}_{p_i,q_i:r}\left[zx^{-2h}y^{-2k} \middle| \begin{matrix} (a_j,\alpha_j)_{1,n}; & (a_{ji},\alpha_{ji})_{n+1,p_i} \\ (b_j,\beta_j)_{1,m}; & (b_{ji},\beta_{ji})_{m+1,q_i} \end{matrix} \right] = 2^{\sigma-2\rho} \sum_{\mu,t=0}^{\infty} \frac{(2t+v+1)}{2u!} H_{2u}(x) J_{v+2t+1}(y)$$

$$I^{m+2,n}_{p_i+1,q_i+3:r}\left[2^{2(h-k)}z \middle| \begin{matrix} (a_j,\alpha_j)_{1,n}; & (a_{ji},\alpha_{ji})_{n+1,p_i}, (\rho-u+1,h), \dots\dots\dots \\ (2\rho+1,2h), \left(\frac{\sigma+v+2t+1}{2},k\right), (b_j,\beta_j)_{1,m}; & (b_{ji},\beta_{ji})_{m+1,q_i}, \left(\frac{\sigma-v-2t-1}{2},k\right) \end{matrix} \right] \tag{4.27}$$

v.

$$M_5 \equiv x^{2\rho} y^{\sigma} I^{m,n}_{p_i,q_i:r}\left[zx^{2h}y^{2k} \middle| \begin{matrix} (a_j,\alpha_j)_{1,n}; & (a_{ji},\alpha_{ji})_{n+1,p_i} \\ (b_j,\beta_j)_{1,m}; & (b_{ji},\beta_{ji})_{m+1,q_i} \end{matrix} \right] = 2^{\sigma-2\rho} \sum_{\mu,t=0}^{\infty} \frac{(2t+v+1)}{2u!} H_{2u}(x) J_{v+2t+1}(y)$$

$$I^{m,n+2}_{p_i+3,q_i+1:r}$$

$$\times\left[2^{2(h-k)}z \middle| \begin{matrix} (-2\rho,2h), \left(\frac{1-\sigma-v-2t}{2},k\right), (a_j,\alpha_j)_{1,n}; & (a_{ji},\alpha_{ji})_{n+1,p_i}, \left(\frac{3-\sigma+v+2t}{2},k\right) \\ (b_j,\beta_j)_{1,m}; & (b_{ji},\beta_{ji})_{m+1,q_i}, (u-\rho,h), \left(\frac{3-\sigma+v+2t}{2},k\right) \dots\dots\dots \end{matrix} \right] \tag{4.28}$$

vi.

$$M_6 \equiv x^{2\rho} y^{\sigma} I^{m,n}_{p_i,q_i:r}\left[zx^{2h}y^{-2k} \middle| \begin{matrix} (a_j,\alpha_j)_{1,n}; & (a_{ji},\alpha_{ji})_{n+1,p_i} \\ (b_j,\beta_j)_{1,m}; & (b_{ji},\beta_{ji})_{m+1,q_i} \end{matrix} \right]$$

$$= 2^{\sigma-2\rho} \sum_{\mu,t=0}^{\infty} \frac{(2t+v+1)}{2u!} H_{2u}(x) J_{v+2t+1}(y)$$

$$I^{m+1,n+1}_{p_i+1,q_i+3:r}$$

$$\times\left[2^{-2(h+k)}z \middle| \begin{matrix} (-2\rho,2h), (a_j,\alpha_j)_{1,n}; & (a_{ji},\alpha_{ji})_{n+1,p_i} \dots\dots\dots \\ \left(\frac{\sigma+v+2t+1}{2},k\right), (b_j,\beta_j)_{1,m}; & (b_{ji},\beta_{ji})_{m+1,q_i}, (u-\rho,h), \left(\frac{\sigma-v-2t-1}{2},k\right) \end{matrix} \right] \tag{4.29}$$

To drive (4.26), we take

$$x^{2\rho}y^{\sigma}I^{m,n}_{p_i,q_i:r}\left[zx^{-2h}y^{2k}\left|\begin{matrix}(a_j,\alpha_j)_{1,n};&(a_{ji},\alpha_{ji})_{n+1,p_i}\\(b_j,\beta_j)_{1,m};&(b_{ji},\beta_{ji})_{m+1,q_i}\end{matrix}\right.\right]=\sum_{u,t=0}^{\infty}A_{u,t}H_{2u}(x)J_{v+2t+1}(y) \tag{4.30}$$

Equation (4.29) is valid since the function on the left is continuous and of bounded variation in the region $-\infty<x<\infty$, $0<y<\infty$.

Multiplying both sides of (4.29) by $y^{-1}J_{v+2t+1}(y)$, integrating with respect to y from 0 to ∞, using (4.38) and the following integral

$$\int_0^{\infty}y^{\sigma-1}J_v(y)x^{2\rho}I^{m,n}_{p_i,q_i:r}\left[zy^{2k}\left|\begin{matrix}(a_j,\alpha_j)_{1,n};&(a_{ji},\alpha_{ji})_{n+1,p_i}\\(b_j,\beta_j)_{1,m};&(b_{ji},\beta_{ji})_{m+1,q_i}\end{matrix}\right.\right]dy$$

$$=z^{\sigma-1}I^{m,n+1}_{p_i+2,q_i+2:r}\left[2^{2k}z\left|\begin{matrix}\left(\frac{2-\sigma-v}{2},k\right),(a_j,\alpha_j)_{1,n};&(a_{ji},\alpha_{ji})_{n+1,p_i},\left(\frac{2-\sigma+v}{2},k\right)\\(b_j,\beta_j)_{1,m};&(b_{ji},\beta_{ji})_{m+1,q_i}\end{matrix}\right.\right]$$

We obtain

$$2^{\sigma-1}x^{2\rho}I^{m,n+1}_{p_i+2,q_i+2:r}$$

$$\times\left[2^{2k}zx^{-2h}\left|\begin{matrix}\left(\frac{1-\sigma-v-2v}{2},k\right),(a_j,\alpha_j)_{1,n};&(a_{ji},\alpha_{ji})_{n+1,p_i},\left(\frac{3-\sigma+v+2v}{2},k\right)\\(b_j,\beta_j)_{1,m};&(b_{ji},\beta_{ji})_{m+1,q_i}\text{.............}\end{matrix}\right.\right]$$

$$=\sum_{u=0}^{\infty}A_{u,v}H_{2u}(x)(4v+2v-2)^{-1} \tag{4.31}$$

Multiplying both sides of (4.30) by $e^{-x^2}H_{2u}(x)$, integrating with respect to x from $-\infty$ to ∞, and using (4.35) and (4.27), we get

$$A_{u,v}=\frac{2^{\sigma-2\sigma}(2v+v+1)}{2u!}\times I^{m+1,n+1}_{p_i+3,q_i+1:r}$$

$$\left[2^{2(h+k)}z\left|\begin{matrix}\left(\frac{1-\sigma-v-2v}{2},k\right),(a_j,\alpha_j)_{1,n};&(a_{ji},\alpha_{ji})_{n+1,p_i}(\rho-u+1,h),\left(\frac{3-\sigma+v+2v}{2},k\right)\\(2\rho+1,2h),(b_j,\beta_j)_{1,m};&(b_{ji},\beta_{ji})_{m+1,q_i}\text{..................................}\end{matrix}\right.\right] \tag{4.32}$$

Using (4.30) and (4.32), formula (4.26) is obtained.

On similar lines, the results (4.27)–(4.29) can also be proved.

4.3 GENERATING FORMULAS OF I-FUNCTION

Here we give various forms of generating formulas of the I-function using Truesdell's ascending and descending F-equations.

Some simple generating formulas will also be added in the later part of the section. Truesdell [36] put forward a theory which yielded a number of results for special functions satisfying the so-called Truesdell F-equations.

4.3.1 Truesdell's *F*-Equations

The function $F(z,\alpha)$ is said to satisfy the ascending F-equation if

a. $D_z^r \mathrm{F}(z,\alpha) = F(z,\alpha + r)$

where $D_z^r = \left(\dfrac{d}{dz}\right)^r$

On the other hand, the function $Y(z,\alpha)$ is said to satisfy the descending F-equation if

b. $Y_z^r \mathrm{F}(z,\alpha) = Y(z,\alpha - r)$

where r is a positive integer.

For $F(z, \alpha)$ satisfying ascending F-equation, Truesdell [36] and Agarwal and Saxena [37] obtained the following generating functions using Taylor's series

c. $F(z+y, \alpha) = \sum_{n=0}^{\infty} y^n F(z,\alpha+n) / n!$

d. $Y(z+y, \alpha) = \sum_{n=0}^{\infty} y^n Y(z,\alpha-n) / n!$

To obtain the main results of this section we will make use of the following well-known results on multiplication formulas for Gamma functions.

e. $\prod_{k=0}^{m-1} \Gamma\left(\dfrac{\alpha + r + k}{m}\right) = m^{-r} (\alpha)_r \prod_{k=0}^{m-1} \Gamma\left(\dfrac{\alpha + k}{m}\right)$

f. $\prod_{k=0}^{m-1} \Gamma\left(\dfrac{\alpha - r - k}{m}\right) = \dfrac{(-m)^{-r}}{(m-\alpha)_r} \prod_{k=0}^{m-1} \Gamma\left(\dfrac{\alpha - k}{m}\right)$

g. $\prod_{k=0}^{m-1}\Gamma\left(\frac{\alpha-r+k}{m}\right)=\frac{(-m)^{-r}}{(1-\alpha)_r}\prod_{k=0}^{m-1}\Gamma\left(\frac{\alpha+k}{m}\right)$

h. $\prod_{k=0}^{m-1}\Gamma\left(\frac{\alpha+r-k}{m}\right)=m^{-r}(\alpha-m+1)_r\prod_{k=0}^{m-1}\Gamma\left(\frac{\alpha-k}{m}\right)$

i. $\Gamma(mz)=m^{mz-\frac{1}{2}}(2\pi)^{-\frac{1}{2}}(m-1)\prod_{k=1}^{m-1}\Gamma\left(z+\frac{k-1}{m}\right)$

j. $\{\Delta(\lambda,\alpha),h\}=\left(\frac{a}{\lambda},h\right),\left(\frac{a+1}{\lambda},h\right),\left(\frac{a+2}{\lambda},h\right),......,\left(\frac{a+\lambda-1}{\lambda},h\right)$

The following forms of I-functions satisfy Truesdell's ascending F-equation

$$\left(z/\lambda\right)^{-\alpha} I^{m,n}_{p_i,q_i:L}\left[\left(\frac{z\beta}{\lambda}\right)^{h\lambda}\left|\begin{matrix}(a_j,\alpha_j)_{1,n}; & (a_{ji},\alpha_{ji})_{n+1,p_i-\rho},\{\Delta(\rho,\alpha),h\}\\ \{\Delta(\lambda,\alpha),h\},(b_j,\beta_j)_{1,m}; & (b_{ji},\beta_{ji})_{m+1,q_i-\rho},\{\Delta(\rho,\alpha),h\}\end{matrix}\right.\right] \tag{4.33}$$

$$\left(z/\lambda\right)^{-\alpha} I^{m,n}_{p_i,q_i:L}\left[\left(\frac{z\beta}{\lambda}\right)^{h\lambda}\left|\begin{matrix}\{\Delta(\lambda,\alpha+1/2),h\},(a_j,\alpha_j)_{\lambda+1,n}; & (a_{ji},\alpha_{ji})_{n+1,p_i}\\ \{\Delta(2\lambda,2\alpha),h\},(b_j,\beta_j)_{2\lambda+1,m}; & (b_{ji},\beta_{ji})_{m+1,q_i}\end{matrix}\right.\right] \tag{4.34}$$

$$\left(z/\lambda\right)^{-\alpha} I^{m,n}_{p_i,q_i:L}$$
$$\times\left[\left(\frac{z\beta}{\lambda}\right)^{h\lambda}\left|\begin{matrix}\{\Delta(\lambda,2\alpha/3),h\},(a_j,\alpha_j)_{\lambda+1,n}; & (a_{ji},\alpha_{ji})_{n+1,p_i-\lambda},\{\Delta(\lambda,\alpha+1/3),h\}\\ \{\Delta(3\lambda,3\alpha),h\},(b_j,\beta_j)_{3\lambda+1,m}; & (b_{ji},\beta_{ji})_{m+1,q_i}\end{matrix}\right.\right] \tag{4.35}$$

$$\left(z/\lambda\right)^{-(\alpha+1)} I^{m,n}_{p_i,q_i:L}$$
$$\times\left[\left(\frac{z\beta}{\lambda}\right)^{h\lambda}\left|\begin{matrix}\{\Delta(2\lambda,2\alpha+1/2),h\},(a_j,\alpha_j)_{2\lambda+1,n}; & (a_{ji},\alpha_{ji})_{n+1,p_i-2\lambda},\{\Delta(2\lambda,2\alpha+1),h\}\\ \{\Delta(4\lambda,4\alpha+1),h\},(b_j,\beta_j)_{4\lambda+1,m}; & (b_{ji},\beta_{ji})_{m+1,q_i-\lambda},\{\Delta(\lambda,\alpha+1),h\}\end{matrix}\right.\right] \tag{4.36}$$

$$\left(\frac{z}{\lambda}\right)^{-\alpha} I^{m,n}_{p_i,q_i:L}\left[\left(\frac{z\beta}{\lambda}\right)^{h\lambda}\middle| \begin{matrix} (a_j,\alpha_j)_{1,n}; & (a_{ji},\alpha_{ji})_{n+1,p_i} \\ \{\Delta(\lambda,\alpha),h\},(b_j,\beta_j)_{\lambda+1,m}; & (b_{ji},\beta_{ji})_{m+1,q_i} \end{matrix}\right] \tag{4.37}$$

$$\left(\frac{z}{\lambda}\right)^{-\alpha} 2^{\alpha} e^{-i\pi\alpha/2} I^{m,n}_{p_i,q_i:L}$$

$$\times\left[\left(\frac{z\beta}{\lambda}\right)^{2h\lambda}\middle| \begin{matrix} (a_j,\alpha_j)_{1,n}; & (a_{ji},\alpha_{ji})_{n+1,p_i} \ldots\ldots\ldots\ldots \\ \{\Delta(\lambda,\alpha/2),h\},(b_j,\beta_j)_{\lambda+1,m}; & (b_{ji},\beta_{ji})_{m+1,q_i-\lambda},\{\Delta(\lambda,\alpha+1/2),h\} \end{matrix}\right] \tag{4.38}$$

$$\left(\frac{z}{\lambda}\right)^{-\alpha} I^{m,n}_{p_i,q_i:L}\left[\left(\frac{z\beta}{\lambda}\right)^{h\lambda}\middle| \begin{matrix} (a_j,\alpha_j)_{1,n}; & (a_{ji},\alpha_{ji})_{n+1,p_i} \\ (b_j,\beta_j)_{\lambda+1,m}; & (b_{ji},\beta_{ji})_{m+1,q_i-\lambda},\{\Delta(\lambda,\alpha),h\} \end{matrix}\right] \tag{4.39}$$

$$\left(\frac{z}{\lambda}\right)^{-\alpha} e^{-i\pi\alpha} I^{m,n}_{p_i,q_i:L}$$

$$\times\left[\left(\frac{z\beta}{\lambda}\right)^{2h\lambda}\middle| \begin{matrix} \{\Delta(\lambda,\alpha),h\},(a_j,\alpha_j)_{\lambda+1,n}; & (a_{ji},\alpha_{ji})_{n+1,p_i} \\ \{\Delta(\rho,\alpha),h\},(b_j,\beta_j)_{\rho+1,m}; & (b_{ji},\beta_{ji})_{m+1,q_i-\lambda},\{\Delta(\lambda,\alpha),h\} \end{matrix}\right] \tag{4.40}$$

Making use of the result (4.33), we obtain

$$D^r[A(z,\alpha)]=$$

$$D^r\left[\left(\frac{z}{\lambda}\right)^{-\alpha} I^{m,n}_{p_i,q_i:L}\left[\left(\frac{z\beta}{\lambda}\right)^{h\lambda}\middle| \begin{matrix} (a_j,\alpha_j)_{1,n}; & (a_{ji},\alpha_{ji})_{n+1,p_i-\rho},\{\Delta(\rho,\alpha),h\} \\ \{\Delta(\lambda,\alpha),h\},(b_j,\beta_j)_{1,m}; & (b_{ji},\beta_{ji})_{m+1,q_i-\rho},\{\Delta(\rho,\alpha),h\} \end{matrix}\right]\right] \tag{4.41}$$

Replacing the I-function by its definition (1.23) and then interchanging the order of integration and differentiation, (4.41) transforms to

$$D^r\left[A(z,\alpha)\right]=\left(\frac{1}{2\pi i}\right)\int_L \frac{\prod_{k=0}^{\lambda-1}\Gamma\left(\frac{\alpha+k}{\lambda}-h\xi\right)\prod_{j=\lambda+1}^{m}\Gamma\left(b_j-\beta_j\xi\right)}{\sum_{i=1}^{L}\prod_{j=m+1}^{q_i-\rho}\Gamma\left(1-b_{ji}-\beta_{ji}\xi\right)\prod_{k=0}^{\rho-1}\Gamma\left(1-\frac{\alpha+k}{\rho}+h\xi\right)}$$

$$\times\frac{\prod_{j=1}^{n}\Gamma\left(1-a_j+\alpha_j\xi\right)\left(\frac{\beta^{\lambda\xi}}{\lambda^{\lambda\xi-\alpha}}\right)\left(\frac{d}{dz}\right)^r z^{h\lambda\xi-\alpha}}{\prod_{j=n+1}^{p_i-\rho}\Gamma\left(a_{ji}-\alpha_{ji}\xi\right)\prod_{k=0}^{\rho-1}\Gamma\left(\frac{\alpha+k}{\rho}-h\xi\right)}d\xi \tag{4.42}$$

Now results (e) and (f) lead to two very interesting identities

$$\prod_{k=0}^{\lambda-1}\Gamma\left(\frac{\alpha+k}{\lambda}-h\xi\right)=\frac{\lambda^r}{(\alpha-h\lambda\xi)}\prod_{k=0}^{\lambda-1}\Gamma\left(\frac{\alpha+r+k}{\lambda}-h\xi\right)$$

and

$$\prod_{k=0}^{\lambda-1}\Gamma\left(1-\frac{\alpha+k}{\lambda}+h\xi\right)=\frac{(\alpha-h\lambda\xi)_r}{(-1)^r\lambda^r}\prod_{k=0}^{\lambda-1}\Gamma\left(1-\frac{\alpha+r+k}{\lambda}+h\xi\right)$$

Using these results (4.42) takes the form

$$D^r\left[A(z,\alpha)\right]=A\left[(z,\alpha+r)\right]$$

which is the Truesdell's ascending F-equation.

Similarly, forms of (4.34)–(4.40) can be shown to satisfy Truesdell's ascending F-equation.

The following are some forms of the I-function which satisfy Truesdell's descending F-equation

$$\left(z/\lambda\right)^{\alpha-1} I^{m,n}_{p_i,q_i:L}\left[\left(\frac{\beta}{\lambda z}\right)^{h\lambda}\middle|\begin{matrix}\{\Delta(\lambda,\alpha),h\},(a_j,\alpha_j)_{\lambda+1,n}; & (a_{ji},\alpha_{ji})_{n+1,p_i-\rho},\{\Delta(\rho,\alpha),h\}\\ (b_j,\beta_j)_{\rho+1,m}; & (b_{ji},\beta_{ji})_{m+1,q_i-\rho},\{\Delta(\rho,\alpha),h\}\end{matrix}\right] \tag{4.43}$$

$$\left(z/\lambda\right)^{\alpha-1} I^{m,n}_{p_i,q_i:L}\left[\left(\frac{\beta}{\lambda z}\right)^{h\lambda}\middle|\begin{matrix}\{\Delta(2\lambda,2\alpha),h\},(a_j,\alpha_j)_{2\lambda+1,n}; & (a_{ji},\alpha_{ji})_{n+1,p_i}\\ \{\Delta(\lambda,\alpha+1/2),h\},(b_j,\beta_j)_{\lambda+1,m}; & (b_{ji},\beta_{ji})_{m+1,q_i}\end{matrix}\right] \tag{4.44}$$

$$\left(z/\lambda\right)^{\alpha-1} I^{m,n}_{p_i,q_i:L}$$

$$\times\left[\left(\frac{\beta}{\lambda z}\right)^{h\lambda}\left|\begin{matrix}\{\Delta(3\lambda,3\alpha),h\},(a_j,\alpha_j)_{3\lambda+1,n}; & (a_{ji},\alpha_{ji})_{n+1,p_i}\\ \{\Delta(\lambda,\alpha+2/3),h\},(b_j,\beta_j)_{\lambda+1,m}; & (b_{ji},\beta_{ji})_{m+1,q_i-\lambda},\{\Delta(\lambda,\alpha+1/3),h\}\end{matrix}\right.\right] \tag{4.45}$$

$$\left(z/\lambda\right)^{\alpha-1} I^{m,n}_{p_i,q_i:L}$$

$$\times\left[\left(\frac{\beta}{\lambda z}\right)^{h\lambda}\left|\begin{matrix}\{\Delta(4\lambda,4\alpha+1),h\},(a_j,\alpha_j)_{4\lambda+1,n}; & (a_{ji},\alpha_{ji})_{n+1,p_i-\lambda},\{\Delta(\lambda,\alpha),h\}\\ \{\Delta(2\lambda,2\alpha+1/2),h\},(b_j,\beta_j)_{2\lambda+1,m}; & (b_{ji},\beta_{ji})_{m+1,q_i-2\lambda},\{\Delta(2\lambda,2\alpha+1),h\}\end{matrix}\right.\right] \tag{4.46}$$

$$\left(z/\lambda\right)^{\alpha-1} e^{\pi i\alpha} I^{m,n}_{p_i,q_i:L}\left[\left(\frac{\beta}{\lambda z}\right)^{h\lambda}\left|\begin{matrix}\{\Delta(\lambda,\alpha),h\},(a_j,\alpha_j)_{\lambda+1,n}; & (a_{ji},\alpha_{ji})_{n+1,p_i}\\ (b_j,\beta_j)_{1,m}; & (b_{ji},\beta_{ji})_{m+1,q_i}\end{matrix}\right.\right] \tag{4.47}$$

$$\left(z/\lambda\right)^{\alpha-1} 2^{-\alpha} e^{i\pi\alpha/2} I^{m,n}_{p_i,q_i:L}$$

$$\times\left[\left(\frac{\beta}{\lambda z}\right)^{2h\lambda}\left|\begin{matrix}\{\Delta(\lambda,\alpha/2),h\},(a_j,\alpha_j)_{\lambda+1,n}; & (a_{ji},\alpha_{ji})_{n+1,p_i-\lambda},\left\{\Delta\left(\lambda,\frac{\alpha+1}{2}\right),h\right\}\\ (b_j,\beta_j)_{1,m}; & (b_{ji},\beta_{ji})_{m+1,q_i},\ldots\ldots\ldots\ldots\end{matrix}\right.\right] \tag{4.48}$$

$$\left(z/\lambda\right)^{\alpha-1} I^{m,n}_{p_i,q_i:L}\left[\left(\frac{\beta}{\lambda z}\right)^{h\lambda}\left|\begin{matrix}(a_j,\alpha_j)_{1,n}; & (a_{ji},\alpha_{ji})_{n+1,p_i-\lambda},\{\Delta(\lambda,\alpha),h\}\\ (b_j,\beta_j)_{1,m}; & (b_{ji},\beta_{ji})_{m+1,q_i}\end{matrix}\right.\right] \tag{4.49}$$

$$\left(z/\lambda\right)^{\alpha-1} e^{\pi i\alpha} I^{m,n}_{p_i,q_i:L}$$

$$\times\left[\left(\frac{\beta}{\lambda z}\right)^{h\lambda}\left|\begin{matrix}\{\Delta(\rho,\alpha),h\},(a_j,\alpha_j)_{\lambda+1,n}; & (a_{ji},\alpha_{ji})_{n+1,p_i-\lambda},\{\Delta(\lambda,\alpha),h\}\\ \{\Delta(\rho,\alpha),h\},(b_j,\beta_j)_{\rho+1,m}; & (b_{ji},\beta_{ji})_{m+1,q_i}\end{matrix}\right.\right] \tag{4.50}$$

As before, taking $\mathrm{B}(z,\alpha)$ equal to this form (4.43), we obtain

$$D^r B(z,\alpha) = \frac{d^r}{dz^r}$$

$$\times\left\{\left(z/\lambda\right)^{\alpha-1} I^{m,n}_{p_i,q_i:L}\left[\left(\frac{\beta}{\lambda z}\right)^{h\lambda}\middle| \begin{matrix} \{\Delta(\lambda,\alpha),h\},(a_j,\alpha_j)_{\lambda+1,n}; & (a_{ji},\alpha_{ji})_{n+1,p_i-\rho},\{\Delta(\rho,\alpha),h\} \\ (b_j,\beta_j)_{\rho+1,m}; & (b_{ji},\beta_{ji})_{m+1,q_i-\rho},\{\Delta(\rho,\alpha),h\} \end{matrix}\right]\right\}$$

Replacing the I-function (1.23) by its definition and then interchanging integration and differentiation, we get

$$D^r\left[\mathrm{B}(z,\alpha)\right] = \left(\frac{1}{2\pi i}\right)\int_L \frac{\prod_{j=1}^{m}\Gamma(b_j-\beta_j\xi)\prod_{k=0}^{\lambda-1}\Gamma\left(1-\frac{\alpha+k}{\lambda}+h\xi\right)}{\sum_{i=1}^{L}\left[\prod_{j=m+1}^{q_i-\rho}\Gamma(1-b_{ji}+\beta_{ji}\xi)\prod_{j=n+1}^{p_i-\rho}\Gamma(a_{ji}-\alpha_{ji}\xi)\right]}$$

$$\times\frac{\prod_{j=1}^{n}\Gamma(1-a_j+\alpha_j\xi)\left(1/\lambda\right)^{\alpha-1}\left(\beta/\lambda\right)^{h\xi\lambda}\left[\frac{d^r}{dz^r}z^{\alpha-1-h\xi\lambda}\right]}{\prod_{k=0}^{\rho-1}\Gamma\left(1-\frac{\alpha+k}{\rho}+h\xi\right)\prod_{k=0}^{\rho-1}\Gamma\left(\frac{\alpha+k}{\rho}-h\xi\right)}d\xi$$

Using the results (*g*) and (*h*), we get the following interesting results

$$\prod_{k=0}^{\lambda-1}\Gamma\left(\frac{\alpha+k}{\lambda}-h\xi\right) = (-\lambda)^{-r}(\lambda\xi h-\alpha+1)_r\prod_{k=0}^{\lambda-1}\Gamma\left(\frac{\alpha-r+k}{\lambda}-h\xi\right)$$

and

$$\prod_{k=0}^{\lambda-1}\Gamma\left(1-\frac{\alpha+k}{\lambda}+h\xi\right) = \frac{(\lambda)^r}{(h\lambda\xi-\alpha+1)_r}\prod_{k=0}^{\lambda-1}\Gamma\left(1-\frac{\alpha-r+k}{\lambda}+h\xi\right)$$

Here function (4.43) takes the above form

Similarly using $D^r\left[B(z,\alpha)\right] = B(z,\alpha+r)$ and the results (g) and (h), it can be shown that (4.44)–(4.50)

satisfy Truesdell's descending *F*-equation

$$D^r\left[B(z,\alpha)\right] = B(z,\alpha+r)$$

Now we present the generating functions which have been obtained by employing forms (4.33)–(4.40).

i.

$$G_1 \equiv \left(1+\frac{h}{\lambda}\right)^{-\alpha} I^{m,n}_{p_i,q_i:L}$$

$$\times\left[\left(1+\frac{h}{\lambda}\right)^{h\lambda} x \,\middle|\, \begin{matrix} (a_j,\alpha_j)_{1,n}; & (a_{ji},\alpha_{ji})_{n+1,p_i-\rho},\{\Delta(\rho,\alpha),h\} \\ \{\Delta(\lambda,\alpha),h\},(b_j,\beta_j)_{\lambda+1,m}; & (b_{ji},\beta_{ji})_{m+1,q_i} \end{matrix}\right]$$

$$= \sum_{r=0}^{\infty}\frac{h^r}{r!} I^{m,n}_{p_i,q_i:L}$$

$$\times\left[x \,\middle|\, \begin{matrix} (a_j,\alpha_j)_{1,n}; & (a_{ji},\alpha_{ji})_{n+1,p_i-\rho},\{\Delta(\rho,\alpha+r),h\} \\ \{\Delta(\lambda,\alpha+r),h\},(b_j,\beta_j)_{\lambda+1,m}; & (b_{ji},\beta_{ji})_{m+1,q_i} \end{matrix}\right] \tag{4.51}$$

ii.

$$G_2 \equiv \left(1+\frac{h}{\lambda}\right)^{-\alpha} I^{m,n}_{p_i,q_i:L}\left[\left(1+\frac{h}{\lambda}\right)^{h\lambda} x \,\middle|\, \begin{matrix} \{\Delta(\lambda,\alpha+1/2),h\},(a_j,\alpha_j)_{\lambda+1,n}; & (a_{ji},\alpha_{ji})_{n+1,p_i} \\ \{\Delta(2\lambda,2\alpha),h\},(b_j,\beta_j)_{2\lambda+1,m}; & (b_{ji},\beta_{ji})_{m+1,q_i} \end{matrix}\right]$$

$$= \sum_{r=0}^{\infty}\frac{h^r}{r!} I^{m,n}_{p_i,q_i:L}\left[x \,\middle|\, \begin{matrix} \{\Delta(\lambda,\alpha+r+1/2),h\},(a_j,\alpha_j)_{\lambda+1,n}; & (a_{ji},\alpha_{ji})_{n+1,p_i} \\ \{\Delta(2\lambda,2(\alpha+r)),h\},(b_j,\beta_j)_{2\lambda+1,m}; & (b_{ji},\beta_{ji})_{m+1,q_i} \end{matrix}\right] \tag{4.52}$$

iii.

$$G_3 \equiv \left(1+\frac{h}{\lambda}\right)^{-\alpha} I^{m,n}_{p_i,q_i:L}$$

$$\times\left[\left(1+\frac{h}{\lambda}\right)^{h\lambda} x \,\middle|\, \begin{matrix} \{\Delta(\lambda,\alpha+2/3),h\},(a_j,\alpha_j)_{\lambda+1,n}; & (a_{ji},\alpha_{ji})_{n+1,p_i-\lambda},\{\Delta(\lambda,\alpha+1/3),h\} \\ \{\Delta(3\lambda,3\alpha),h\},(b_j,\beta_j)_{3\lambda+1,m}; & (b_{ji},\beta_{ji})_{m+1,q_i} \end{matrix}\right]$$

$$= \sum_{r=0}^{\infty}\frac{h^r}{r!} I^{m,n}_{p_i,q_i:L}$$

$$\times\left[x \,\middle|\, \begin{matrix} \{\Delta(\lambda,\alpha+r+2/3),h\},(a_j,\alpha_j)_{\lambda+1,n}; & (a_{ji},\alpha_{ji})_{n+1,p_i-\lambda},\{\Delta(\lambda,\alpha+r+1/3),h\} \\ \{\Delta(3\lambda,3(\alpha+r)),h\},(b_j,\beta_j)_{3\lambda+1,m}; & (b_{ji},\beta_{ji})_{m+1,q_i} \end{matrix}\right] \tag{4.53}$$

iv.

$$G_4 \equiv \left(1+\frac{h}{\lambda}\right)^{-\alpha} I^{m,n}_{p_i,q_i:L}$$

$$\times\left[\left(1+\frac{h}{\lambda}\right)^{h\lambda} x \left| \begin{matrix} \{\Delta(2\lambda,2\alpha+\tfrac{1}{2}),h\},(a_j,\alpha_j)_{2\lambda+1,n}; & (a_{ji},\alpha_{ji})_{n+1,p_i-2\lambda},\{\Delta(2\lambda,2\alpha+1),h\} \\ \{\Delta(4\lambda,4\alpha+1),h\},(b_j,\beta_j)_{4\lambda+1,m}; & (b_{ji},\beta_{ji})_{m+1,q_i-\lambda},\{\Delta(\lambda,\alpha+1),h\} \end{matrix}\right.\right]$$

$$=\sum_{r=0}^{\infty}\frac{\boldsymbol{h}^r}{\boldsymbol{r}!} I^{m,n}_{p_i,q_i:L}$$

$$\times\left[x \left| \begin{matrix} \{\Delta(2\lambda,2\alpha+2r+\tfrac{1}{2}),h\},(a_j,\alpha_j)_{2\lambda+1,n}; & (a_{ji},\alpha_{ji})_{n+1,p_i-2\lambda},\{\Delta(2\lambda,2\alpha+2r+1),h\} \\ \{\Delta(4\lambda,4(\alpha+r)+1),h\},(b_j,\beta_j)_{4\lambda+1,m}; & (b_{ji},\beta_{ji})_{m+1,q_i-\lambda},\{\Delta(\lambda,4+r+1),h\} \end{matrix}\right.\right] \tag{4.54}$$

v.

$$G_5 \equiv \left(1+\frac{h}{\lambda}\right)^{-\alpha} I^{m,n}_{p_i,q_i:L}\left[\left(1+\frac{h}{\lambda}\right)^{h\lambda} x \left| \begin{matrix} (a_j,\alpha_j)_{1,n}; & (a_{ji},\alpha_{ji})_{n+1,p_i} \\ \{\Delta(\lambda,\alpha),h\},(b_j,\beta_j)_{\lambda+1,m}; & (b_{ji},\beta_{ji})_{m+1,q_i} \end{matrix}\right.\right]$$

$$=\sum_{r=0}^{\infty}\frac{\boldsymbol{h}^r}{\boldsymbol{r}!}(-1)^r I^{m,n}_{p_i,q_i:L}\left[x \left| \begin{matrix} (a_j,\alpha_j)_{1,n}; & (a_{ji},\alpha_{ji})_{n+1,p_i} \\ \{\Delta(\lambda,(\alpha+r)),h\},(b_j,\beta_j)_{\lambda+1,m}; & (b_{ji},\beta_{ji})_{m+1,q_i} \end{matrix}\right.\right] \tag{4.55}$$

vi.

$$G_6 \equiv \left(1+\frac{h}{\lambda}\right)^{-\alpha} I^{m,n}_{p_i,q_i:L}$$

$$\times\left[\left(1+\frac{h}{\lambda}\right)^{2h\lambda} x \left| \begin{matrix} (a_j,\alpha_j)_{1,n}; & (a_{ji},\alpha_{ji})_{n+1,p_i},\ldots\ldots\ldots\ldots \\ \{\Delta(\lambda,\alpha/2),h\},(b_j,\beta_j)_{\lambda+1,m}; & (b_{ji},\beta_{ji})_{m+1,q_i-\lambda},\{\Delta(\lambda,\alpha+\tfrac{1}{2}),h\} \end{matrix}\right.\right]$$

$$=\sum_{r=0}^{\infty}\frac{(2\boldsymbol{h})^r}{\boldsymbol{r}!}(-1)^{r/2} I^{m,n}_{p_i,q_i:L}$$

$$\times\left[x^2 \left| \begin{matrix} (a_j,\alpha_j)_{1,n}; & (a_{ji},\alpha_{ji})_{n+1,p_i} \\ \{\Delta(\lambda,(\alpha+r/2)),h\},(b_j,\beta_j)_{\lambda+1,m}; & (b_{ji},\beta_{ji})_{m+1,q_i-\lambda},\{\Delta(\lambda,\alpha+r+\tfrac{1}{2}),h\} \end{matrix}\right.\right] \tag{4.56}$$

vii.

$$G_7 \equiv \left(1+\frac{h}{\lambda}\right)^{-\alpha} I^{m,n}_{p_i,q_i:L}\left[\left(1+\frac{h}{\lambda}\right)^{h\lambda} x \left| \begin{matrix} (a_j,\alpha_j)_{1,n}; & (a_{ji},\alpha_{ji})_{n+1,p_i} \\ (b_j,\beta_j)_{1,m}; & (b_{ji},\beta_{ji})_{m+1,q_i-\lambda}, \{\Delta(\lambda,\alpha),h\} \end{matrix}\right.\right]$$

$$= \sum_{r=0}^{\infty} \frac{(h)^r}{r!} I^{m,n}_{p_i,q_i:L}\left[x \left| \begin{matrix} (a_j,\alpha_j)_{1,n}; & (a_{ji},\alpha_{ji})_{n+1,p_i} \\ (b_j,\beta_j)_{1,m}; & (b_{ji},\beta_{ji})_{m+1,q_i-\lambda}, \{\Delta(\lambda,\alpha+r),h\} \end{matrix}\right.\right] \tag{4.57}$$

viii.

$$G_8 \equiv \left(1+\frac{h}{\lambda}\right)^{-\alpha} I^{m,n}_{p_i,q_i:L}$$

$$\times\left[\left(1+\frac{h}{\lambda}\right)^{h\lambda} x \left| \begin{matrix} \{\Delta(\rho,\alpha),h\},(a_j,\alpha_j)_{\rho+1,n}; & (a_{ji},\alpha_{ji})_{n+1,p_i} \\ \{\Delta(\rho,\alpha+r),h\},(b_j,\beta_j)_{\rho+1,m}; & (b_{ji},\beta_{ji})_{m+1,q_i-\lambda}, \{\Delta(\lambda,\alpha),h\} \end{matrix}\right.\right]$$

$$= \sum_{r=0}^{\infty} \frac{(h)^r}{r!}(-1)^r I^{m,n}_{p_i,q_i:L}$$

$$\times\left[x \left| \begin{matrix} \{\Delta(\rho,\alpha+r),h\},(a_j,\alpha_j)_{\rho+1,n}; & (a_{ji},\alpha_{ji})_{n+1,p_i} \\ \{\Delta(\rho,\alpha+r),h\},(b_j,\beta_j)_{\rho+1,m}; & (b_{ji},\beta_{ji})_{m+1,q_i-\lambda}, \{\Delta(\lambda,\alpha+r),h\} \end{matrix}\right.\right] \tag{4.58}$$

To prove G_1 we employ form (4.33) in Truesdell's ascending F-equation (c) to obtain

$$\left(1+\frac{h}{\lambda}\right)^{-\alpha} I^{m,n}_{p_i,q_i:L}$$

$$\times\left[\left(1+\frac{h}{\lambda}\right)^{h\lambda} x \left| \begin{matrix} (a_j,\alpha_j)_{1,n}; & (a_{ji},\alpha_{ji})_{n+1,p_i-\rho}, \{\Delta(\rho,\alpha),h\} \\ \{\Delta(\rho,\alpha),h\},(b_j,\beta_j)_{\rho+1,m}; & (b_{ji},\beta_{ji})_{m+1,q_i-\lambda}, \{\Delta(\lambda,\alpha),h\} \end{matrix}\right.\right]$$

$$= \sum_{r=0}^{\infty} \frac{(h)^r}{r!} I^{m,n}_{p_i,q_i:L}$$

$$\times\left[\left(1+\frac{h}{\lambda}\right)^{h\lambda} x \left| \begin{matrix} (a_j,\alpha_j)_{1,n}; & (a_{ji},\alpha_{ji})_{n+1,p_i-\rho}, \{\Delta(\rho,\alpha+r),h\} \\ \{\Delta(\rho,\alpha+r),h\},(b_j,\beta_j)_{\lambda+1,m}; & (b_{ji},\beta_{ji})_{m+1,q_i} \end{matrix}\right.\right]$$

On replacing z by $\frac{y\lambda}{h}$ and $\left(y\beta/h\right)^{h\lambda}$ by x in succession we get the formula. The proofs for the remaining results G_2 to G_8 are similar.

Now the forms (4.43)–(4.50) have been employed to establish the following generating functions for I-functions using Truesdell's descending F-equation technique

ix.

$$G_9 \equiv \left(1+\frac{h}{\lambda}\right)^{\alpha-1} I^{m,n}_{p_i,q_i:L}$$

$$\times\left[\left(1+\frac{h}{\lambda}\right)^{-h\lambda} x \left| \begin{matrix} \{\Delta(\lambda,\alpha),h\},(a_j,\alpha_j)_{\lambda+1,n}; & (a_{ji},\alpha_{ji})_{n+1,p_i-\rho},\{\Delta(\rho,\alpha),h\} \\ (b_j,\beta_j)_{1,m}; & (b_{ji},\beta_{ji})_{m+1,q_i-\rho},\{\Delta(\rho,\alpha),h\} \end{matrix} \right.\right] =$$

$$= \sum_{r=0}^{\infty} \frac{(\boldsymbol{h})^r}{\boldsymbol{r}!} I^{m,n}_{p_i,q_i:L}$$

$$\times\left[x \left| \begin{matrix} \{\Delta(\lambda,\alpha-r),h\},(a_j,\alpha_j)_{\lambda+1,n}; & (a_{ji},\alpha_{ji})_{n+1,p_i-\rho},\{\Delta(\rho,\alpha-r),h\} \\ (b_j,\beta_j)_{1,m}; & (b_{ji},\beta_{ji})_{m+1,q_i-\rho},\{\Delta(\rho,\alpha-r),h\} \end{matrix} \right.\right]$$

(4.59)

x.

$$G_{10} \equiv \left(1+\frac{h}{\lambda}\right)^{\alpha-1} I^{m,n}_{p_i,q_i:L}$$

$$\times\left[\left(1+\frac{h}{\lambda}\right)^{-h\lambda} x \left| \begin{matrix} \{\Delta(2\lambda,2\alpha),h\},(a_j,\alpha_j)_{2\lambda+1,n}; & (a_{ji},\alpha_{ji})_{n+1,p_i} \\ \{\Delta(\lambda,\alpha+1/2),h\},\ (b_j,\beta_j)_{\lambda+1,m}; & (b_{ji},\beta_{ji})_{m+1,q_i} \end{matrix} \right.\right]$$

$$= \sum_{r=0}^{\infty} \frac{(\boldsymbol{h})^r}{\boldsymbol{r}!} I^{m,n}_{p_i,q_i:L}$$

(4.60)

$$\times\left[x \left| \begin{matrix} \{\Delta(2\lambda,2\alpha-2r),h\},(a_j,\alpha_j)_{2\lambda+1,n}; & (a_{ji},\alpha_{ji})_{n+1,p_i} \\ \{\Delta(\lambda,\alpha-r+1/2),h\},\ (b_j,\beta_j)_{\lambda+1,m}; & (b_{ji},\beta_{ji})_{m+1,q_i} \end{matrix} \right.\right]$$

xi.

$$G_{11} \equiv \left(1+\frac{h}{\lambda}\right)^{\alpha-1} I^{m,n}_{p_i,q_i:L}$$

$$\times\left[\left(1+\frac{h}{\lambda}\right)^{-h\lambda} x \left| \begin{matrix} \{\Delta(3\lambda,3\alpha),h\},(a_j,\alpha_j)_{3\lambda+1,n}; & (a_{ji},\alpha_{ji})_{n+1,p_i} \\ \{\Delta(\lambda,\alpha+2/3),h\},\ (b_j,\beta_j)_{\lambda+1,m}; & (b_{ji},\beta_{ji})_{m+1,q_i},\{\Delta(\lambda,\alpha+1/3),h\} \end{matrix}\right.\right]$$

$$=\sum_{r=0}^{\infty}\frac{(h)^r}{r!} I^{m,n}_{p_i,q_i:L}$$

$$\times\left[x \left| \begin{matrix} \{\Delta(3\lambda,\ 3\alpha-3r),h\},(a_j,\alpha_j)_{3\lambda+1,n}; & (a_{ji},\alpha_{ji})_{n+1,p_i} \\ \{\Delta(\lambda,\alpha-r+2/3),h\},\ (b_j,\beta_j)_{\lambda+1,m}; & (b_{ji},\beta_{ji})_{m+1,q_i-\lambda},\{\Delta(\lambda,\alpha-r+1/3),h\} \end{matrix}\right.\right] \tag{4.61}$$

xii.

$$G_{12} \equiv \left(1+\frac{h}{\lambda}\right)^{\alpha-1} I^{m,n}_{p_i,q_i:L}$$

$$\times\left[\left(1+\frac{h}{\lambda}\right)^{-h\lambda} x \left| \begin{matrix} \{\Delta(4\lambda,4\alpha+1),h\},(a_j,\alpha_j)_{4\lambda+1,n}; & (a_{ji},\alpha_{ji})_{n+1,p_i-\lambda},\{\Delta(\lambda,\ \alpha),h\} \\ \{\Delta(2\lambda,2\alpha+1/2),h\},\ (b_j,\beta_j)_{2\lambda+1,m}; & (b_{ji},\beta_{ji})_{m+1,q_i-2\lambda},\{\Delta(2\lambda,2\alpha+1),h\} \end{matrix}\right.\right]$$

$$\sum_{r=0}^{\infty}\frac{(h)^r}{r!} I^{m,n}_{p_i,q_i:L}$$

$$\times\left[x \left| \begin{matrix} \{\Delta(4\lambda,\ 4\alpha-4r+1),h\},(a_j,\alpha_j)_{4\lambda+1,n}; & (a_{ji},\alpha_{ji})_{n+1,p_i-\lambda},\{\Delta(\lambda,\ \alpha-r),h\} \\ \{\Delta(2\lambda,2\alpha-2r+1/2),h\},\ (b_j,\beta_j)_{2\lambda+1,m}; & (b_{ji},\beta_{ji})_{m+1,q_i-2\lambda},\{\Delta(2\lambda,2\alpha-2r+1),h\} \end{matrix}\right.\right] \tag{4.62}$$

xiii.

$$G_{13} \equiv \left(1+\frac{h}{\lambda}\right)^{\alpha-1} I^{m,n}_{p_i,q_i:L}\left[\left(1+\frac{h}{\lambda}\right)^{-h\lambda} x \left| \begin{matrix} \{\Delta(\lambda,\alpha),h\},(a_j,\alpha_j)_{\lambda+1,n}; & (a_{ji},\alpha_{ji})_{n+1,p_i} \\ (b_j,\beta_j)_{1,m}; & (b_{ji},\beta_{ji})_{m+1,q_i} \end{matrix}\right.\right]$$

$$=\sum_{r=0}^{\infty}\frac{(h)^r(-1)^r}{r!} I^{m,n}_{p_i,q_i:L}\left[x \left| \begin{matrix} \{\Delta(\lambda,\ \alpha-r),h\},(a_j,\alpha_j)_{\lambda+1,n}; & (a_{ji},\alpha_{ji})_{n+1,p_i} \\ (b_j,\beta_j)_{1,m}; & (b_{ji},\beta_{ji})_{m+1,q_i} \end{matrix}\right.\right] \tag{4.63}$$

xiv.

$$G_{14} \equiv \left(1+\frac{h}{\lambda}\right)^{\alpha-1} I^{m,n}_{p_i,q_i:L}$$

$$\times\left[\left(1+\frac{h}{\lambda}\right)^{-2h\lambda} x^2 \left| \begin{matrix} \left\{\Delta\left(\lambda,\alpha/2\right),h\right\},\left(a_j,\alpha_j\right)_{\lambda+1,n}; & \left(a_{ji},\alpha_{ji}\right)_{n+1,p_i-\lambda},\left\{\Delta\left(\lambda,\alpha/2\right),h\right\} \\ \left(b_j,\beta_j\right)_{1,m}; & \left(b_{ji},\beta_{ji}\right)_{m+1,q_i} \end{matrix} \right.\right]$$

$$= \sum_{r=0}^{\infty} \frac{(2h)^r}{r!}(-1)^{r/2} I^{m,n}_{p_i,q_i:L}$$

$$\times\left[x^2 \left| \begin{matrix} \left\{\Delta\left(\lambda,\alpha-r/2\right),h\right\},\left(a_j,\alpha_j\right)_{\lambda+1,n}; & \left(a_{ji},\alpha_{ji}\right)_{n+1,p_i-\lambda},\left\{\Delta\left(\lambda,\,\alpha-r+1/2\right),h\right\} \\ \left(b_j,\beta_j\right)_{1,m}; & \left(b_{ji},\beta_{ji}\right)_{m+1,q_i} \end{matrix} \right.\right] \tag{4.64}$$

xv.

$$G_{15} \equiv \left(1+\frac{h}{\lambda}\right)^{\alpha-1} I^{m,n}_{p_i,q_i:L}\left[\left(1+\frac{h}{\lambda}\right)^{-h\lambda} x \left| \begin{matrix} \left(a_j,\alpha_j\right)_{1,n}; & \left(a_{ji},\alpha_{ji}\right)_{n+1,p_i-\lambda},\left\{\Delta(\lambda,\alpha),h\right\} \\ \left(b_j,\beta_j\right)_{1,m}; & \left(b_{ji},\beta_{ji}\right)_{m+1,q_i} \end{matrix} \right.\right]$$

$$= \sum_{r=0}^{\infty} \frac{(h)^r}{r!} I^{m,n}_{p_i,q_i:L}\left[x \left| \begin{matrix} \left(a_j,\alpha_j\right)_{1,n}; & \left(a_{ji},\alpha_{ji}\right)_{n+1,p_i-\lambda},\left\{\Delta(\lambda,\,\alpha-r),h\right\} \\ \left(b_j,\beta_j\right)_{1,m}; & \left(b_{ji},\beta_{ji}\right)_{m+1,q_i} \end{matrix} \right.\right] \tag{4.65}$$

xvi.

$$G_{16} \equiv \left(1+\frac{h}{\lambda}\right)^{\alpha-1} I^{m,n}_{p_i,q_i:L}$$

$$\times\left[\left(1+\frac{h}{\lambda}\right)^{-h\lambda} x \left| \begin{matrix} \left\{\Delta(\rho,\alpha),h\right\},\left(a_j,\alpha_j\right)_{\rho+1,n}; & \left(a_{ji},\alpha_{ji}\right)_{n+1,p_i-\lambda},\left\{\Delta(\lambda,\alpha),h\right\} \\ \left\{\Delta(\rho,\alpha),h\right\},\left(b_j,\beta_j\right)_{\rho+1,m}; & \left(b_{ji},\beta_{ji}\right)_{m+1,q_i} \end{matrix} \right.\right]$$

$$= \sum_{r=0}^{\infty} \frac{(h)^r}{r!}(-1)^r I^{m,n}_{p_i,q_i:L}$$

$$\times\left[x \left| \begin{matrix} \left\{\Delta(\rho,\alpha-r),h\right\},\left(a_j,\alpha_j\right)_{\rho+1,n}; & \left(a_{ji},\alpha_{ji}\right)_{n+1,p_i-\lambda},\left\{\Delta(\lambda,\,\alpha-r),h\right\} \\ \left\{\Delta(\rho,\alpha-r),h\right\},\left(b_j,\beta_j\right)_{\rho+1,m}; & \left(b_{ji},\beta_{ji}\right)_{m+1,q_i} \end{matrix} \right.\right] \tag{4.66}$$

The formulas from G_9 to G_{16} can be obtained by the technique applied earlier and making use of the results (4.43)–(4.50), respectively,

Finally, we write some linear generating functions which can be obtained by using the following simple relations

$$(\alpha)_n = \frac{\Gamma(\alpha+n)}{\Gamma(\alpha)} \quad ; \quad \frac{\Gamma(1-\alpha-n)}{\Gamma(1-\alpha)} = \frac{(-1)^n}{(\alpha)_n}$$

$$\sum_{n=0}^{\infty} \frac{(\alpha)_n t^n}{n!} = (1-t)^{-\alpha} \quad ; \quad \sum_{n=0}^{\infty} \frac{(\alpha)_n (-t)^n}{n!} = (1+t)^{-\alpha}$$

xvii.

$$G_{17} \equiv \sum_{r=0}^{\infty} \frac{t^r}{r!} I^{m,n+1}_{p_i+1,q_i:r}\left[x \left| \begin{matrix} (\lambda-r,\alpha),(a_j,\alpha_j)_{1,n}; & (a_{ji},\alpha_{ji})_{n+1,p_i} \\ (b_j,\beta_j)_{1,m}; & (b_{ji},\beta_{ji})_{m+1,q_i} \end{matrix} \right. \right]$$

$$= (1-t)^{\lambda-1} I^{m,n+1}_{p_i+1,q_i:r}\left[x(1-t)^{-\alpha} \left| \begin{matrix} (\lambda,r),(a_j,\alpha_j)_{1,n}; & (a_{ji},\alpha_{ji})_{n+1,p_i}, \{\Delta(\lambda,\alpha-r),h\} \\ (b_j,\beta_j)_{1,m}; & (b_{ji},\beta_{ji})_{m+1,q_i} \end{matrix} \right. \right] \tag{4.67}$$

xviii.

$$G_{18} \equiv \sum_{r=0}^{\infty} \frac{t^r}{r!} I^{m,n+1}_{p_i+1,q_i:r}\left[x \left| \begin{matrix} (a_j,\alpha_j)_{1,n}; & (a_{ji},\alpha_{ji})_{n+1,p_i} \\ (\lambda+r,\alpha),(b_j,\beta_j)_{1,m}; & (b_{ji},\beta_{ji})_{m+1,q_i} \end{matrix} \right. \right] \tag{4.68}$$

$$= (1-t)^{-\lambda} I^{m+1,n}_{p_i,q_i+1:r}\left[x(1-t)^{\alpha} \left| \begin{matrix} (a_j,\alpha_j)_{1,n}; & (a_{ji},\alpha_{ji})_{n+1,p_i} \\ (\lambda,r),(b_j,\beta_j)_{1,m}; & (b_{ji},\beta_{ji})_{m+1,q_i} \end{matrix} \right. \right]$$

xix.

$$G_{19} \equiv \sum_{r=0}^{\infty} \frac{t^r}{r!} I^{m,n}_{p_i+1,q_i:r}\left[x \left| \begin{matrix} (a_j,\alpha_j)_{1,n}; & (a_{ji},\alpha_{ji})_{n+1,p_i}, (1-\lambda-r,\alpha) \\ (b_j,\beta_j)_{1,m}; & (b_{ji},\beta_{ji})_{m+1,q_i} \end{matrix} \right. \right] \tag{4.69}$$

$$= (1-t)^{-\lambda} I^{m,n}_{p_i+1,q_i:r}\left[x(1+t)^{-\alpha} \left| \begin{matrix} (a_j,\alpha_j)_{1,n}; & (a_{ji},\alpha_{ji})_{n+1,p_i}, (1-\lambda,\alpha) \\ (b_j,\beta_j)_{1,m}; & (b_{ji},\beta_{ji})_{m+1,q_i} \end{matrix} \right. \right]$$

xx.

$$G_{20} \equiv \sum_{r=0}^{\infty} \frac{t^r}{r!} I_{p_i,q_i+1:r}^{m,n}\left[x \left| \begin{matrix} (a_j,\alpha_j)_{1,n}; & (a_{ji},\alpha_{ji})_{n+1,p_i} \\ (b_j,\beta_j)_{1,m}; & (b_{ji},\beta_{ji})_{m+1,q_i},(\lambda+r,\alpha) \end{matrix} \right.\right]$$

$$= (1+t)^{-\lambda} I_{p_i,q_i+1:r}^{m,n}\left[x(1+t)^{\alpha} \left| \begin{matrix} (a_j,\alpha_j)_{1,n}; & (a_{ji},\alpha_{ji})_{n+1,p_i} \\ (b_j,\beta_j)_{1,m}; & (b_{ji},\beta_{ji})_{m+1,q_i},(\lambda,\alpha) \end{matrix} \right.\right] \tag{4.70}$$

xxi.

$$G_{21} \equiv \sum_{r=0}^{\infty} \frac{(-t)^r}{r!} I_{p_i+1,q_i:r}^{m,n+1}\left[x \left| \begin{matrix} (\lambda-r,\alpha),(a_j,\alpha_j)_{1,n}; & (a_{ji},\alpha_{ji})_{n+1,p_i} \\ (b_j,\beta_j)_{1,m}; & (b_{ji},\beta_{ji})_{m+1,q_i} \end{matrix} \right.\right]$$

$$= (1+t)^{-\lambda} I_{p_i+1,q_i:r}^{m,n+1}\left[x(1+t)^{-\alpha} \left| \begin{matrix} (\lambda,\alpha),(a_j,\alpha_j)_{1,n}; & (a_{ji},\alpha_{ji})_{n+1,p_i} \\ (b_j,\beta_j)_{1,m}; & (b_{ji},\beta_{ji})_{m+1,q_i},(\lambda,\alpha) \end{matrix} \right.\right] \tag{4.71}$$

xxii.

$$G_{22} \equiv \sum_{r=0}^{\infty} \frac{(-t)^r}{r!} I_{p_i,q_i+1:r}^{m+1,n}\left[x \left| \begin{matrix} (a_j,\alpha_j)_{1,n}; & (a_{ji},\alpha_{ji})_{n+1,p_i} \\ (\lambda+r,\alpha),(b_j,\beta_j)_{1,m}; & (b_{ji},\beta_{ji})_{m+1,q_i} \end{matrix} \right.\right]$$

$$= (1+t)^{-\lambda} I_{p_i,q_i+1:r}^{m+1,n}\left[x(1+t)^{\alpha} \left| \begin{matrix} (a_j,\alpha_j)_{1,n}; & (a_{ji},\alpha_{ji})_{n+1,p_i} \\ (\lambda,\alpha),(b_j,\beta_j)_{1,m}; & (b_{ji},\beta_{ji})_{m+1,q_i} \end{matrix} \right.\right] \tag{4.72}$$

xxiii.

$$G_{23} \equiv \sum_{r=0}^{\infty} \frac{(-t)^r}{r!} I_{p_i+1,q_i:r}^{m,n}\left[x \left| \begin{matrix} (a_j,\alpha_j)_{1,n}; & (a_{ji},\alpha_{ji})_{n+1,p_i},(1-\lambda-r,\alpha) \\ (b_j,\beta_j)_{1,m}; & (b_{ji},\beta_{ji})_{m+1,q_i} \end{matrix} \right.\right]$$

$$= (1-t)^{-\lambda} I_{p_i+1,q_i:r}^{m,n}\left[x(1-t)^{-\alpha} \left| \begin{matrix} (a_j,\alpha_j)_{1,n}; & (a_{ji},\alpha_{ji})_{n+1,p_i},(1-\lambda,\alpha) \\ (b_j,\beta_j)_{1,m}; & (b_{ji},\beta_{ji})_{m+1,q_i} \end{matrix} \right.\right] \tag{4.73}$$

xxiv.

$$G_{24} \equiv \sum_{r=0}^{\infty} \frac{(-t)^r}{r!} I_{p_i,q_i+1:r}^{m,n} \left[x \left| \begin{matrix} (a_j,\alpha_j)_{1,n}; & (a_{ji},\alpha_{ji})_{n+1,p_i} \\ (b_j,\beta_j)_{1,m}; & (b_{ji},\beta_{ji})_{m+1,q_i}, (\lambda+r,\alpha) \end{matrix} \right. \right]$$

$$= (1-t)^{-\lambda} I_{p_i,q_i+1:r}^{m,n} \left[x(1-t)^{\alpha} \left| \begin{matrix} (a_j,\alpha_j)_{1,n}; & (a_{ji},\alpha_{ji})_{n+1,p_i} \\ (b_j,\beta_j)_{1,m}; & (b_{ji},\beta_{ji})_{m+1,q_i}, (\lambda,\alpha) \end{matrix} \right. \right] \quad (4.74)$$

In all the above results, it is assumed that all the parameters satisfy the conditions necessary for the existence of the I-functions involved.

4.3.2 Special Cases

These results yield certain interesting generating functions for Fox's H-function. If we take $L=1$, then the series (4.51) reduces to an interesting series of Fox's H-function.

$$\left(1+\frac{h}{\lambda}\right)^{-\alpha} H_{p,q}^{m,n} \left[\left(1+\frac{h}{\lambda}\right)^{-h\lambda} x \left| \begin{matrix} (a_j,\alpha_j)_{1,p-\rho}, & \{\Delta(\rho,\alpha),h\} \\ \{\Delta(\lambda,\alpha),h\}, & (b_j,\beta_j)_{\lambda+1,q} \end{matrix} \right. \right]$$

$$= \sum_{r=0}^{\infty} \frac{h^r}{r!} H_{p,q}^{m,n} \left[x \left| \begin{matrix} (a_j,\alpha_j)_{1,p-\rho}, & \{\Delta(\rho,\alpha+r),h\} \\ \{\Delta(\lambda,\alpha+r),h\}, & (b_j,\beta_j)_{\lambda+1,q} \end{matrix} \right. \right] \quad (4.75)$$

Similarly, putting $L=1$ in results (4.59), we get the corresponding results for Fox's H-function

$$\left(1+\frac{h}{\lambda}\right)^{\alpha-1} H_{p,q}^{m,n} \left[\left(1+\frac{h}{\lambda}\right)^{-h\lambda} x \left| \begin{matrix} \{\Delta(\lambda,\alpha),h\},(a_j,\alpha_j)_{\lambda+1,n}; & (a_{ji},\alpha_{ji})_{n+1,p-\rho},\{\Delta(\rho,\alpha),h\} \\ (b_j,\beta_j)_{1,m}; & (b_{ji},\beta_{ji})_{m+1,q-\rho},\{\Delta(\rho,\alpha),h\} \end{matrix} \right. \right]$$

$$= \sum_{r=0}^{\infty} \frac{(h)^r}{r!} H_{p,q}^{m,n} \left[x \left| \begin{matrix} \{\Delta(\lambda,\alpha-r),h\},(a_j,\alpha_j)_{\lambda+1,n}; & (a_{ji},\alpha_{ji})_{n+1,p-\rho},\{\Delta(\rho,\alpha-r),h\} \\ (b_j,\beta_j)_{1,m}; & (b_{ji},\beta_{ji})_{m+1,q-\rho},\{\Delta(\rho,\alpha-r),h\} \end{matrix} \right. \right] \quad (4.76)$$

4.4 FINITE SERIES

A number of finite summation formulas for the I-function based on certain identities for Gauss hypergeometric functions are given in this section. Proofs of some selected formulas are given at the end.

The following are finite series involving I-functions

i.

$$FS_1 \equiv \sum_{k=0}^{n} \frac{1}{k!}\left\{\frac{(d-b)_k}{(d+1)_k} I^{m+1,n+1}_{p_i+2,q_i+2:r}\left[z\left|\begin{matrix}(-1-a,h),(a_j,\alpha_j)_{1,n}; & (a_{ji},\alpha_{ji})_{n+1,p_i},(d-a-1,h)\\(d-a-1+k,h),\ (b_j,\beta_j)_{1,m}; & (b_{ji},\beta_{ji})_{m+1,q_i},(-a,h)\end{matrix}\right.\right]\right.$$

$$\left.-(1+b)\frac{(d-b-1)_k}{(d+1)_k} I^{m+1,n}_{p_i+1,q_i+1:r}\left[z\left|\begin{matrix}(a_j,\alpha_j)_{1,n}; & (a_{ji},\alpha_{ji})_{n+1,p_i},(d-a,h)\\(d-a+k,h),\ (b_j,\beta_j)_{1,m}; & (b_{ji},\beta_{ji})_{m+1,q_i}\end{matrix}\right.\right]\right\}$$

$$=\frac{(d-b)_n}{n!(d+1)_k} I^{m+1,n+1}_{p_i+2,q_i+2:r}\left[z\left|\begin{matrix}(b-a,h),(a_j,\alpha_j)_{1,n}; & (a_{ji},\alpha_{ji})_{n+1,p_i},(d-a,h)\\(d-a+n,h),\ (b_j,\beta_j)_{1,m}; & (b_{ji},\beta_{ji})_{m+1,q_i},(b-a+1,h)\end{matrix}\right.\right] \tag{4.77}$$

provided $|\arg z| < \frac{1}{2}B_i\pi,\ A_i\langle 0, B_i\rangle 0$ where

$$A_i = \sum_{j=1}^{p_i} a_j - \sum_{j=1}^{q_j} b_j,$$

$$B_i = \sum_{j=1}^{n}\alpha_j - \sum_{j=n+1}^{p_i}\alpha_{ji} + \sum_{j=1}^{m}\beta_j - \sum_{j=m+1}^{q_i}\beta_{ji}\ ; i = 1,2,\ldots,r \text{ and } h \text{ is a positive number.}$$

ii.

$$FS_2 \equiv \sum_{k=0}^{n} \frac{1}{k!}\left\{a\, I^{m+2,n}_{p_i+2,q_i+2:r}\left[z\left|\begin{matrix}(a_j,\alpha_j)_{1,n};(a_{ji},\alpha_{ji})_{n+1,p_i}, & (d+k,h),(d-a-1,h)\\(d-b+k,h),(d-a-1+k,h),\ (b_j,\beta_j)_{1,m}; & (b_{ji},\beta_{ji})_{m+1,q_i}\end{matrix}\right.\right]\right.$$

$$\left.-I^{m+2,n+1}_{p_i+3,q_i+3:r}\left[z\left|\begin{matrix}(d-a+b,h),(a_j,\alpha_j)_{1,n}; & (a_{ji},\alpha_{ji})_{n+1,p_i},(d+k,h),(d-a,h)\\(d-b+k,h),\ (b_j,\beta_j)_{1,m}; & (b_{ji},\beta_{ji})_{m+1,q_i},(1-a-b+d,h)\end{matrix}\right.\right]\right\}$$

$$=\frac{1}{n!} I^{m+2,n}_{p_i+2,q_i+2:r}\left[z\left|\begin{matrix}(a_j,\alpha_j)_{1,n}; & (a_{ji},\alpha_{ji})_{n+1,p_i},(d+n,h),(d-a,h)\\(d+n-b+1,h),(d-a+n,h),\ (b_j,\beta_j)_{1,m}; & (b_{ji},\beta_{ji})_{m+1,q_i}\end{matrix}\right.\right] \tag{4.78}$$

iii.

$$\mathrm{FS}_3 \equiv \sum_{k=0}^{n} \frac{1}{k!}\left\{\frac{1}{\Gamma(d+k)} I^{m+1,n+1}_{p_i+2,q_i+2:r}\left[z\left|\begin{matrix}(1-d+b-k,h),(a_j,\alpha_j)_{1,n};(a_{ji},\alpha_{ji})_{n+1,p_i}, & (d-a,h)\\ (d-a+k,h),\ (b_j,\beta_j)_{1,m};(b_{ji},\beta_{ji})_{m+1,q_i}, & (1-d+b,h)\end{matrix}\right.\right]\right.$$

$$\left.+\frac{d-a-b+1}{\Gamma(d+k+1)} I^{m+1,n+1}_{p_i+2,q_i+2:r}\left[z\left|\begin{matrix}(b-d-k,h),(a_j,\alpha_j)_{1,n}; & (a_{ji},\alpha_{ji})_{n+1,p_i},(d+a+1,h)\\ (d-a+k+1,h),\ (b_j,\beta_j)_{1,m}; & (b_{ji},\beta_{ji})_{m+1,q_i},(b-d,h)\end{matrix}\right.\right]\right\}$$

$$=\frac{2d-a-b+n-1}{n!}\frac{1}{\Gamma(d+n+1)} I^{m+1,n+1}_{p_i+2,q_i+2:r}$$

$$\times\left[z\left|\begin{matrix}(b-d-n,h),(a_j,\alpha_j)_{1,n};(a_{ji},\alpha_{ji})_{n+1,p_i}, & (1+d-a,h)\\ (d-a+n+1,h),\ (b_j,\beta_j)_{1,m};(b_{ji},\beta_{ji})_{m+1,q_i}, & (b-d,h)\end{matrix}\right.\right] \tag{4.79}$$

Conditions for the convergence of results (4.78) and (4.79) are the same as in (4.77).

iv.

$$\mathrm{FS}_4 \equiv \sum_{k=0}^{n} \frac{1}{k!}$$

$$\times\left\{(1+a) I^{m+2,n}_{p_i+3,q_i+2:r}\left[z\left|\begin{matrix}(a_j,\alpha_j)_{1,n};(a_{ji},\alpha_{ji})_{n+1,p_i},(d-b,h),(d+1+k,h), & (d-a-1,h)\\ (d-b+k,h),(d-a-1+k,h),\ (b_j,\beta_j)_{1,m}; & (b_{ji},\beta_{ji})_{m+1,q_i}\end{matrix}\right.\right]\right.$$

$$\left.-(1+b) I^{m+2,n}_{p_i+3,q_i+2:r}\left[z\left|\begin{matrix}(a_j,\alpha_j)_{1,n};(a_{ji},\alpha_{ji})_{n+1,p_i},(d-b-1,h),(d+1+k,h), & (d-a,h)\\ (d-b+k,h),(d-a+k,h),\ (b_j,\beta_j)_{1,m}; & (b_{ji},\beta_{ji})_{m+1,q_i}\end{matrix}\right.\right]\right\}$$

$$=\frac{a-b}{n!} I^{m+2,n}_{p_i+3,q_i+2:r}\left[z\left|\begin{matrix}(a_j,\alpha_j)_{1,n};(a_{ji},\alpha_{ji})_{n+1,p_i},(d-b,h),(d-a,h), & (d+1+n,h)\\ (d-a+n,h),(d-b+n,h),\ (b_j,\beta_j)_{1,m}; & (b_{ji},\beta_{ji})_{m+1,q_i}\end{matrix}\right.\right] \tag{4.80}$$

v.

$$\mathrm{FS}_5 \equiv \sum_{k=0}^{n} \frac{1}{k!(1+d+k)}$$

$$\times\left\{I^{m+2,n+1}_{p_i+3,q_i+3:r}\left[z\left|\begin{matrix}(-1-a,h),(a_j,\alpha_j)_{1,n};(a_{ji},\alpha_{ji})_{n+1,p_i},(d-b,h'), & (d-a-1,h)\\ (d-b+k,h'),(d-a-1+k,h),\ (b_j,\beta_j)_{1,m}; & (b_{ji},\beta_{ji})_{m+1,q_i},(-a,h)\end{matrix}\right.\right]\right.$$

$$-I^{m+2,n+1}_{p_i+3,q_i+3:r}\left[z\left|\begin{matrix}(-1-a,h),(a_j,\alpha_j)_{1,n};(a_{ji},\alpha_{ji})_{n+1,p_i},(d-b-1,h'), & (d-a,h)\\ (d-b+k-1,h'),(d-a+k,h),\ (b_j,\beta_j)_{1,m}; & (b_{ji},\beta_{ji})_{m+1,q_i},(-b,h)\end{matrix}\right.\right]\Bigg\}$$

$$=\frac{1}{n!\Gamma(1+d+n)}$$

$$\times\left\{I^{m+2,n+1}_{p_i+3,q_i+3:r}\left[z\left|\begin{matrix}(-a,h),(a_j,\alpha_j)_{1,n};(a_{ji},\alpha_{ji})_{n+1,p_i},(d-b,h'), & (d-a,h)\\ (d-a+n,h),(d-b+n,h'),\ (b_j,\beta_j)_{1,m}; & (b_{ji},\beta_{ji})_{m+1,q_i},(1-a,h)\end{matrix}\right.\right]\right.$$

$$\left.-I^{m+2,n+1}_{p_i+3,q_i+3:r}\left[z\left|\begin{matrix}(-b,h'),(a_j,\alpha_j)_{1,n};(a_{ji},\alpha_{ji})_{n+1,p_i},(d-b,h'), & (d-a,h)\\ (d-a+n,h),(d-b+n,h'),\ (b_j,\beta_j)_{1,m}; & (b_{ji},\beta_{ji})_{m+1,q_i},(1-a,h')\end{matrix}\right.\right]\right\} \quad (4.81)$$

vi.

$$FS_6\equiv\sum_{k=0}^{n}\frac{\Gamma(d-b+k)}{k!\Gamma(d+k)}\left\{I^{m+1,n+1}_{p_i+2,q_i+2:r}\left[z\left|\begin{matrix}(-a,h)(a_j,\alpha_j)_{1,n};(a_{ji},\alpha_{ji})_{n+1,p_i}, & (d-a-1,h)\\ (d-a-1+k,h),\ (b_j,\beta_j)_{1,m}; & (b_{ji},\beta_{ji})_{m+1,q_i},(1-a,h)\end{matrix}\right.\right]\right.$$

$$\left.-I^{m+1,n+1}_{p_i+3,q_i+2:r}\left[z\left|\begin{matrix}(d-a-b,h),(a_j,\alpha_j)_{1,n};(a_{ji},\alpha_{ji})_{n+1,p_i}, & (d-a,h)\\ (d-a+k,h),\ (b_j,\beta_j)_{1,m}; & (b_{ji},\beta_{ji})_{m+1,q_i},(d-a-b+1,h)\end{matrix}\right.\right]\right\}$$

$$=\frac{1\Gamma(1+d+n-b)}{n!\Gamma(d+n)}I^{m+1,n}_{p_i+1,q_i+1:r}\left[z\left|\begin{matrix}(a_j,\alpha_j)_{1,n};(a_{ji},\alpha_{ji})_{n+1,p_i} & (d-a,h)\\ (d-a+n,h),\ (b_j,\beta_j)_{1,m}; & (b_{ji},\beta_{ji})_{m+1,q_i}\end{matrix}\right.\right] \quad (4.82)$$

vii.

$$FS_7\equiv\sum_{k=0}^{n}\frac{\Gamma(d-a-1k)}{k!\Gamma(d+k)}\left\{aI^{m,n+1}_{p_i+1,q_i+1:r}\left[z\left|\begin{matrix}(1+b-d-k,h)(a_j,\alpha_j)_{1,n}; & (a_{ji},\alpha_{ji})_{n+1,p_i}\\ (b_j,\beta_j)_{1,m}; & (b_{ji},\beta_{ji})_{m+1,q_i},(1+b-d,h)\end{matrix}\right.\right]\right.$$

$$\left.-\frac{(d-a+k-1)}{(d-a,1)}I^{m+1,n+1}_{p_i+2,q_i+2:r}\left[z\left|\begin{matrix}(1+b-d-k,h),(a_j,\alpha_j)_{1,n};(a_{ji},\alpha_{ji})_{n+1,p_i}, & (a+b-d,h)\\ (a+b+1-d,h),\ (b_j,\beta_j)_{1,m}; & (b_{ji},\beta_{ji})_{m+1,q_i},(1+b-d,h)\end{matrix}\right.\right]\right\}$$

$$=\frac{\Gamma(d+n-b)}{n!\Gamma(d+n)\Gamma(d-a-1)}I^{m,n+1}_{p_i+1,q_i+1:r}\left[z\left|\begin{matrix}(b-d-n,h)(a_j,\alpha_j)_{1,n}; & (a_{ji},\alpha_{ji})_{n+1,p_i}\\ (b_j,\beta_j)_{1,m}; & (b_{ji},\beta_{ji})_{m+1,q_i},(b-d+1,h)\end{matrix}\right.\right] \quad (4.83)$$

viii.

$$FS_8 \equiv \sum_{k=0}^{n} \frac{1}{k!\Gamma(d+k)} \left\{ I^{m+2,n+1}_{p_i+3,q_i+3:r} \left[z \middle| \begin{matrix} (-a,h)(a_j,\alpha_j)_{1,n};(a_{ji},\alpha_{ji})_{n+1,p_i}, & (d-a-1,h),(d-b,h') \\ (d-b+k,h'),(d-a+k,h)(b_j,\beta_j)_{1,m}; & (b_{ji},\beta_{ji})_{m+1,q_i},(1-a,h) \end{matrix} \right] \right.$$

$$\left. - I^{m+2,n+1}_{p_i+3,q_i+3:r} \left[z \middle| \begin{matrix} (d-a-b,h'),(a_j,\alpha_j)_{1,n};(a_{ji},\alpha_{ji})_{n+1,p_i}, & (d-b,h)(d-a,h) \\ (d-b+k,h'),\ (d-a+k,h)(b_j,\beta_j)_{1,m}; & (b_{ji},\beta_{ji})_{m+1,q_i},(1+d-a-b,h+h') \end{matrix} \right] \right\}$$

$$= \frac{1}{n!\Gamma(d+n)} I^{m+2,n}_{p_i+2,q_i+2:r} \left[z \middle| \begin{matrix} (a_j,\alpha_j)_{1,n};(a_{ji},\alpha_{ji})_{n+1,p_i}, & (d-b,h'),(d-a,h) \\ (d+n-b+1,h'),(d-a+n,h)(b_j,\beta_j)_{1,m}; & (b_{ji},\beta_{ji})_{m+1,q_i}, \end{matrix} \right] \tag{4.84}$$

ix.

$$FS_{10} \equiv \sum_{k=0}^{n} \frac{1}{k!} \left\{ I^{m,n+2}_{p_i+2,q_i+3:r} \left[z \middle| \begin{matrix} (1-d+k+b,h),(1+a-d-k,h),(a_j,\alpha_j)_{1,n}; & (a_{ji},\alpha_{ji})_{n+1,p_i} \\ (b_j,\beta_j)_{1,m};(b_{ji},\beta_{ji})_{m+1,q_i},(1-d-k,h),(1-d+b,h), & (1-d+a,h) \end{matrix} \right] \right.$$

$$\left. - I^{m+1,n+2}_{p_i+3,q_i+4:r} \left[z \middle| \begin{matrix} (b-d-k,h),(a-d-k,h)(a_j,\alpha_j)_{1,n}; & (a_{ji},\alpha_{ji})_{n+1,p_i},(a-b-d-1,h) \\ (a+b-d,h),\ (b_j,\beta_j)_{1,m};(b_{ji},\beta_{ji})_{m+1,q_i}, & (-d-k,h),(b-d,h),(a-d,h) \end{matrix} \right] \right\}$$

$$= \frac{2}{n!} I^{m,n+2}_{p_i+2,q_i+3:r} \left[z \middle| \begin{matrix} \left(\frac{a+b-n-1-2d}{2},h\right),(a-d-n,h),(a_j,\alpha_j)_{1,n}; & (a_{ji},\alpha_{ji})_{n+1,p_i} \\ (b_j,\beta_j)_{1,m};(b_{ji},\beta_{ji})_{m+1,q_i},\left(\frac{a+b+1-n-2d}{2},h\right),(a-d,h), & (b-d,h) \end{matrix} \right] \tag{4.85}$$

xi.

$$FS_{11} \equiv \sum_{k=0}^{n} \frac{1}{k!} \left\{ \frac{1}{\Gamma(d+k)} I^{m+2,n}_{p_i+2,q_i+2:r} \left[z \middle| \begin{matrix} (a_j,\alpha_j)_{1,n};(a_{ji},\alpha_{ji})_{n+1,p_i},(d-a,h), & (d-b,h) \\ (d-a+k,h),(d-b+k,h),(b_j,\beta_j)_{1,m}; & (b_{ji},\beta_{ji})_{m+1,q_i} \end{matrix} \right] \right.$$

$$\left. - \frac{2}{\Gamma(d+k+1)} I^{m+2,n+1}_{p_i+3,q_i+3:r} \left[z \middle| \begin{matrix} \left(\frac{d-a-b+1}{2},h\right)(a_j,\alpha_j)_{1,n};(a_{ji},\alpha_{ji})_{n+1,p_i},(d-a+1,h), & (d-b+1,h) \\ (1+d-a+k,h),\ (1+d+k-b,h),(b_j,\beta_j)_{1,m};(b_{ji},\beta_{ji})_{m+1,q_i}, & \left(\frac{d-a-b+3}{2},h\right) \end{matrix} \right] \right\}$$

$$= \frac{4}{n!\Gamma(d+n+1)} I^{m+3,n}_{p_i+3,q_i+3:r}$$

$$\times \left[z \middle| \begin{matrix} (a_j,\alpha_j)_{1,n};(a_{ji},\alpha_{ji})_{n+1,p_i},(1+d-a,h),(1+d-b,h), & \left(\frac{2d-a+n-b+1}{2},h\right) \\ (1+d-a+n,h),(1+d-b-n,h),\left(\frac{2d-a-b+3}{2},h\right)(b_j,\beta_j)_{1,m}; & (b_{ji},\beta_{ji})_{m+1,q_i} \end{matrix} \right] \tag{4.86}$$

The restrictions for (4.80)–(4.87) are the same as indicated in earlier results.

xii.

$$\mathrm{FS}_{12} \equiv \sum_{k=0}^{n} \frac{\Gamma(b+k)}{k!\Gamma(d+k)} \left\{ I_{p_i+2,q_i+1:r}^{m+1,n+1}\left[z \left| \begin{matrix} (2-a-k,h),(a_j,\alpha_j)_{1,n};(a_{ji},\alpha_{ji})_{n+1,p_i}, & (d-a,h) \\ (d-a+1,h),(b_j,\beta_j)_{1,m}; & (b_{ji},\beta_{ji})_{m+1,q_i},(2-a,h) \end{matrix} \right. \right] \right.$$

$$\left. -I_{p_i+2,q_i+2:r}^{m+1,n+1}\left[z \left| \begin{matrix} (1-a-k,h)(a_j,\alpha_j)_{1,n};(a_{ji},\alpha_{ji})_{n+1,p_i}, & (d-a-b,h) \\ (d-a-b+1,h),(b_j,\beta_j)_{1,m};(b_{ji},\beta_{ji})_{m+1,q_i}, & (1-a,h) \end{matrix} \right. \right] \right\}$$

$$= \frac{\Gamma(b+n+1)}{n!\Gamma(d+n)} I_{p_i+1,q_i+1:r}^{m,n+1}\left[z \left| \begin{matrix} (1-a-n,h)(a_j,\alpha_j)_{1,n}; & (a_{ji},\alpha_{ji})_{n+1,p_i} \\ (b_j,\beta_j)_{1,m}; & (b_{ji},\beta_{ji})_{m+1,q_i},(1-a,h) \end{matrix} \right. \right] \tag{4.88}$$

xiii.

$$\mathrm{FS}_{13} \equiv \sum_{k=0}^{n} \frac{\Gamma(b+k)}{k!\Gamma(d+k)} \left\{ bI_{p_i+1,q_i+1:r}^{m,n+1}\left[z \left| \begin{matrix} (1-a-k,h),(a_j,\alpha_j)_{1,n}; & (a_{ji},\alpha_{ji})_{n+1,p_i} \\ (b_j,\beta_j)_{1,m}; & (b_{ji},\beta_{ji})_{m+1,q_i},(1-a,h) \end{matrix} \right. \right] \right.$$

$$\left. +\frac{b+k}{d+k} I_{p_i+2,q_i+2:r}^{m,n+2}\left[z \left| \begin{matrix} (d-a-b-1,h),(-a-k,h)(a_j,\alpha_j)_{1,n}; & (a_{ji},\alpha_{ji})_{n+1,p_i} \\ (b_j,\beta_j)_{1,m};(b_{ji},\beta_{ji})_{m+1,q_i},(d-a-b,h) & (-a,h) \end{matrix} \right. \right] \right\}$$

$$= \frac{\Gamma(b+n+1)}{n!\Gamma(d+n+1)} I_{p_i+2,q_i+2:r}^{m,n+2}$$

$$\times \left[z \left| \begin{matrix} (-a-b-n-1,h),(a-n,h),(a_j,\alpha_j)_{1,n}; & (a_{ji},\alpha_{ji})_{n+1,p_i} \\ (b_j,\beta_j)_{1,m}; & (b_{ji},\beta_{ji})_{m+1,q_i},(-a-b-n,h),(-a,h) \end{matrix} \right. \right] \tag{4.89}$$

The conditions for convergence for the results (4.88) and (4.89) are same as that for (4.77).

The following finite summation formulas can also be established using the following results.

xiv.

$$\mathrm{FS}_{14} \equiv \sum_{k=0}^{n} \frac{\Gamma(b+k)}{\Gamma(k)} \left\{ (a-1)_k \, I_{p_i+2,q_i+2:r}^{m,n+2}\left[z \middle| \begin{matrix} (a-d,h),(1-d,h), & (a_j,\alpha_j)_{1,n};(a_{ji},\alpha_{ji})_{n+1,p_i} \\ (b_j,\beta_j)_{1,m}; & (b_{ji},\beta_{ji})_{m+1,q_i},(1+a-d,h),(1-d-k,h) \end{matrix} \right] \right.$$

$$\left. -(a)_k \, I_{p_i+2,q_i+2:r}^{m,n+2}\left[z \middle| \begin{matrix} (a+b-d,h),(1-d,h)(a_j,\alpha_j)_{1,n}; & (a_{ji},\alpha_{ji})_{n+1,p_i} \\ (b_j,\beta_j)_{1,m};(b_{ji},\beta_{ji})_{m+1,q_i},(1+a+b-d,h) & (1-d-k,h) \end{matrix} \right] \right\}$$

$$= \frac{\Gamma(b+n+1)}{n!}(a)_n \, I_{p_i+1,q_i+1:r}^{m,n+1}\left[z \middle| \begin{matrix} (1-d,h),(a-n,h),(a_j,\alpha_j)_{1,n}; & (a_{ji},\alpha_{ji})_{n+1,p_i} \\ (b_j,\beta_j)_{1,m}; & (b_{ji},\beta_{ji})_{m+1,q_i},(1-d-n,h) \end{matrix} \right] \tag{4.90}$$

xv.

$$\mathrm{FS}_{15} \equiv \sum_{k=0}^{n} \frac{1}{k!(d)_k} \left\{ I_{p_i+3,q_i+2:r}^{m+1,n+2}\left[z \middle| \begin{matrix} (1-b-k,h),(2-a-k,h), & (a_j,\alpha_j)_{1,n};(a_{ji},\alpha_{ji})_{n+1,p_i} \\ (d-a+1,h),(b_j,\beta_j)_{1,m}; & (b_{ji},\beta_{ji})_{m+1,q_i},(2-a,h) \end{matrix} \right] \right.$$

$$\left. -2I_{p_i+3,q_i+2:r}^{m+1,n+2}\left[z \middle| \begin{matrix} (1-b-k,h),(1-a-k,h)(a_j,\alpha_j)_{1,n}; & (a_{ji},\alpha_{ji})_{n+1,p_i},\left(\frac{d-a-b}{2},h\right) \\ \left(\frac{d-a-b+2}{2},h\right),(b_j,\beta_j)_{1,m};(b_{ji},\beta_{ji})_{m+1,q_i}, & (1-a,h) \end{matrix} \right] \right\}$$

$$= \frac{1}{n!(d)_n} I_{p_i+2,q_i+1:r}^{m,n+2}\left[z \middle| \begin{matrix} (-b-n,h),(1-a-n,h),(a_j,\alpha_j)_{1,n}; & (a_{ji},\alpha_{ji})_{n+1,p_i} \\ (b_j,\beta_j)_{1,m}; & (b_{ji},\beta_{ji})_{m+1,q_i},(1-a,h) \end{matrix} \right] \tag{4.91}$$

xvi.

$$\mathrm{FS}_{16} \equiv \sum_{k=0}^{n} \left\{ I_{p_i+3,q_i+3:r}^{m+2,n+1}\left[z \middle| \begin{matrix} (2-a-k,h), & (a_j,\alpha_j)_{1,n};(a_{ji},\alpha_{ji})_{n+1,p_i},(d-a,h),(b,h) \\ (d-a+1,h),(b+k,h)(b_j,\beta_j)_{1,m}; & (b_{ji},\beta_{ji})_{m+1,q_i},(2-a,h) \end{matrix} \right] \right.$$

$$\left. -(d-a-b)\, I_{p_i+2,q_i+2:r}^{m+1,n+1}\left[z \middle| \begin{matrix} (1-a-k,h)(a_j,\alpha_j)_{1,n}; & (a_{ji},\alpha_{ji})_{n+1,p_i},(b,h) \\ (b+k,h),(b_j,\beta_j)_{1,m};(b_{ji},\beta_{ji})_{m+1,q_i}, & (1-a,h) \end{matrix} \right] \right\}$$

$$= \frac{1}{n!(d)_n} I_{p_i+2,q_i+2:r}^{m+1,n+1}\left[z \middle| \begin{matrix} (1-a-n,h),(a_j,\alpha_j)_{1,n}; & (a_{ji},\alpha_{ji})_{n+1,p_i},(b,h) \\ (b+n+1,h)(b_j,\beta_j)_{1,m}; & (b_{ji},\beta_{ji})_{m+1,q_i},(1-a,h) \end{matrix} \right] \tag{4.92}$$

xvii.

$$FS_{17} \equiv \sum_{k=0}^{n} \frac{\Gamma(a+k)}{k!\Gamma(d+k)} \left\{ a I^{m+1,n}_{p_i+1,q_i+1:r}\left[z \left| \begin{matrix} (a_j,\alpha_j)_{1,n}; & (a_{ji},\alpha_{ji})_{n+1,p_i}, (b,h) \\ (b+k),(b_j,\beta_j)_{1,m}; & (b_{ji},\beta_{ji})_{m+1,q_i} \end{matrix} \right. \right] \right.$$

$$\left. + \frac{a+k}{d+k} I^{m+2,n}_{p_i+2,q_i+2:r}\left[z \left| \begin{matrix} (a_j,\alpha_j)_{1,n};(a_{ji},\alpha_{ji})_{n+1,p_i}, & (a+b-d+1,h),(1+b,h) \\ (a+b-d+2,h),(1+b+k,h),(b_j,\beta_j)_{1,m};, & (b_{ji},\beta_{ji})_{m+1,q_i} \end{matrix} \right. \right] \right\}$$

$$= \frac{\Gamma(a+n+1)}{n!\Gamma(d+n+1)} I^{m+2,n}_{p_i+2,q_i+2:r}\left[z \left| \begin{matrix} (a_j,\alpha_j)_{1,n};(a_{ji},\alpha_{ji})_{n+1,p_i},(a+b+n+1,h), & (b+1,h) \\ (a+b+n+2,h),(b+n+1,h)(b_j,\beta_j)_{1,m}; & (b_{ji},\beta_{ji})_{m+1,q_i}, \end{matrix} \right. \right] \tag{4.93}$$

xviii.

$$FS_{18} \equiv \sum_{k=0}^{n} \frac{1}{k!} \left\{ \frac{\Gamma(a+1)}{\Gamma(d+k)} I^{m,n+2}_{p_i+1,q_i+2:r}\left[z \left| \begin{matrix} (1-b-k,h'),(1-a-k,h),(a_j,\alpha_j)_{1,n}; & (a_{ji},\alpha_{ji})_{n+1,p_i} \\ (b_j,\beta_j)_{1,m};(b_{ji},\beta_{ji})_{m+1,q_i} & (1-b,h'),(1-a,h) \end{matrix} \right. \right] \right.$$

$$\left. + I^{m,n+3}_{p_i+3,q_i+3:r}\left[z \left| \begin{matrix} (d-a-b-1,h+h'),(-b-k,h'),(-a-k,h)(a_j,\alpha_j)_{1,n}; & (a_{ji},\alpha_{ji})_{n+1,p_i} \\ (b_j,\beta_j)_{1,m};(b_{ji},\beta_{ji})_{m+1,q_i},(d-a-b,h+h'),(-b,h'), & (-a,h) \end{matrix} \right. \right] \right\}$$

$$= \frac{1}{n!} I^{m,n+3}_{p_i+3,q_i+3:r}\left[z \left| \begin{matrix} (-a-b-n-1,h+h'),(-b-n,h'),(-a-n,h)(a_j,\alpha_j)_{1,n}; & (a_{ji},\alpha_{ji})_{n+1,p_i} \\ (b_j,\beta_j)_{1,m};(b_{ji},\beta_{ji})_{m+1,q_i},(-a-b-n,h+h'),(-b,h'), & (-a,h) \end{matrix} \right. \right] \tag{4.94}$$

Conditions for the existence of results (4.90)–(4.94) are same as those for (4.77).

xix.

$$FS_{19} \equiv \sum_{k=0}^{n} \binom{n}{k} (\beta)_k \, I^{m,n+1}_{p_i+1,q_i+1:r}\left[z \left| \begin{matrix} (1-\alpha-n+k,h),(a_j,\alpha_j)_{1,n}; & (a_{ji},\alpha_{ji})_{n+1,p_i} \\ (b_j,\beta_j)_{1,m}; & (b_{ji},\beta_{ji})_{m+1,q_i},(1-\alpha,h) \end{matrix} \right. \right]$$

$$= I^{m,n+1}_{p_i+1,q_i+1:r}\left[z \left| \begin{matrix} (1-\alpha-\beta-n,h),(a_j,\alpha_j)_{1,n}; & (a_{ji},\alpha_{ji})_{n+1,p_i} \\ (b_j,\beta_j)_{1,m}; & (b_{ji},\beta_{ji})_{m+1,q_i},(1-\alpha-\beta,h) \end{matrix} \right. \right] \tag{4.95}$$

xx.

$$FS_{20} \equiv \sum_{k=0}^{n} \binom{n}{k} \left(\frac{1}{\beta}\right)_k I_{p_i+1,q_i+1:r}^{m,n+1}\left[z \middle| \begin{array}{cc} (1-\alpha,h),\left(a_j,\alpha_j\right)_{1,n}; & \left(a_{ji},\alpha_{ji}\right)_{n+1,p_i} \\ \left(b_j,\beta_j\right)_{1,m}; & \left(b_{ji},\beta_{ji}\right)_{m+1,q_i},(1-\alpha-n+k,h) \end{array} \right]$$

$$= \left(\frac{1}{\beta}\right)_n I_{p_i+2,q_i+2:r}^{m,n+2}\left[z \middle| \begin{array}{cc} (2-\alpha-\beta-2n,h),(1-\alpha,h),\left(a_j,\alpha_j\right)_{1,n}; & \left(a_{ji},\alpha_{ji}\right)_{n+1,p_i} \\ \left(b_j,\beta_j\right)_{1,m};\left(b_{ji},\beta_{ji}\right)_{m+1,q_i},(2-\alpha-\beta-n,h), & (1-\alpha-n,h) \end{array} \right] \tag{4.96}$$

Under the condition of (4.77).

xxi.

$$FS_{21} \equiv \sum_{k=0}^{n} \binom{n}{k} (\alpha)_{n-k} I_{p_i+1,q_i+1:r}^{m,n+1}\left[z \middle| \begin{array}{cc} (1-\beta-k,h),\left(a_j,\alpha_j\right)_{1,n}; & \left(a_{ji},\alpha_{ji}\right)_{n+1,p_i} \\ \left(b_j,\beta_j\right)_{1,m}; & \left(b_{ji},\beta_{ji}\right)_{m+1,q_i},(1-\beta,h) \end{array} \right]$$

$$= I_{p_i+1,q_i+1:r}^{m,n+1}\left[z \middle| \begin{array}{cc} (1-\alpha-\beta-n,h),\left(a_j,\alpha_j\right)_{1,n}; & \left(a_{ji},\alpha_{ji}\right)_{n+1,p_i} \\ \left(b_j,\beta_j\right)_{1,m};\left(b_{ji},\beta_{ji}\right)_{m+1,q_i}, & (1-\alpha-\beta,h) \end{array} \right] \tag{4.97}$$

xxii.

$$FS_{22} \equiv \sum_{k=0}^{n} \binom{n}{k} (\beta)_k I_{p_i+1,q_i+1:r}^{m+1,n}\left[z \middle| \begin{array}{cc} \left(a_j,\alpha_j\right)_{1,n}; & \left(a_{ji},\alpha_{ji}\right)_{n+1,p_i},(\alpha,h) \\ (\alpha+n-k,h)\left(b_j,\beta_j\right)_{1,m}; & \left(b_{ji},\beta_{ji}\right)_{m+1,q_i} \end{array} \right]$$

$$= I_{p_i+1,q_i+1:r}^{m+1,n}\left[z \middle| \begin{array}{cc} \left(a_j,\alpha_j\right)_{1,n}; & \left(a_{ji},\alpha_{ji}\right)_{n+1,p_i},(\alpha+\beta,h) \\ (\alpha+\beta+n,h),\left(b_j,\beta_j\right)_{1,m}; & \left(b_{ji},\beta_{ji}\right)_{m+1,q_i} \end{array} \right] \tag{4.98}$$

xxiii.

$$FS_{23} \equiv \sum_{k=0}^{n} \binom{n}{k} (\beta)_k I_{p_i+2,q_i+2:r}^{m+1,n+1}\left[z \middle| \begin{array}{cc} (1-\alpha-n+k,h)\left(a_j,\alpha_j\right)_{1,n}; & \left(a_{ji},\alpha_{ji}\right)_{n+1,p_i},(\beta,h) \\ (\beta+k,h)\left(b_j,\beta_j\right)_{1,m}; & \left(b_{ji},\beta_{ji}\right)_{m+1,q_i},(1-\alpha,h) \end{array} \right]$$

$$= (\alpha+\beta)_n I_{p_i,q_i:r}^{m,n}\left[z \middle| \begin{array}{cc} \left(a_j,\alpha_j\right)_{1,n}; & \left(a_{ji},\alpha_{ji}\right)_{n+1,p_i} \\ \left(b_j,\beta_j\right)_{1,m}; & \left(b_{ji},\beta_{ji}\right)_{m+1,q_i} \end{array} \right] \tag{4.99}$$

xxiv.

$$\mathrm{FS}_{24} \equiv \sum_{k=0}^{n} \binom{n}{k} I^{m,n+2}_{p_i+2,q_i+2:r}\left[z \left| \begin{matrix} (1-\beta-k,h),(1-\alpha-n+k,h),(a_j,\alpha_j)_{1,n}; & (a_{ji},\alpha_{ji})_{n+1,p_i} \\ (b_j,\beta_j)_{1,m};(b_{ji},\beta_{ji})_{m+1,q_i},(1-\beta,h), & (1-\alpha,h) \end{matrix} \right. \right]$$

$$= 2^n I^{m,n}_{p_i+2,q_i+2:r}\left[z \left| \begin{matrix} \left(\frac{2-\alpha-\beta-n}{2},h\right),\left(\frac{1-\alpha-\beta-n}{2},h\right),(a_j,\alpha_j)_{1,n}; & (a_{ji},\alpha_{ji})_{n+1,p_i} \\ (b_j,\beta_j)_{1,m};(b_{ji},\beta_{ji})_{m+1,q_i},\left(\frac{2-\alpha-\beta}{2},h\right), & \left(\frac{1-\alpha-\beta}{2},h\right) \end{matrix} \right. \right] \tag{4.100}$$

xxv.

$$\mathrm{FS}_{25} \equiv \sum_{k=0}^{n} \binom{n}{k} I^{m+1,n}_{p_i+1,q_i+1:r}\left[z \left| \begin{matrix} (a_j,\alpha_j)_{1,n};(a_{ji},\alpha_{ji})_{n+1,p_i}, & (\alpha+n-k,h) \\ (\alpha,h),(b_j,\beta_j)_{1,m}; & (b_{ji},\beta_{ji})_{m+1,q_i} \end{matrix} \right. \right]$$

$$= \left(\frac{1}{\beta}\right)_n I^{m+2,n}_{p_i+2,q_i+2:r}\left[z \left| \begin{matrix} (a_j,\alpha_j)_{1,n};(a_{ji},\alpha_{ji})_{n+1,p_i},(\alpha+\beta+n-1,h), & (\alpha+n,h) \\ (\alpha+\beta+2n-1,h),(\alpha,h),(b_j,\beta_j)_{1,m}; & (b_{ji},\beta_{ji})_{m+1,q_i} \end{matrix} \right. \right] \tag{4.101}$$

xxvi.

$$\mathrm{FS}_{26} \equiv \sum_{k=0}^{n} \binom{n}{k} I^{m+1,n+1}_{p_i+2,q_i+2:r}\left[z \left| \begin{matrix} (1-\alpha,h),(a_j,\alpha_j)_{1,n};(a_{ji},\alpha_{ji})_{n+1,p_i}, & (\beta+k,h) \\ (\beta,h),(b_j,\beta_j)_{1,m};(b_{ji},\beta_{ji})_{m+1,q_i}, & (1-\alpha-n+k,h) \end{matrix} \right. \right]$$

$$= (\alpha+\beta+n-1)_n I^{m+1,n+1}_{p_i+2,q_i+2:r}\left[z \left| \begin{matrix} (1-\alpha,h),(a_j,\alpha_j)_{1,n};(a_{ji},\alpha_{ji})_{n+1,p_i}, & (\beta+n,h) \\ (\beta,h),(\alpha,h),(b_j,\beta_j)_{1,m}; & (b_{ji},\beta_{ji})_{m+1,q_i},(1-\alpha-n,h) \end{matrix} \right. \right] \tag{4.102}$$

xxvii.

$$\mathrm{FS}_{27} \equiv \sum_{k=0}^{n} \binom{n}{k} I^{m,n+2}_{p_i+2,q_i+2:r}\left[z \left| \begin{matrix} (1-\beta,h),(1-\alpha,h),(a_j,\alpha_j)_{1,n}; & (a_{ji},\alpha_{ji})_{n+1,p_i} \\ (b_j,\beta_j)_{1,m};(b_{ji},\beta_{ji})_{m+1,q_i},(1-\beta-k,h) & ,(1-\alpha-n+k,h) \end{matrix} \right. \right]$$

$$= 2^n I^{m,n+4}_{p_i+4,q_i+4:r}$$

$$\left[z \left| \begin{matrix} \left(\frac{3-\alpha-\beta-n}{2},h\right),\left(\frac{2-\alpha-\beta-2n}{2},h\right),(1-\alpha,h),(1-\beta,h)(a_j,\alpha_j)_{1,n}; & (a_{ji},\alpha_{ji})_{n+1,p_i} \\ (b_j,\beta_j)_{1,m};(b_{ji},\beta_{ji})_{m+1,q_i},\left(\frac{3-\alpha-\beta-n}{2},h\right),\left(\frac{2-\alpha-\beta-n}{2},h\right),(1-\alpha-n,h), & (1-\beta-n,h) \end{matrix} \right. \right] \tag{4.103}$$

Conditions for the convergence of results from (4.97) to (4.103) are same as those in (4.77).

xxviii.

$$\mathrm{FS}_{28} \equiv \sum_{k=0}^{n} \binom{n}{k} I_{p_i,q_i:r}^{m+1,n}\left[z \,\middle|\, \begin{matrix} (a_j,\alpha_j)_{1,n}; & (a_{ji},\alpha_{ji})_{n+1,p_i} \\ (0,k)(b_j,\beta_j)_{1,m}; & (b_{ji},\beta_{ji})_{m+1,q_i} \end{matrix} \right]$$

$$= \pi^{\rho} 2^{\rho} \sum_{t=1}^{k} \left\{ I_{p_i+n+3,q_i+m+5:r}^{2m+3,2n+1}\left[\frac{e^{-\pm i(k+1/2)\pi} z^2}{4^{\delta}} \,\middle|\, \begin{matrix} \left(1,\frac{k}{2}\right),\left(\frac{a_j}{2},\alpha_j\right)_{1,n}, \left(\frac{1}{2}+\frac{a_j}{2},\alpha_j\right)_{1,n};(a_{ji},2\alpha_{ji})_{n+1,p_i},\left(\frac{1}{2},t\right),\left(1,\frac{1}{2}\right) \\ (0,k),\left(\frac{1}{2},k\right),\left(1,\frac{k}{2}\right), \left(\frac{b_j}{2},\beta_j\right)_{1,m},\left(\frac{1}{2}+\frac{b_j}{2},\beta_j\right)_{1,m},(b_{ji},2\beta_{ji})_{m+1,q_i},\left(\frac{1}{2},t\right),\left(1,\frac{1}{2}\right) \end{matrix} \right] \right. \tag{4.104}$$

$$\pm i I_{p_i+n+3,q_i+m+5:r}^{2m+3,2n+1}\left[\frac{e^{-\pm i(k+1/2)\pi} z^2}{4^{\delta}} \,\middle|\, \begin{matrix} \left(1,\frac{k}{2}\right),\left(\frac{a_j}{2},\alpha_j\right)_{1,n}, \left(\frac{1}{2}+\frac{a_j}{2},\alpha_j\right)_{1,n};(a_{ji},2\alpha_{ji})_{n+1,p_i},(0,t),\left(1,\frac{1}{2}\right) \\ (0,k),\left(\frac{1}{2},k\right),\left(1,\frac{k}{2}\right),\left(\frac{b_j}{2},\beta_j\right)_{1,m}, \left(\frac{1}{2}+\frac{b_j}{2},\beta_j\right)_{1,m},(b_{ji},2\beta_{ji})_{m+1,q_i},(0,t),\left(1,\frac{1}{2}\right) \end{matrix} \right] \left. \right\}$$

where $\sigma = \frac{1}{2} - \frac{m}{2} - \frac{n}{2}$, $\rho = \sum_{j=1}^{m} b_j - \sum_{j=1}^{n} a_j - m$,

$$\delta = \sum_{j=1}^{m} \beta_j - \sum_{j=1}^{n} \alpha_j - k.$$

To obtain some of the results of this section the following well-known identities for Gauss hypergeometric functions were employed [10]

a.
$$(1+a)_2 F_1\left[\begin{matrix} d-b, d-a-1 \\ 1+d \end{matrix}\right]_n - (1+b)_2 F_1\left[\begin{matrix} d-b-1, d-a \\ 1+d \end{matrix}\right]_n$$
$$= \frac{a-b}{n!} \times \frac{(d-b)_n (d-a)_n}{(1+d)_n}$$

b.
$$a\ {}_2F_1\left[\begin{matrix} d-c, d-a-1 \\ d \end{matrix}\right]_n - (a+c-d)_2 F_1\left[\begin{matrix} d-c, d-a \\ d \end{matrix}\right]_n$$
$$= \frac{a+n-c}{n!} \times \frac{(d-c)_n (d-a)_n}{(d)_n}$$

c.
$$d\ {}_2F_1\left[\begin{matrix} d-c, d-a \\ d \end{matrix}\right]_n - (a+c-d-1)_2 F_1\left[\begin{matrix} d-c+1, d-a+1 \\ d+1 \end{matrix}\right]_n$$
$$= \frac{2d-c-a+n+1}{n!} \times \frac{(d-c+1)_n (d-a+1)_n}{(d+1)_n}$$

d.
$$(d-a)_2 F_1\left[\begin{matrix} b, a-1 \\ d \end{matrix}\right]_{n+1} - (d-a-b)_2 F_1\left[\begin{matrix} b, a \\ d \end{matrix}\right]_{n+1} = \frac{b+n}{n!} \times \frac{(a)_n (b)_n}{(d)_n}$$

e.
$$d\ {}_2F_1\left[\begin{matrix} b, a \\ d \end{matrix}\right]_{n+1} + (a+b-d-1)_2 F_1\left[\begin{matrix} b+1, a+1 \\ d+1 \end{matrix}\right]_{n+1}$$
$$= \frac{(a+b+n+1)}{n!} \times \frac{(b+1)_n (a+1)_n}{(d+1)_n}$$

where suffix $(n+1)$ on the left-hand side indicates that only first $(n+1)$ terms of the Gauss series have been taken.

f. $$\sum_{k=0}^{n} \binom{n}{k} (\beta)_k (\alpha)_{n-k} = (\beta+\alpha)_n$$

g. $$\sum_{k=0}^{n} \binom{n}{k} \frac{1}{(\beta)_k (\alpha)_{n-k}} = \frac{(\beta+\alpha+n-1)_n}{(\alpha)_n (\beta)_n}$$

Here we give brief derivations of some selected results. Other derivations are left as the same can be obtained on similar lines.

To establish (4.41) we substitute value of I-functions from definition on the left-hand side of (4.4.1), so that

$$\mathrm{FS}_1 = \sum_{k=0}^{n} \frac{1}{k!}\frac{1}{(2\pi i)}\int_L \phi(\xi)$$

$$\times\left[\frac{(d-b)_k\,\Gamma(d-a-1+k-h\xi)(a+1+h\xi)}{(d+1)_k\,\Gamma(d-a-1-h\xi)}\frac{(d-b-1)_k\,\Gamma(d-a-k-h\xi)(1+b)}{(d+1)_k\,\Gamma(d-a-h\xi)}\right]z^{\xi}d\xi$$

Changing the order of summation and integration, which is readily justified as the integral is finite, the aforementioned expression takes the form

$$\frac{1}{(2\pi i)}\int_L \phi(\xi)$$

$$\times\left\{\frac{1}{(d+1)_k}\left[\frac{(d-b)_k(d-a-1-h\xi)_k(a+1+h\xi)}{(d+1)_k\,\Gamma(d-a-1-h\xi)}\frac{(d-b-1)_k(d-a-h\xi)(1+b)}{(d+1)_k\,\Gamma(d-a-h\xi)}\right]z^{\xi}d\xi\right\}$$

Now, using (a), we get the right-hand side of (4.77).

Considering the summation (4.78) we have

$$FS_2 = \sum_{k=0}^{n}\frac{1}{k!}\left\{a\, I^{m+2,n}_{p_i+2,q_i+2:r}\left[z\left|\begin{matrix}(a_j,\alpha_j)_{1,n};(a_{ji},\alpha_{ji})_{n+1,p_i}, & (d+k,h),(d-a-1,h)\\ (d-b+k,h),(d-a-1+k,h),\ (b_j,\beta_j)_{1,m}; & (b_{ji},\beta_{ji})_{m+1,q_i}\end{matrix}\right.\right]\right.$$

$$\left.-I^{m+2,n+1}_{p_i+3,q_i+3:r}\left[z\left|\begin{matrix}(d-a+b,h),(a_j,\alpha_j)_{1,n}; & (a_{ji},\alpha_{ji})_{n+1,p_i},(d+k,h),(d-a,h)\\ (d-b+k,h),\ (b_j,\beta_j)_{1,m}; & (b_{ji},\beta_{ji})_{m+1,q_i},(1-a-b+d,h)\end{matrix}\right.\right]\right\}$$

Replacing the I-function by the definition of (1.23), we get

$$\mathrm{FS}_2 = \sum_{k=0}^{n}\frac{1}{k!}\frac{1}{(2\pi i)}\int_L \phi(\xi)\left[\frac{\Gamma(d-b+k-h\xi)\Gamma(d-a-1+k-h\xi)\Gamma(d-b+k-h\xi)}{\Gamma(d+k-h\xi)\Gamma(d-a-1-h\xi)\Gamma(a+b-d+h\xi)}\right.$$

$$\left.\frac{\Gamma(d-a+k-h\xi)\Gamma(1+a+b-d+h\xi)}{\Gamma(d+k-h\xi)\Gamma(d-a-h\xi)}\right]z^{\xi}d\xi$$

Interchanging the order of summation and integration we obtain

$$\mathrm{FS}_2 = \frac{1}{(2\pi i)} \int_L \phi(\xi) \left\{ \sum_{k=0}^{n} \frac{1}{k!} \left[a \frac{(d-b-h\xi)_k (d-a-1-h\xi)_k}{(d-h\xi)_k} \right.\right. \tag{4.105}$$

$$\left.\left. - \frac{(a+b-d-h\xi)(d-b-h\xi)_k (d-a-h\xi)_k}{(d-h\xi)_k} \right] \right\} z^{\xi} d\xi$$

Now using (*b*), we get the right-hand side of (4.78)

To establish relation (4.79) we again use the definition of I-function on the left-hand side, so that

$$\mathrm{FS}_3 = \sum_{k=0}^{n} \frac{1}{k!} \frac{1}{(2\pi i)} \int_L \phi(\xi) \left[\frac{\Gamma(d-a+k-h\xi)\Gamma(d-b+k+h\xi)}{\Gamma(d+k)\Gamma(d-b+h\xi)\Gamma(d-a-h\xi)} + \frac{(d-a-b+1)}{\Gamma(d+k+1)} \right.$$

$$\left. \times \frac{\Gamma(d-a+1+k-h\xi)\Gamma(1-b+d+k+h\xi)}{\Gamma(1-b+d+h\xi)\Gamma(d-a+1-h\xi)} \right] z^{\xi} d\xi$$

Changing the order of summation and integration and using (c), we get the right-hand side of FS_3

To establish (4.88) we employ (1.23) to the left side of (4.88) so that

$$\mathrm{FS}_{12} = \sum_{k=0}^{n} \frac{1}{k!} \frac{1}{(2\pi i)} \int_L \phi(\xi) \left\{ \frac{\Gamma(b+k)}{\Gamma(d+k)} \left[\frac{\Gamma(a-1+k-h\xi)\Gamma(d-a+1-h\xi)}{\Gamma(d-a-h\xi)\Gamma(a-1+h\xi)} \right.\right.$$

$$\left.\left. - \frac{\Gamma(a+k+h\xi)\Gamma(d-a-b+1-h\xi)}{\Gamma(a+h\xi)\Gamma(d-a-b-h\xi)} \right] \right\} z^{\xi} d\xi$$

Changing the order of summation and integration, the above expression takes the form

$$\frac{1}{(2\pi i)} \int_L \phi(\xi) \left\{ \sum_{k=0}^{n} \frac{1}{k!} \frac{\Gamma(b+k)}{\Gamma(d+k)} \left[(a-1+h\xi)_k (d-a-h\xi) - (a+h\xi)_k (d-a-h\xi) \right] \right\} z^{\xi} d\xi \tag{4.106}$$

Now using (d), we get the right-hand side of (4.88).

Considering the summation (4.89) we have

$$FS_{13}=\sum_{k=0}^{n}\frac{\Gamma(b+k)}{k!\Gamma(d+k)}\left\{bI^{m,n+1}_{p_i+1,q_i+1:r}\left[z\left|\begin{matrix}(1-a-k,h),(a_j,\alpha_j)_{1,n}; & (a_{ji},\alpha_{ji})_{n+1,p_i}\\ (b_j,\beta_j)_{1,m}; & (b_{ji},\beta_{ji})_{m+1,q_i},(1-a,h)\end{matrix}\right.\right]\right.$$

$$\left.+\frac{b+k}{d+k}I^{m,n+2}_{p_i+2,q_i+2:r}\left[z\left|\begin{matrix}(d-a-b-1,h),(-a-k,h)(a_j,\alpha_j)_{1,n}; & (a_{ji},\alpha_{ji})_{n+1,p_i}\\ (b_j,\beta_j)_{1,m};(b_{ji},\beta_{ji})_{m+1,q_i},(d-a-b,h) & (-a,h)\end{matrix}\right.\right]\right\}$$

Replacing the I-function by the definition we get,

$$FS_{13}=\sum_{k=0}^{n}\frac{1}{k!}\frac{1}{(2\pi i)}\int_L \phi(\xi)\frac{\Gamma(b+k)}{k!\Gamma(d+k)}$$

$$\left\{b\frac{\Gamma(a+k+h\xi)}{\Gamma(a+h\xi)(d+k)}+\frac{(b+k)}{(d+k)}\frac{\Gamma(2+a-b-d+h\xi)}{\Gamma(1+a+h\xi)}\right\}z^{\xi}d\xi \tag{4.107}$$

Then interchanging the order of summation and integration and using (*e*) we get the right-hand side of (4.89).

To establish FS_{19} we employ (1.23) and the formulas (f) and (g) so that

$$FS_{19}\equiv\sum_{k=0}^{n}\binom{n}{k}(\beta)_k\frac{1}{(2\pi i)}\int_L \phi(\xi)\left[\frac{\Gamma(\alpha-n+k+h\xi)}{\Gamma(\alpha+hk)}\right]z^{\xi}d\xi$$

.

Changing the order of summation and integration, which is readily justified, the above expression takes the form

$$\frac{1}{(2\pi i)}\int_L \phi(\xi)\left[\sum_{k=0}^{n}\binom{n}{k}(\beta)_k(\alpha+h\xi)_{n-k}\right]z^{\xi}\,d\xi$$

Now, using (f), we get the right-hand side of (4.95).

Similarly

$$FS_{20}\equiv\sum_{k=0}^{n}\binom{n}{k}\frac{1}{(\beta)_k}\frac{1}{(2\pi i)}\int_L \phi(\xi)\left[\frac{\Gamma(\alpha+h\xi)}{\Gamma(\alpha+n+hk)}\right]z^{\xi}d\xi \tag{4.108}$$

Justifiably, changing the order of summation (finite) and integration, we arrive at the result.

All the remaining results up to (4.103) can be established along the same line. To drive (4.104) we use the result

$$\Gamma(-k\xi)\Gamma\left(\frac{1}{2}-k\xi\right)=\frac{\sin\left(\frac{\pi\xi}{2}\right)}{\sin\left(\frac{k\pi\xi}{2}\right)}\pi e^{\pm i\left(k+\frac{1}{2}\right)\pi\xi}\sum_{t=1}^{k}\frac{\Gamma(-k\xi)\Gamma\left(\frac{1}{2}-k\xi\right)}{\Gamma\left(\frac{1}{2}+t\xi\right)\Gamma\left(\frac{1}{2}-t\xi\right)}\pm i\frac{\Gamma(-k\xi)\Gamma\left(\frac{1}{2}-k\xi\right)}{\Gamma(1+t\xi)\Gamma(-t\xi)} \tag{4.109}$$

And the formula

$$\pi\text{cosec}\,\pi\theta=\Gamma(\theta)\Gamma(1-\theta) \tag{4.110}$$

We consider

$$\text{FS}_{28}=I_{p_i,q_i:r}^{m+1,n}\left[z\left|\begin{matrix}(a_j,\alpha_j)_{1,n}; & (a_{ji},\alpha_{ji})_{n+1,p_i}\\ (0,k)(b_j,\beta_j)_{1,m}; & (b_{ji},\beta_{ji})_{m+1,q_i}\end{matrix}\right.\right]$$

$$=\frac{1}{(2\pi i)}\int_L\frac{\Gamma(-k\xi)\prod_{j=1}^{m}\Gamma(b_j-\beta_j\xi)\prod_{j=1}^{n}\Gamma(1-a_j-\alpha_j)z^{\xi}d\xi}{\sum_{i=1}^{r}\prod_{j=m+1}^{q_i}\Gamma(1-b_{ji}+\beta_{ji}\xi)\prod_{j=n+1}^{p_i}\Gamma(a_{ji}-\alpha_{ji}\xi)}$$

Putting $\xi=2\theta$ and applying the Legendre duplication formula $\Gamma(2z)=\pi^{-\frac{1}{2}}2^{2z-1}\Gamma(z)\Gamma\left(z+\frac{1}{2}\right)$ we obtain

$$\text{FS}_{28}=\frac{1}{(2\pi i)}\int_L\frac{\pi^{-\frac{1}{2}}2^{-2k\theta-1}\Gamma(k\theta)\Gamma\left(-k\theta+\frac{1}{2}\right)\prod_{j=1}^{m}\left\{\pi^{-\frac{1}{2}}2^{b_j-2\beta_j\theta-1}\Gamma\left(\frac{b_j}{2}-\beta_j\theta\right)\right.}{\sum_{i=1}^{r}\prod_{j=m+1}^{q_i}\Gamma(1-b_{ji}+2\beta_{ji}\theta)\prod_{j=n+1}^{p_i}\Gamma(a_{ji}-2\alpha_{ji}\theta)}$$

$$\times\frac{\left.\Gamma\left(\frac{b_j}{2}-\beta_j\theta+\frac{1}{2}\right)\right\}\prod_{j=1}^{p}\left\{\left\{\pi^{-\frac{1}{2}}2^{-a_j+2\alpha_j\xi}\Gamma\left(\frac{1}{2}-\frac{a_j}{2}+\alpha_j\theta\right)\Gamma\left(1-\frac{a_j}{2}+\alpha_j\theta\right)\right\}z^{2\theta}2d\theta}{1}$$

$$=\frac{\pi^{\frac{1}{2}-\frac{m}{2}-\frac{n}{2}}2^{\rho}}{2\pi i}\int_L\frac{\Gamma(-k\theta)\Gamma\left(-k\theta+\frac{1}{2}\right)\prod_{j=1}^{m}\Gamma\left(\frac{b_j}{2}-\beta_j\theta\right)\Gamma\left(\frac{b_j}{2}-\beta_j\theta+\frac{1}{2}\right)}{\sum_{i=1}^{r}\prod_{j=m+1}^{q_i}\Gamma(1-b_{ji}+2\beta_{ji}\theta)\prod_{j=n+1}^{p_i}\Gamma(a_{ji}-2\alpha_{ji}\theta)}$$

$$\times\frac{\prod_{j=1}^{n}\Gamma\left(\frac{1}{2}-\frac{a_j}{2}+\alpha_j\theta\right)\prod_{j=1}^{n}\Gamma\left(1-\frac{a_j}{2}+\alpha_j\theta\right)z^{2\theta}2^{-2\left(k+\sum_{j=1}^{m}\beta_j-\sum_{j=1}^{n}\alpha_j\right)^{\theta}}}{1}d\theta$$

Accordingly (4.109) and (4.110), we get

$$FS_{28} = \pi^{\frac{1}{2}-\frac{m}{2}-\frac{n}{2}}2^{\rho}\int_L \frac{\prod_{j=1}^{m}\left\{\Gamma\left(\frac{b_j}{2}-\beta_j\theta\right)\Gamma\left(\frac{b_j}{2}-\beta_j\theta+\frac{1}{2}\right)\right\}\prod_{j=1}^{n}\left\{\Gamma\left(\frac{1}{2}-\frac{a_j}{2}+\alpha_j\theta\right)\prod_{j=1}^{n}\Gamma\left(1-\frac{a_j}{2}+\alpha_j\theta\right)\right\}}{\sum_{i=1}^{r}\prod_{j=m+1}^{q_i}\Gamma\left(1-b_{ji}+2\beta_{ji}\theta\right)\prod_{j=n+1}^{p_i}\Gamma\left(a_{ji}-2\alpha_{ji}\theta\right)}$$

$$\times\frac{z^2}{4^{\delta}}\frac{\sin(\pi\theta/2)}{\sin(k\pi\theta/2)}\pi e^{\pm i(k+1/2)\pi\theta}\sum_{t=1}^{k}\left[\frac{\Gamma(-k\theta)\Gamma\left(-k\theta+\frac{1}{2}\right)}{\Gamma\left(\frac{1}{2}-t\theta\right)\Gamma\left(t\theta+\frac{1}{2}\right)}\pm i\frac{\Gamma(-k\theta)\Gamma\left(-k\theta+\frac{1}{2}\right)}{\Gamma(1+t\theta)\Gamma(-t\theta)}\right]d\theta$$

$$=\frac{\pi^{\frac{1}{2}-\frac{m}{2}-\frac{n}{2}}2^{\rho}}{2\pi i}\int_L\frac{\Gamma(k\theta/2)\Gamma(1-k\theta/2)}{\Gamma(\theta/2)\Gamma(1-\theta/2)}\left(\frac{e^{\pm i(k+1/2)\pi\theta}z^2}{4^{\delta}}\right)\theta$$

$$\sum_{t=1}^{k}\left[\frac{\Gamma(-k\theta)\Gamma\left(-k\theta+\frac{1}{2}\right)}{\Gamma\left(\frac{1}{2}-t\theta\right)\Gamma\left(t\theta+\frac{1}{2}\right)}\pm i\frac{\Gamma(-k\theta)\Gamma\left(-k\theta+\frac{1}{2}\right)}{\Gamma(1+t\theta)\Gamma(-t\theta)}\right]$$

$$\times\frac{\prod_{j=1}^{m}\left\{\Gamma\left(\frac{b_j}{2}-\beta_j\theta\right)\Gamma\left(\frac{b_j}{2}-\beta_j\theta+\frac{1}{2}\right)\right\}\prod_{j=1}^{n}\left\{\Gamma\left(\frac{1}{2}-\frac{a_j}{2}+\alpha_j\theta\right)\prod_{j=1}^{n}\Gamma\left(1-a_j+\alpha_j\theta\right)\right\}}{\sum_{i=1}^{r}\prod_{j=m+1}^{q_i}\Gamma\left(1-b_{ji}+2\beta_{ji}\theta\right)\prod_{j=n+1}^{p_i}\Gamma\left(a_{ji}-2\alpha_{ji}\theta\right)}$$

$$=\frac{\pi^{\sigma}2^{\rho}}{2\pi i}\int_L\sum_{t=1}^{k}\left[\frac{\Gamma(1-k\theta/2)\Gamma(-k\theta)\Gamma\left(-k\theta+\frac{1}{2}\right)\Gamma(k\theta/2)}{\Gamma(\theta/2)\Gamma\left(t\theta+\frac{1}{2}\right)+\Gamma(1-\theta/2)\Gamma\left(\frac{1}{2}-t\theta\right)}\right.$$

$$\left.\pm i\frac{\Gamma(1-k\theta/2)\Gamma(-k\theta)\Gamma\left(-k\theta+\frac{1}{2}\right)\Gamma(k\theta/2)}{\Gamma(1+t\theta)\Gamma(\theta/2)\Gamma(-t\theta)\Gamma(1-t\theta/2)}\right]$$

$$\times\frac{\prod_{j=1}^{m}\left\{\Gamma\left(\frac{b_j}{2}-\beta_j\theta\right)\Gamma\left(\frac{b_j}{2}-\beta_j\theta+\frac{1}{2}\right)\right\}\prod_{j=1}^{n}\left\{\Gamma\left(1-a_j+\alpha_j\theta\right)\prod_{j=1}^{n}\Gamma\left(\frac{1}{2}-\frac{a_j}{2}+\alpha_j\theta\right)\right\}}{\sum_{i=1}^{r}\prod_{j=m+1}^{q_i}\Gamma\left(1-b_{ji}+2\beta_{ji}\theta\right)\prod_{j=n+1}^{p_i}\Gamma\left(a_{ji}-2\alpha_{ji}\theta\right)}$$

$$\left\{e^{\pm i(k+1/2)\pi}\frac{z^2}{4^{\delta}}\right\}\theta d\theta$$

$$=\pi^{\sigma}2^{\rho}\sum_{t=1}^{h}\left\{I^{2m+3,2n+1}_{p_i+n+3,q_i+m+5:r}\left[\frac{e^{+i(k+1/2)\pi}z^2}{4^{\delta}}\left|\begin{matrix}(1,k/2),(a_j,\alpha_j)_{1,n}\\(0,k),(1/2,k),(1,k/2),\end{matrix}\right.\right.\right.$$

$$\left.\begin{matrix}(1/2+a_j/2,\alpha_j)_{1,n};(a_{ji},2\alpha_{ji})_{n+1,p_i},(1/2,t),(1,1/2)\\(b_j/2,\beta_j)_{1,m},(1/2+b_j/2,\beta_j)_{1,m},(b_{ji},2\beta_{ji})_{m+1,q_i},(1/2,t),(1,1/2)\end{matrix}\right]\pm i$$

$$\times I^{2m+3,2n+1}_{p_i+n+3,q_i+m+5:r}\left[\frac{e^{\pm i(k+1/2)\pi}z^2}{4^{\delta}}\left|\begin{matrix}(1,k/2),(a_j/2,\alpha_j)_{1,n},(1/2+a_j/2,\alpha_j)_{1,n};\\(0,k),(1/2,k),(1,k/2),(b_j/2,\beta_j)_{1,m};\end{matrix}\right.\right.$$

$$\left.\left.\begin{matrix}(a_{ji},2\alpha_{ji})_{n+1,p_i},(0,t),(1,1/2)\\(1/2+b_j/2,\beta_j)_{1,m};(b_{ji},2\beta_{ji})_{m+1,p_i},(0,t),(1,1/2)\end{matrix}\right]\right\}\tag{4.111}$$

4.4.1 Special Cases

Here we present special cases of certain finite series of the I-function.

i. If $r = 1$ the result (4.77) reduces to an interesting result of Fox's H-function.

$$\sum_{k=0}^{n}\frac{1}{k!}\left\{\frac{(d-b)_k}{(d+1)_k}H_{p+2,q+2}^{m+1,n+1}\left[z\left|\begin{matrix}(-a-1,h),(a_j,\alpha_j)_{1,p}, & (d-a-1,h)\\ (d-a-1+k,h),\ (b_j,\beta_j)_{1,q}, & (-a,h)\end{matrix}\right.\right]\right.$$

$$\left.-(1+b)\frac{(d-b+1)_k}{(d+1)_k}H_{p+1,q+1}^{m+1,n}\left[z\left|\begin{matrix}(a_j,\alpha_j)_{1,p}, & (d-a,h)\\ (d-a+k,h), & (b_j,\beta_j)_{1,q}\end{matrix}\right.\right]\right\}$$

$$=\frac{(d-b)_n}{n!(d+1)_k}H_{p+2,q+2}^{m+1,n+1}\left[z\left|\begin{matrix}(b-a,h),(a_j,\alpha_j)_{1,p}, & (d-a,h)\\ (d-a+n,h),\ (b_j,\beta_j)_{1,q}, & (b-a+1,h)\end{matrix}\right.\right] \tag{4.112}$$

ii. If $r = 1$ the result (4.78) we get the corresponding result for Fox's H-function.

$$\sum_{k=0}^{n}\frac{1}{k!}\left\{a\ H_{p+2,q+2}^{m+2,n}\left[z\left|\begin{matrix}(a_j,\alpha_j)_{1,p},(d+k,h), & (d-a-1,h)\\ (d-b+k,h),(d-a-1+k,h), & (b_j,\beta_j)_{1,q}\end{matrix}\right.\right]\right.$$

$$\left.-H_{p+3,q+3}^{m+2,n+1}\left[z\left|\begin{matrix}(d-a+b,h),(a_j,\alpha_j)_{1,p}, & (d+k,h),(d-a,h)\\ (d-b+k,h),\ (b_j,\beta_j)_{1,q}, & (1-a-b+d,h)\end{matrix}\right.\right]\right\}$$

$$=\frac{1}{n!}H_{p+2,q+2}^{m+2,n}\left[z\left|\begin{matrix}(a_j,\alpha_j)_{1,p},(d+n,h), & (d-a,h)\\ (d+n-b+1,h),(d-a+n,h), & (b_j,\beta_j)_{1,m}\end{matrix}\right.\right] \tag{4.113}$$

iii. If $r = 1$ the result (4.79) takes the form of the following result for Fox's H-function.

$$\sum_{k=0}^{n}\frac{1}{k!}\left\{\frac{1}{\Gamma(d+k)}H_{p+2,q+2}^{m+1,n+1}\left[z\left|\begin{matrix}(1-d+b-k,h),(a_j,\alpha_j)_{1,p}, & (d-a,h)\\ (d-a+k,h),\ (b_j,\beta_j)_{1,q}, & (1-d+b,h)\end{matrix}\right.\right]\right.$$

$$\left.+\frac{d-a-b+1}{\Gamma(d+k+1)}H_{p+2,q+2}^{m+1,n+1}\left[z\left|\begin{matrix}(b-d-k,h),(a_j,\alpha_j)_{1,p}, & (d+a+1,h)\\ (d-a+k+1,h),\ (b_j,\beta_j)_{1,q}, & (b-d,h)\end{matrix}\right.\right]\right\}$$

$$=\frac{2d-a-b+n-1}{n!}\frac{1}{\Gamma(d+n+1)}H_{p+2,q+2}^{m+1,n+1}\left[z\left|\begin{matrix}(b-d-n,h),(a_j,\alpha_j)_{1,p}, & (1+d-a,h)\\ (d-a+n+1,h),\ (b_j,\beta_j)_{1,q}, & (b-d,h)\end{matrix}\right.\right] \tag{4.114}$$

Similarly, results (4.80) and (4.87) can be employed to yield interesting results for Fox's H-functions.

iv. If $r = 1$ the result (4.88) is reduced to the following result for Fox's H-function.

$$\sum_{k=0}^{n}\frac{\Gamma(b+k)}{k!\Gamma(d+k)}\left\{H_{p+2,q+2}^{m+1,n+1}\left[z\left|\begin{matrix}(2-a-k,h),(a_j,\alpha_j)_{1,p}, & (d-a,h)\\ (d-a+1,h),(b_j,\beta_j)_{1,q}, & (2-a,h)\end{matrix}\right.\right]\right.$$

$$\left.-H_{p+2,q+2}^{m+1,n+1}\left[z\left|\begin{matrix}(1-a-k,h)(a_j,\alpha_j)_{1,p}, & (d-a-b,h)\\ (d-a-b+1,h),\ (b_j,\beta_j)_{1,q}, & (1-a,h)\end{matrix}\right.\right]\right\} \tag{4.115}$$

$$=\frac{\Gamma(b+n+1)}{n!\Gamma(d+n)}H_{p+1,q+1}^{m,n+1}\left[z\left|\begin{matrix}(1-a-n,h) & (a_j,\alpha_j)_{1,p}\\ (b_j,\beta_j)_{1,q}, & (1-a,h)\end{matrix}\right.\right]$$

v. If $r = 1$ the result (4.89) reduces to the corresponding result for Fox's H-function.

$$\sum_{k=0}^{n}\frac{\Gamma(b+k)}{k!\Gamma(d+k)}\left\{bH_{p+1,q+1}^{m,n+1}\left[z\left|\begin{matrix}(1-a-k,h), & (a_j,\alpha_j)_{1,p}\\ (b_j,\beta_j)_{1,q}, & (1-a,h)\end{matrix}\right.\right]\right.$$

$$\left.+\frac{b+k}{d+k}H_{p+2,q+2}^{m,n+2}\left[z\left|\begin{matrix}(d-a-b-1,h),(-a-k,h) & (a_j,\alpha_j)_{1,p}\\ (b_j,\beta_j)_{1,q},(d-a-b,h) & (-a,h)\end{matrix}\right.\right]\right\}$$

$$=\frac{\Gamma(b+n+1)}{n!\Gamma(d+n+1)}H_{p+2,q+2}^{m,n+2}\left[z\left|\begin{matrix}(-a-b-n-1,h),(a-n,h), & (a_j,\alpha_j)_{1,p}\\ (b_j,\beta_j)_{1,q}, & (-a-b-n,h),(-a,h)\end{matrix}\right.\right]$$

(4.116)

Similarly, results (4.90) and (4.94) can be employed to yield interesting and new results for Fox's H-functions.

vi. If $r=1$ (4.96) reduces to a new result for Fox's H-function.

$$\sum_{k=0}^{n}\binom{n}{k}\left(\frac{1}{\beta}\right)_k H_{p+1,q+1}^{m,n+1}\left[z\left|\begin{matrix}(1-\alpha-n+k,h), & (a_j,\alpha_j)_{1,p}\\ (b_j,\beta_j)_{1,q}, & (1-\alpha,h)\end{matrix}\right.\right]$$

(4.117)

$$=H_{p+1,q+1}^{m,n+1}\left[z\left|\begin{matrix}(1-\alpha-\beta-n,h), & (a_j,\alpha_j)_{1,p}\\ (b_j,\beta_j)_{1,q}, & (1-\alpha-\beta,h)\end{matrix}\right.\right]$$

vii. If $r=1$ (4.97) reduces to a result for Fox's H-function.

$$\sum_{k=0}^{n}\binom{n}{k}\left(\frac{1}{\beta}\right)_k H_{p+1,q+1}^{m,n+1}\left[z\left|\begin{matrix}(1-\alpha,h), & (a_j,\alpha_j)_{1,p}\\ (b_j,\beta_j)_{1,q}, & (1-\alpha-n+k,h)\end{matrix}\right.\right]$$

(4.118)

$$=\left(\frac{1}{\beta}\right)_n H_{p+2,q+2}^{m,n+2}\left[z\left|\begin{matrix}(2-\alpha-\beta-2n,h),(1-\alpha,h), & (a_j,\alpha_j)_{1,p}\\ (b_j,\beta_j)_{1,q},(2-\alpha-\beta-n,h), & (1-\alpha-n,h)\end{matrix}\right.\right]$$

Similarly, results (4.98)–(4.103) can be employed to yield apparently new results for Fox's functions.

viii. If $r = 1$ the disjunction formula of the I-function (4.104) reduces to the following disjunction formula of H-function.

$$H_{p,q+1}^{m+1,n}\left[z \left| \begin{matrix} (a_j,\alpha_j)_{1,n}; & (a_j,\alpha_j)_{n+1,p} \\ (0,k)(b_j,\beta_j)_{1,m}; & (b_j,\beta_j)_{m+1,q} \end{matrix} \right.\right]$$

$$= \pi^{\rho} 2^{\rho} \sum_{t=1}^{k} \left\{ H_{p+n+3,q+m+5}^{2m+3,2n+1}\left[\frac{e^{-\pm i(k+1/2)\pi} z^2}{4^{\delta}} \left| \begin{matrix} \left(1,\frac{k}{2}\right),\left(\frac{a_j}{2},\alpha_j\right)_{1,n}, \\ (0,k),\left(\frac{1}{2},k\right),\left(1,\frac{k}{2}\right), \left(\frac{b_j}{2},\beta_j\right)_{1,m} \end{matrix} \right.\right.\right.$$

$$\left.\begin{matrix} \left(\frac{1}{2}+\frac{a_j}{2},\alpha_j\right)_{1,n};(a_j,2\alpha_j)_{n+1,p},\left(\frac{1}{2},t\right),\left(1,\frac{1}{2}\right) \\ \left(\frac{1}{2}+\frac{b_j}{2},\beta_j\right)_{1,m},(b_j,2\beta_j)_{m+1,q},\left(\frac{1}{2},t\right),\left(1,\frac{1}{2}\right) \end{matrix}\right]$$

$$\pm iH_{p+n+3,q+m+5}^{2m+3,2n+1}\left[\frac{e^{-\pm i(k+1/2)\pi} z^2}{4^{\delta}} \left| \begin{matrix} \left(1,\frac{k}{2}\right),\left(\frac{a_j}{2},\alpha_j\right)_{1,n}, \\ (0,k),\left(\frac{1}{2},k\right),\left(1,\frac{k}{2}\right), \end{matrix} \right.\right.$$

$$\left.\left.\begin{matrix} \left(\frac{1}{2}+\frac{a_j}{2},\alpha_j\right)_{1,n};(a_j,2\alpha_j)_{n+1,p},\left(\frac{1}{2},t\right),\left(1,\frac{1}{2}\right) \\ \left(\frac{b_j}{2},\beta_j\right)_{1,m},\left(\frac{1}{2}+\frac{b_j}{2},\beta_j\right)_{1,m},(b_j,2\beta_j)_{m+1,q},(0,t),\left(1,\frac{1}{2}\right) \end{matrix}\right]\right\} \tag{4.119}$$

ix. Using Legendre duplication formula given in proof of FS_{28} also reduces to the disjunction formula of H-function

$$H_{p,q+1}^{m+1,n}\left[z\left|\begin{matrix}(a_j,\alpha_j)_{1,n}; & (a_j,\alpha_j)_{n+1,p}\\ (0,k)(b_j,\beta_j)_{1,m}; & (b_j,\beta_j)_{m+1,q}\end{matrix}\right.\right]$$

$$=\pi^{\rho'}2^{\rho'}\sum_{t=1}^{k}\left\{H_{p+n+3,q+m+5}^{2m+3,2n+1}\left[\frac{e^{-\pm i(k+1/2)\pi}z^2}{4^{\delta}}\left|\begin{matrix}\left(1,\frac{k}{2}\right),\left(\frac{a_j}{2},\alpha_j\right)_{1,n},\left(\frac{1}{2}+\frac{a_j}{2},\alpha_j\right)_{1,n};\\ (0,k),\left(\frac{1}{2},k\right),\left(1,\frac{k}{2}\right),\left(\frac{b_j}{2},\beta_j\right)_{1,m};\end{matrix}\right.\right.\right.$$

$$\left.\left.\begin{matrix}\left(\frac{a_j}{2},\alpha_j\right)_{n+1,p},\left(\frac{1}{2}+\frac{a_j}{2},\alpha_j\right)_{n+1,p},\left(\frac{1}{2},t\right),\left(1,\frac{1}{2}\right)\\ \left(\frac{1}{2}+\frac{b_j}{2},\beta_j\right)_{1,m},\left(\frac{b_j}{2},\beta_j\right)_{m+1,q},\left(\frac{1}{2}+\frac{b_j}{2},\beta_j\right)_{m+1,q},\left(\frac{1}{2},t\right),\left(1,\frac{1}{2}\right)\end{matrix}\right.\right]$$

$$\pm iH_{p+n+3,q+m+5}^{2m+3,2n+1}\left[\frac{e^{-\pm i(k+1/2)\pi}z^2}{4^{\delta}}\left|\begin{matrix}\left(1,\frac{k}{2}\right),\left(\frac{a_j}{2},\alpha_j\right)_{1,n},\left(\frac{1}{2}+\frac{a_j}{2},\alpha_j\right)_{1,n};\\ (0,k),\left(\frac{1}{2},k\right),\left(1,\frac{k}{2}\right),\left(\frac{b_j}{2},\beta_j\right)_{1,m};\end{matrix}\right.\right.$$

$$\left.\left.\left.\begin{matrix}\left(\frac{a_j}{2},\alpha_j\right)_{n+1,p},\left(\frac{1}{2}+\frac{a_j}{2},\alpha_j\right)_{n+1,p},(0,t),\left(1,\frac{1}{2}\right)\\ \left(\frac{1}{2}+\frac{b_j}{2},\beta_j\right)_{1,m},\left(\frac{b_j}{2},\beta_j\right)_{m+1,q},\left(\frac{1}{2}+\frac{b_j}{2},\beta_j\right)_{m+1,q},(0,t),\left(1,\frac{1}{2}\right)\end{matrix}\right.\right]\right\}\quad(4.120)$$

where $\sigma'=\frac{1}{2}(p+q+1)-m-n;\quad \rho'=\sum_{j=1}^{q}b_j-\sum_{j=1}^{q}a_j+p-m-n$

and $\delta'=\sum_{j=1}^{q}\beta_j-\sum_{j=1}^{p}\alpha_j+k.$

4.5 INFINITE SERIES

In this section, certain I-functions are expressed in terms of single and double infinite series involving the I-function and other functions.

i.

$$\mathrm{IS}_1 \equiv \pi^{-1/2} I^{m,n}_{p_i,q_i:r}\left[z\cos^2(\theta/2) \left| \begin{matrix} (a_j,\alpha_j)_{1,n}; & (a_{ji},\alpha_{ji})_{n+1,p_i} \\ (b_j,\beta_j)_{1,m}; & (b_{ji},\beta_{ji})_{m+1,q_i} \end{matrix} \right. \right]$$

$$= I^{m,n+1}_{p_i+2,q_i+1:r}\left[z \left| \begin{matrix} (1/2,1),(a_j,\alpha_j)_{1,n}; & (a_{ji},\alpha_{ji})_{n+1,p_i} \\ (b_j,\beta_j)_{1,m}; & (b_{ji},\beta_{ji})_{m+1,q_i},(0,1) \end{matrix} \right. \right] + 2\sum_{k=1}^{\infty}\cos(k\theta)$$

$$\times I^{m,n+2}_{p_i+2,q_i+2:r}\left[z \left| \begin{matrix} (0,1),(1/2,1),(a_j,\alpha_j)_{1,n}; & (a_{ji},\alpha_{ji})_{n+1,p_i} \\ (b_j,\beta_j)_{1,m}; & (b_{ji},\beta_{ji})_{m+1,q_i},(k,1),(-k,1) \end{matrix} \right. \right] \tag{4.121}$$

provided $0 \le \theta \le \pi$, $|\arg z| < \frac{1}{2}B_i\pi$, B_i and $A_i \le 0$, where

$$A_i = \sum_{j=1}^{p_i} a_{ji} - \sum_{j=1}^{q_i} b_{ji},\ B_i = \sum_{j=1}^{n}\alpha_j - \sum_{j=n+1}^{p_i}\alpha_{ji} + \sum_{j=1}^{m}\beta_j - \sum_{j=m+1}^{q_i}\beta_{ji} < \quad i = 1,2,\ldots,r$$

We know that

$$\sqrt{\pi}\,\frac{\Gamma(s+1)}{\Gamma(s+1/2)}\left(\cos(\theta/2)\right)^{2s} = 1 + 2\sum_{k=1}^{\infty}\frac{(-1)^k(-s)_k}{(s+1)_k}\cos k\theta \tag{4.122}$$

where $\mathrm{Re}(s) > \frac{-1}{2}$, $0 \le \theta < \pi$ so that

$$\mathrm{IS}_1 = \frac{1}{2\pi i}\int_L \phi(\xi) z^{\xi}\,\frac{\Gamma\left(\xi+\frac{1}{2}\right)}{\Gamma(\xi+1)}\left\{1 + 2\sum_{k=1}^{\infty}\frac{(-1)^k(\ \xi)_k}{(\xi+1)_k}\cos k\theta\right\} d\xi$$

$$= \frac{1}{2\pi i}\int_L \phi(\xi)\frac{\Gamma\left(\xi+\frac{1}{2}\right)}{\Gamma(\xi+1)} z^{\xi} d\xi + 2\sum_{k=1}^{\infty}\cos(k\theta)\frac{1}{2\pi i}$$

$$\int_L \phi(\xi)\frac{\Gamma\left(\xi+\frac{1}{2}\right)\Gamma(\xi+1)}{\Gamma(\xi+1+k)\Gamma(\xi+1-k)} z^{\xi}\ d\xi$$

where $\phi(\xi)$ is defined in (1.23).

Hence,

$$\mathrm{IS}_1 = I^{m,n+1}_{p_i+1,q_i+1:r}\left[z\left|\begin{matrix}(1/2,1),(a_j,\alpha_j)_{1,n}; & (a_{ji},\alpha_{ji})_{n+1,p_i}\\ (b_j,\beta_j)_{1,m}; & (b_{ji},\beta_{ji})_{m+1,q_i},(0,1)\end{matrix}\right.\right]+2\sum_{k=1}^{\infty}\cos(k\theta)$$

$$\times I^{m,n+2}_{p_i+2,q_i+2:r}\left[z\left|\begin{matrix}(0,1),(1/2,1),(a_j,\alpha_j)_{1,n}; & (a_{ji},\alpha_{ji})_{n+1,p_i}\\ (b_j,\beta_j)_{1,m}; & (b_{ji},\beta_{ji})_{m+1,q_i},(k,1),(-k,1)\end{matrix}\right.\right] \tag{4.123}$$

ii.

$$\mathrm{IS}_2 \equiv e^{ix}x^{\delta}I^{m,n}_{p_i,q_i:r}\left[zx^h\left|\begin{matrix}(a_j,\alpha_j)_{1,n}; & (a_{ji},\alpha_{ji})_{n+1,p_i}\\ (b_j,\beta_j)_{1,m}; & (b_{ji},\beta_{ji})_{m+1,q_i}\end{matrix}\right.\right]$$

$$= \frac{1}{2^{\lambda-1}\sqrt{\pi}}\sum_{k=0}^{\infty}\mathrm{Re}^{(1/2)\pi i(\delta+R)} \quad \times I^{m+,n+1}_{p_i+2,q_i+1:r}$$

$$\left[z\left(\frac{e^{(1/2)\pi t}}{2}\right)^h\left|\begin{matrix}(1-\lambda-R,h),(a_j,\alpha_j)_{1,n}; & (a_{ji},\alpha_{ji})_{n+1,p_i},(R-\delta+1,h)\\ \left(\frac{1}{2}-\lambda,h\right),(b_j,\beta_j)_{1,m}; & (b_{ji},\beta_{ji})_{m+1,q_i}\end{matrix}\right.\right]J_R(x)$$

$$= \frac{1}{2^{\lambda-1}\sqrt{\pi}}\sum_{k=0}^{\infty}\mathrm{Re}^{(1/2)\pi i(\delta+R)} \quad \times I^{m+,n+1}_{p_i+2,q_i+1:r}$$

$$\left[z\left(\frac{e^{(1/2)\pi t}}{2}\right)^h\left|\begin{matrix}(1-\lambda-R,h),(a_j,\alpha_j)_{1,n}; & (a_{ji},\alpha_{ji})_{n+1,p_i},(R-\delta+1,h)\\ \left(\frac{1}{2}-\lambda,h\right),(b_j,\beta_j)_{1,m}; & (b_{ji},\beta_{ji})_{m+1,q_i}\end{matrix}\right.\right]J_R(x) \tag{4.124}$$

J_R denotes Bessel's function; $R = u + tk + 1$,
$\mathrm{Re}\left(\mu + h\min\left(b_j/\beta_j\right)\right) > 0\ (j = 1,2,\ldots,m)$; $\mathrm{Re}(\mu) < \frac{1}{2}, |\arg z|\left\langle \frac{1}{2}B_i\pi,\quad B_i\right\rangle 0,\ A_i \le 0.$

iii.

$$\mathrm{IS}_3 \equiv x^{\delta} J_{\mu}(x) I^{m,n}_{p_i,q_i:r}\left[zx^{h} \left| \begin{matrix} (a_j,\alpha_j)_{1,n}; & (a_{ji},\alpha_{ji})_{n+1,p_i} \\ (b_j,\beta_j)_{1,m}; & (b_{ji},\beta_{ji})_{m+1,q_i} \end{matrix} \right. \right]$$

$$= \frac{1}{\sqrt{\pi}} J_R(x)$$

$$\times \sum_{k=0}^{\infty} J_R(x) R I^{m+2,n+1}_{p_i+4,q_i+2:r}\left[z \left| \begin{matrix} \left(\frac{2+\delta+\mu-R}{2},h\right),(a_j,\alpha_j)_{1,n}; & (a_{ji},\alpha_{ji})_{n+1,p_i}, \\ \left(\frac{\delta+1}{2},h\right),\left(\frac{\delta+2}{2},h\right),(b_j,\beta_j)_{1,m}; & (b_{ji},\beta_{ji})_{m+1,q_i}, \end{matrix} \right. \right.$$

$$\left. \begin{matrix} \left(\frac{2+\delta+\mu+R}{2},h\right),\left(\frac{2+\delta-\mu+R}{2},h\right), & \left(\frac{2+\delta+\mu-R}{2},h\right) \\ \ldots\ldots\ldots\ldots & \ldots\ldots\ldots\ldots\ldots \end{matrix} \right] \tag{4.125}$$

where
$R = u + 2k + 1$, $\mathrm{Re}\left((\mu-\delta+2h)\min\left(b_j / \beta_j\right)\right) > 0$ $(j = 1,2,\ldots,m)$; $|\arg z|\left\langle \frac{1}{2} B_i\pi, B_i \right\rangle 0$, $A_i \le 0$.

To obtain (4.124) we put

$$f(x) = e^{ix} x^{\delta} I^{m,n}_{p_i,q_i:r}\left[zx^{h} \left| \begin{matrix} (a_j,\alpha_j)_{1,n}; & (a_{ji},\alpha_{ji})_{n+1,p_i} \\ (b_j,\beta_j)_{1,m}; & (b_{ji},\beta_{ji})_{m+1,q_i} \end{matrix} \right. \right] \tag{4.126}$$

$$= \sum_{k=0}^{\infty} b_k J_R(x)$$

where $R = u + 2k + 1$ and $0 < x < \infty$.

Equation (4.126) is valid since $f(x)$ is continuous and of bounded variation in the open interval $(0,\infty)$ when $\delta \ge 0$.

Multiplying both the sides of (4.126) by $x^{-1}J_T(x)$, putting $T = u + 2t + 1$ and integrating from 0 to ∞, with respect to x, we get

$$\int_0^\infty e^{ix}x^{\delta-1}J_T(x)I_{p_i,q_i:r}^{m,n}\left[zx^h\left|\begin{matrix}(a_j,\alpha_j)_{1,n}; & (a_{ji},\alpha_{ji})_{n+1,p_i}\\(b_j,\beta_j)_{1,m}; & (b_{ji},\beta_{ji})_{m+1,q_i}\end{matrix}\right.\right]$$

$$=\sum_{k=0}^\infty b_k\left\{\int_L^\infty x^{-1}J_R(x)J_T(x)dx\right\} \tag{4.127}$$

Evaluating the left-hand side of (4.127) and applying orthogonality property of Bessel functions on the right-hand side we get

$$\int_0^\infty t^{-1}J_{u+2m+1}(t)J_{u+2n+1}(t)dt=0,\ \ m\neq n$$

$$=(4m+2u+2)^{-1}\ \text{ if } m=n$$

$$\mathrm{Re}(u+m+n)>-1.$$

We have

$$b_t=\frac{2T}{2^\lambda\sqrt{\pi}}e^{1/2\pi i(\delta+T)}$$

$$\times I_{p_i+2,q_i+1:r}^{m+1,n+1}\left[z\left(\frac{e^{(1/2)\pi t}}{2}\right)^h\left|\begin{matrix}(-\delta-T+1,h),(a_j,\alpha_j)_{1,n}; & (a_{ji},\alpha_{ji})_{n+1,p_i},(T-\delta+1,h)\\(-\delta+1/2,h)(b_j,\beta_j)_{1,m}; & (b_{ji},\beta_{ji})_{m+1,q_i}\end{matrix}\right.\right] \tag{4.128}$$

where $T=u+2t+1$

Substituting the value of b_k in (4.126) we get the result (4.124). Proceeding on similar lines the result (4.125) can also be obtained.

iv.

$$\mathrm{IS}_4\equiv x^\delta(1+x)^\delta I_{p_i,q_i:r}^{m,n}\left[z(1+x)^h\left|\begin{matrix}(a_j,\alpha_j)_{1,n}; & (a_{ji},\alpha_{ji})_{n+1,p_i}\\(b_j,\beta_j)_{1,m}; & (b_{ji},\beta_{ji})_{m+1,q_i}\end{matrix}\right.\right]$$

$$=2^\delta\sum_{N=0}^\infty\sum_{K=0}^\infty\frac{(-1)^{N2k}(-\delta)_k(\alpha+\beta+2N+1)\Gamma(\alpha+\beta+N+1)}{k!\Gamma(\beta+N+1)} \tag{4.129}$$

$$\times I_{p_i+2,q_i+2:r}^{m,n+2}\left[2^hz\left|\begin{matrix}(-\delta-k,h), & (-\beta-\delta-k,h), & (a_j,\alpha_j)_{1,n}; & (a_{ji},\alpha_{ji})_{n+1,p_i}\\(b_j,\beta_j)_{1,m}; & (b_{ji},\beta_{ji})_{m+1,q_i}, & (-\delta+N-k,h), & (\alpha-\beta-\delta-N-1,h)\end{matrix}\right.\right]P_N^{\alpha,\beta}(x)$$

where, $\delta, h > 0$, m, n, $p_i, q_i \,(i = 1,2,\ldots.,r)$ are integers such that $\mathrm{Re}(\delta) > -1, \mathrm{Re}\left(\delta + h \min\left(b_j / \beta_j\right)\right) > -1,\ j = 1,2,\ldots\ldots,m.$

v.

$$\mathrm{IS}_5 \equiv (1-x)^{\delta} P_m^{(\rho,\sigma)}(x) I_{p_i,q_i:r}^{m,n}\left[z(1-x)^h \middle| \begin{matrix} (a_j,\alpha_j)_{1,n}; & (a_{ji},\alpha_{ji})_{n+1,p_i} \\ (b_j,\beta_j)_{1,m}; & (b_{ji},\beta_{ji})_{m+1,q_i} \end{matrix}\right]$$

$$= 2^{\delta} \sum_{K=0}^{\infty}\sum_{N=0}^{\infty} \frac{(-m)_k (\rho+\sigma+m+1)(\alpha+\beta+2N+1)\Gamma(\alpha+\beta+N+1)}{k!m!\Gamma(\rho+k+1)}$$

$$\left(\frac{\Gamma(\sigma+m+1)}{\Gamma(\alpha+N+1)}\right) P_N^{(\alpha,\beta)}(x) \tag{4.130}$$

$$\times I_{p_i+4,q_i+4:r}^{m+1,n+3}\left[2^h z \middle| \begin{matrix} (-\delta-k,h),(N-\delta,h), & (-\beta-\delta-k,h), & (a_j,\alpha_j)_{1,n}; & (a_{ji},\alpha_{ji})_{n+1,p_i},(-\delta,h) \\ (-\delta+N,h)(b_j,\beta_j)_{1,m}; & (b_{ji},\beta_{ji})_{m+1,q_i}, & (-\beta-\alpha-\delta-N-1,h),(-\delta,h) & (N-\delta-k,h) \end{matrix}\right]$$

vi.

$$\mathrm{IS}_6 \equiv (1-x)^{\rho}(1-x)^{\sigma}\, I_{p_i,q_i:r}^{m,n}\left[z(1-x)^h(1+x)^t \middle| \begin{matrix} (a_j,\alpha_j)_{1,n}; & (a_{ji},\alpha_{ji})_{n+1,p_i} \\ (b_j,\beta_j)_{1,m}; & (b_{ji},\beta_{ji})_{m+1,q_i} \end{matrix}\right]$$

$$= 2^{\rho+\sigma} \sum_{N=0}^{\infty}\sum_{K=0}^{\infty} \frac{(-N)_k\, n!(\alpha+\beta+2N+1)\Gamma(\alpha+\beta+N+k+1)}{k!(\alpha+1)_k\,\Gamma(\alpha+N+1)\Gamma(\beta+N+1)} P_N^{(\alpha,\beta)}(x)$$

$$\times I_{p_i+2,q_i+1:r}^{m,n+2}\left[2^{h+t} z \middle| \begin{matrix} (-\rho-\alpha-k,h)(-\beta-\sigma,h),(a_j,\alpha_j)_{1,n}; & (a_{ji},\alpha_{ji})_{n+1,p_i} \\ (b_j,\beta_j)_{1,m};(b_{ji},\beta_{ji})_{m+1,q_i}, & (-1-\rho-\sigma-\alpha-\beta-k,h+t) \end{matrix}\right] \tag{4.131}$$

ix.

$$IS_8 \equiv (1-x)^{\rho-M/2}(1+x)^{\sigma+N/2}\ I^{m,n}_{p_i,q_i:r}\left[z(1-x)^h(1+x)^t \left| \begin{matrix} (a_j,\alpha_j)_{1,n}; & (a_{ji},\alpha_{ji})_{n+1,p_i} \\ (b_j,\beta_j)_{1,m}; & (b_{ji},\beta_{ji})_{m+1,q_i} \end{matrix} \right.\right]$$

$$= 2^{\rho+\sigma-1}\sum_{R=0}^{\infty}\sum_{P=0}^{R}\frac{(2R+N-M+1)\Gamma(R-M+1)(-R)_P\,\Gamma(1+P+R-M+N)}{R!\Gamma(R+N+1)P!\Gamma(1-M+P)}$$

$$I^{m,n+3}_{p_i+3,q_i+2:r}\left[2^{h+t}z \left| \begin{matrix} (-M+\rho-P,h),(M+\sigma-\rho-1,h+t) & (-\sigma-N,t), \\ (b_j,\beta_j)_{1,m},(b_{ji},\beta_{ji})_{m+1,q_i}, & (M+\sigma-\rho-1-P,h-t) \end{matrix} \right.\right.$$

$$\left.\begin{matrix} (a_j,\alpha_j)_{1,n};(a_{ji},\alpha_{ji})_{n+1,p_i} \\ (M-\rho-\sigma-N-1,h+t) \end{matrix}\right] P^{M,N}_{R-(M-N/2)}(x) \tag{4.136}$$

To prove this, we take

$$(1-x)^{\rho-M/2}(1+x)^{\sigma+N/2}\ I^{m,n}_{p_i,q_i:r}\left[z(1-x)^h(1+x)^t \left| \begin{matrix} (a_j,\alpha_j)_{1,n}; & (a_{ji},\alpha_{ji})_{n+1,p_i} \\ (b_j,\beta_j)_{1,m}; & (b_{ji},\beta_{ji})_{m+1,q_i} \end{matrix} \right.\right]$$

$$= \sum_{R=0}^{\infty} A_R P^{M,N}_{R-(M-N/2)}(x) \tag{4.137}$$

Multiplying both the sides of (4.137) by $P^{M,N}_{R-(M-N/2)}(x)$ and integrate with respect to x from −1 to +1. Evaluating the left-hand side with the help of (4.16) and interchanging the order of summation on the right-hand side. Integrating and applying the orthogonality property of generalized associated Legendre's function [32], we get

$$\int_{-1}^{+1} P^{M,N}_{K-(M-N/2)}(x)P^{M,N}_{R-(M-N/2)}(x)\,dx = 0 \quad \text{if } K \neq R$$

$$= \frac{2^{-M+N+1}K!\Gamma(K+M+1)}{(2K-M+N+1)\Gamma(K-M+1)\Gamma(K-M+N+1)} \quad \text{if } K = R$$

provided that $\mathrm{Re}(M) < 1$, $\mathrm{Re}(N) > -1$, we obtain

$$A_K = \frac{2^{\rho+\sigma-1}(2K-M+N+1)\Gamma(K-M+1)}{K!\Gamma(K+N+1)}$$

$$\sum_{P=0}^{\infty}\frac{(-K)_P\,\Gamma(1+K+N-M+P)}{P!\Gamma(1-M+P)}$$

$$\times I^{m,n+3}_{p_i+3,q_i+2:r}\left[2^{h+t}z\left|\begin{array}{l}(-M+\rho-P,h),(M+\sigma-\rho-1,h+t) \quad (-\sigma-N,t),\\ (b_j,\beta_j)_{1,m},(b_{ji},\beta_{ji})_{m+1,q_i}, \quad (M+\sigma-\rho-1-P,h-t)\end{array}\right.\right.$$

$$\left.\begin{array}{l}(a_j,\alpha_j)_{1,n};(a_{ji},\alpha_{ji})_{n+1,p_i}\\ (M-\rho-\sigma-N-1,h+t)\end{array}\right] \tag{4.138}$$

Finally, substituting the value of $A_K$in (4.137) the result (4.136) follows.

4.5.2 Special Cases

i. In a special case when $r=1$, $M=m$, $N=n$, $h=v$, $t=\delta$, (4.136) gives rise to a well-known expansion of H-function involving generalized Legendre functions.
ii. If we replace δ by $\lambda-1$ and put $\alpha=\beta=\rho=\sigma=0$ the infinite expansion (4.130) reduces to the following infinite expansion theorem involving the product of Legendre polynomials

$$(1-x)^{\lambda-1}P_m(x)I^{m,n}_{p_i,q_i:r}\left[z(1-x)^h\left|\begin{array}{ll}(a_j,\alpha_j)_{1,n}; & (a_{ji},\alpha_{ji})_{n+1,p_i}\\ (b_j,\beta_j)_{1,m}; & (b_{ji},\beta_{ji})_{m+1,q_i}\end{array}\right.\right]$$

$$=2^{\lambda-1}\sum_{K=0}^{\infty}\sum_{N=0}^{\infty}\frac{(-m)_k(m+1)_k(2N+1)}{(k!)^2}I^{m+1,n+3}_{p_i+4,q_i+4:r}$$

$$\left[2^hz\left|\begin{array}{ll}(1-\lambda-k,h),(1-\lambda-k,h), & (N+1-\lambda,h),\\ (1-\lambda+N,h)(b_j,\beta_j)_{1,m}; & (b_{ji},\beta_{ji})_{m+1,q_i},\end{array}\right.\right. \tag{4.139}$$

$$\left.\begin{array}{ll}(a_j,\alpha_j)_{1,n}; & (a_{ji},\alpha_{ji})_{n+1,p_i},(1-\lambda,h)\\ (-\lambda-k-N,h),(1-\lambda,h) & (1+N-\lambda-k,h)\end{array}\right]P_N(x)$$

iii. If $\alpha=\beta=0, \rho$ is replaced by $\rho-1$ and σ is replaced by $\sigma-1$,(4.131) transforms into a finite expansion theorem involving Legendre polynomials of the I-function

$$(1-x)^{\rho-1}(1+x)^{\sigma-1} I^{m,n}_{p_i,q_i:r}\left[z(1-x)^h(1+x)^t \left| \begin{matrix} (a_j,\alpha_j)_{1,n}; & (a_{ji},\alpha_{ji})_{n+1,p_i} \\ (b_j,\beta_j)_{1,m}; & (b_{ji},\beta_{ji})_{m+1,q_i} \end{matrix} \right.\right]$$

$$=\sum_{N=0}^{\infty}\sum_{K=0}^{\infty} \frac{2^{\rho+\sigma-2}(-N)_k(2N+1)\Gamma(N+k+1)}{(k!)^2 N!} I^{m,n+2}_{p_i+2,q_i+1:r}$$

$$\left[2^{h+t}z \left| \begin{matrix} (1-\rho-k,h)(1-\sigma,t),(a_j,\alpha_j)_{1,n}; & (a_{ji},\alpha_{ji})_{n+1,p_i} \\ (b_j,\beta_j)_{1,m};(b_{ji},\beta_{ji})_{m+1,q_i}, & (1-\sigma-\rho-k,h+t) \end{matrix} \right.\right] P_N(x) \tag{4.140}$$

iv. Replacing $\rho$by $\rho-1,\sigma$ by $\sigma-1$ and putting $\alpha=\beta=0$ the infinite expansion theorem (4.133) takes the form of the following infinite expansion for the I-function

$$(1-x)^{\rho-1}(1+x)^{\sigma-1} I^{m,n}_{p_i,q_i:r}\left[z(1-x)^h(1+x)^{-t} \left| \begin{matrix} (a_j,\alpha_j)_{1,n}; & (a_{ji},\alpha_{ji})_{n+1,p_i} \\ (b_j,\beta_j)_{1,m}; & (b_{ji},\beta_{ji})_{m+1,q_i} \end{matrix} \right.\right]$$

$$=\sum_{N=0}^{\infty}\sum_{K=0}^{\infty} \frac{2^{\rho+\sigma-2}(-N)_k(2N+1)\Gamma(N+k+1)}{(k!)^2 N!} I^{m+1,n+1}_{p_i+2,q_i+1:r}$$

$$\left[2^{h-t}z \left| \begin{matrix} (-\rho-k,h),(a_j,\alpha_j)_{1,n}; & (a_{ji},\alpha_{ji})_{n+1,p_i};(1-k-\rho-\sigma,h-t) \\ (1+\sigma,t) & (b_j,\beta_j)_{1,m};(b_{ji},\beta_{ji})_{m+1,q_i} \end{matrix} \right.\right] P_N(x) \tag{4.141}$$

These infinite expansions for the I-function can also be proved without considering them as a special case of infinite expansions involving Jacobi polynomials.

v. If all the $\alpha's$ and $\beta's$ are unity, then expansion formula (4.121) reduces to the following expansion formula

$$\pi^{-1/2} I_{p_i,q_i:r}^{m,n}\left[x\cos^2(\theta/2)\left|\begin{matrix}(a_j,\alpha_j)_{1,n}; & (a_{ji},\alpha_{ji})_{n+1,p_i}\\ (b_j,\beta_j)_{1,m}; & (b_{ji},\beta_{ji})_{m+1,q_i}\end{matrix}\right.\right]$$

$$= I_{p_i+1,q_i+1:r}^{m,n+1}\left[x\left|\begin{matrix}(1/2,1),(a_j,\alpha_j)_{1,n}; & (a_{ji},\alpha_{ji})_{n+1,p_i}\\ (b_j,\beta_j)_{1,m}; & (b_{ji},\beta_{ji})_{m+1,q_i},(0,1)\end{matrix}\right.\right]+2\sum_{k=1}^{\infty}\cos(k\theta)$$

$$\times I_{p_i+2,q_i+2:r}^{m,n+2}\left[x\left|\begin{matrix}(0,1),(1/2,1),(a_j,\alpha_j)_{1,n}; & (a_{ji},\alpha_{ji})_{n+1,p_i}\\ (b_j,\beta_j)_{1,m}; & (b_{ji},\beta_{ji})_{m+1,q_i},(k,1),(-k,1)\end{matrix}\right.\right] \tag{4.142}$$

provided $0 \le \theta \le \pi$.

vi. If $r = 1$, the expansion formula (4.131) reduces to the expansion formula of the H-function

$$\pi^{-1/2} H_{p,q}^{m,n}\left[x\cos^2(\theta/2)\left|\begin{matrix}(a_j,\alpha_j)_{1,n}; & (a_{ji},\alpha_{ji})_{n+1,p_i}\\ (b_j,\beta_j)_{1,m}; & (b_{ji},\beta_{ji})_{m+1,q_i}\end{matrix}\right.\right]$$

$$= H_{p+1,q+1}^{m,n+1}\left[x\left|\begin{matrix}(1/2,1),(a_j,\alpha_j)_{1,n}; & (a_{ji},\alpha_{ji})_{n+1,p}\\ (b_j,\beta_j)_{1,m}; & (b_{ji},\beta_{ji})_{m+1,q},(0,1)\end{matrix}\right.\right]+2\sum_{k=1}^{\infty}\cos(k\theta)$$

$$\times H_{p+2,q+2:r}^{m,n+2}\left[x\left|\begin{matrix}(0,1),(1/2,1),(a_j,\alpha_j)_{1,n}; & (a_{ji},\alpha_{ji})_{n+1,p}\\ (b_j,\beta_j)_{1,m}; & (b_{ji},\beta_{ji})_{m+1,q},(k,1),(-k,1)\end{matrix}\right.\right] \tag{4.143}$$

provided $0 \le \theta \le \pi$.

5 Single and Dual Integral Equations Involving I-Function

In this chapter, an attempt has been made to solve integral equations having I-function as kernel by using fractional integral operators using generalized Hardy–Tichmarch theorem [37].

5.1 I-FUNCTION AS KERNEL

We consider the integral equation

$$g(x)=\int_0^\infty I\left[ux\left|\begin{matrix}\{(1-a_j,e_j),_{1\,q}\};\{(a_{ji},e_{ji})_{q+1,t_i}\}\\ \{(b_j,f_j)_{1,p}\};\{1-b_{ji},f_{ji})_{p+1,s_i}\}\end{matrix}\right.\right]F(u)\,du \tag{5.1}$$

Here $g(x)$ is known and $f(x)$ is to be determined. The notation of the I-function, used as a kernel function, which is in different form of equation (1.22).

And Mellin transform of I-function is given by

$$\mathrm{M}\left[I\left[u\left|\begin{matrix}\{(1-a_j,e_j)_{1,q}\};\{(a_{ji},e_{ji})_{q+1,t_i}\}\\ \{(b_j,f_j)_{1,p}\};\{(1-b_{ji},f_{ji})_{p+1,s_i}\}\end{matrix}\right.\right]\right]$$

$$=\frac{\prod_{j=1}^{p}\Gamma(b_j+f_j\xi)\prod_{j=1}^{q}\Gamma(1-a_j-e_j\xi)}{\sum_{i=1}^{r}\left\{\prod_{j=p+1}^{s_i}\Gamma(1-b_{ji}-f_{ji}\xi)\prod_{j=q+1}^{t_i}\Gamma(a_{ji}+e_{ji}\xi)\right\}} \tag{5.2}$$

The Mellin transform of a function $f(x)$ is defined as

$$F(\xi)=M[f(u)]=\int_0^\infty u^{\xi-1}f(u)\,du \tag{5.3}$$

The Parseval's theorem of the transform is

$$\int_0^\infty g(ux)f(u)\,du=\lim_{T\to\infty}\frac{1}{2\pi i}\int_{\sigma_0-iT}^{\sigma_0+iT}x^{-\xi}g(\xi)f(1-\xi)\,d\xi \tag{5.4}$$

DOI: 10.1201/9781003517757-5

From Equations (5.2) and (5.4), we obtain

$$\frac{1}{2\pi i}\int_L \frac{\prod_{j=1}^{p}\Gamma\left(b_j+f_j\xi\right)\prod_{j=1}^{q}\Gamma\left(1-a_j-e_j\xi\right)}{\sum_{i=1}^{r}\left\{\prod_{p+1}^{s_i}\Gamma\left(1-b_j-f_j\xi\right)\prod_{q+1}^{t_i}\Gamma\left(a_j+e_j\xi\right)\right.} x^{-\xi}F(1-\xi)\,d\xi \tag{5.5}$$

$$= g(x)$$

where $F(\xi) = M[f(u)]$

Now we shall use the following definition of Beta function

$$\int_x^{\infty}\left(v^{1/c}-x^{1/c}\right)^{b-\beta-1} v^{\frac{1}{c}-\frac{b}{c}-s-1}\,dv = \frac{c\,\Gamma(b-\beta)\Gamma(\beta+cs)}{\Gamma(b+cs)}x^{-\frac{\beta}{c}-s} \tag{5.6}$$

Replace x with v in (5.6) and multiply by

$$\left(v^{\frac{1}{f_1}}-x^{\frac{1}{f_1}}\right)^{b_1-\mu_1-1} v^{\frac{1}{f_1}-\frac{b_1}{f_1}-1},$$

and integrate under the integral sign with respect to v from x to ∞ where $x > 1$ we obtain

$$\int_x^{\infty}\left(v^{\frac{1}{f_1}}-x^{\frac{1}{f_1}}\right)^{b_1-\mu_1-1} v^{\frac{1}{f_1}-\frac{b_1}{f_1}-1}\,dv$$

$$= \frac{f_1\Gamma(b_1-\mu_1)\Gamma(\mu_1+f_1\xi)}{\Gamma(b_1+f_1\xi)}x^{-\frac{\mu_1}{f_1}}$$

From Equation (5.5) we have

$$\frac{1}{2\pi i}\int_L \frac{\prod_{j=2}^{P}\Gamma\left(b_j+f_j\xi\right)\Gamma\left(b_1+f_1\xi\right)\Gamma\left(\mu_1+f_1\xi\right)\prod_{j=1}^{q}\Gamma\left(1-a_j-e_j\xi\right)}{\sum_{i=1}^{r}\left\{\prod_{j=p+1}^{s_i}\Gamma\left(1-b_{ji}-f_{ji}\xi\right)\prod_{j=q+1}^{t_i}\Gamma\left(a_{ji}+e_{ji}\xi\right)\right\}}$$

$$\times\frac{1}{\Gamma\left(b_1+f_1\xi\right)}x^{-\xi}f(1-\xi)d\xi \tag{5.7}$$

$$=\frac{1}{f_1\Gamma(b_1-\mu_1)}x^{\frac{\mu_1}{f_1}}\int_x^{\infty}\left(v^{\frac{1}{f_1}}-x^{\frac{1}{f_1}}\right)^{b_1-\mu_1-1} v^{\frac{1}{f_1}-\frac{b_1}{f_1}-1}\,g(v)\,dv$$

Let us denote right-hand side of above equations by $g_1(x)$, Then it is evident that

$$g_1(x) = \Re\left[b_1 - \mu_1, \frac{\mu_1}{f_1} : \frac{1}{f_1} \right] g(x) \tag{5.8}$$

where $\Re$ is second operator of fractional integration defined in (5.6) and (1.26) again, if we apply the operator, we get

$$\frac{1}{2\pi i}\int_L \frac{\prod_{j=3}^{p}\Gamma(b_j+f_j\xi)\Gamma(\mu_1+f_1\xi)\Gamma(\mu_2+f_2\xi)\prod_{j=1}^{q}\Gamma(1-a_j-e_j\xi)}{\sum_{i=1}^{r}\left[\prod_{p+1}^{s_i}\Gamma(1-b_{ji}-f_{ji}\xi)\prod_{q+1}^{t_i}\Gamma(a_{ji}+e_{ji}\xi)\right]} x^{-\xi}F(1-\xi)d\xi \tag{5.9}$$

$$= g_2(x)$$

where

$$g_2(x) = \Re\left[b_2 - \mu_2, \frac{\mu_2}{f_2} : \frac{1}{f_2} \right] g_1(x) \tag{5.10}$$

In a similar manner, we have

$$\frac{1}{2\pi i}\int_L \frac{\prod_{j=3}^{P}\Gamma(b_j+f_j\xi)\ \Gamma(\mu_1+f_1\xi)\ \Gamma(\mu_2+f_2\xi)\prod_{j=1}^{q}\Gamma(1-a_j-e_j\xi)}{\sum_{i=1}^{r}\left[\prod_{p+1}^{s_i}\Gamma(1-b_{ji}-f_{ji}\xi)\prod_{q+1}^{t_i-1}\Gamma(a_{ji}+e_{ji}\xi)\right]\Gamma(v_L+e_L\xi)} \tag{5.11}$$

$$\times x^{-\xi}F(1-\xi)d\xi = g_3(x)$$

$$g_3(x) = \left[v_L - a_L, \frac{a_L}{e_L} : \frac{1}{e_L} \right] g_2(x)$$

Here, subscripts, $L = t_r$.

Continuing in this manner, and making use of Perseval's theorem [38] we obtain from (5.7), (5.9), (5.11) respectively, that

$$g_1(x) = \int_0^\infty I\left[ux \middle| \begin{matrix} \{(1-a_j, e_j)_{1,q}\} : \{(a_{ji}, e_{ji})_{q+1,t_i}\}(v_L, e_L) \\ \{(\mu_1, f_1)(b_j, f_j)_{2,p}\}((1-b_{ji}, f_{ji})_{p+1,s_i}) \end{matrix} \right] f(u)\,du \tag{5.12}$$

$$g_2(x) = \int_0^\infty I\left[ux \middle| \begin{matrix} \{(1-a_j, e_j)_{1q}\} : \{(a_{ji}, e_{ji})_{q+1,t_i}\}(v_L, e_L) \\ \{(\mu_1, f_1),(\mu_2, f_2)\}, \{(b_j, f_j)_{3,P}\}\{(1-b_{ji}, f_{ji})_{p+1,s_i}\} \end{matrix} \right] f(u)\,du \tag{5.13}$$

$$g_3(x) = \int_0^\infty I\left[ux \left| \begin{matrix} \{(1-a_j, e_j)_{1,q}\}, \{(a_{ji}, e_{ji})_{q+1,t_i-1})(v_L, e_L) \\ \{(\mu_1, f_1)\{(\mu_2, f_2)\}, \{b_j, f_j)_{3,P}\}\{(1-b_{ji}, f_{ji})_{p+1,s_i}\} \end{matrix} \right.\right] f(u)du \tag{5.14}$$

where the functions $g_1(x)$, $g_2(x)$ and $g_3(x)$ are given in (5.8), (5.10), and (5.11).

Now, we split the equation either of the form (5.12)– (5.14), or in any suitable form, into the pair of known integral equations. We have already indicated that the choice of the form depends on the type and number of parameters in (5.1). The criteria for the selection will be clearer when we arrive at the final solution.

First, we assure that any one of f_j ($j = 1, 2 \ldots p$) is equal to unity or reducible to unity, say $f_1 = 1$. In this case we consider the integral equation (5.12) may be written in the form, (where $\mu_1 = 0$)

$$\int_o^\infty I\left[ux \left| \begin{matrix} \{(1-a_j, e_j)_{1,q}\}: \{(a_{ji}, e_{ji})_{q+1,t_i}\} \\ \{(0,1)\}, \{(b_j, f_j)_{2,p}\}\{1-b_{ji}, f_{ji})_{p+1,s_i}\} \end{matrix} \right.\right] f(u)du = g_1(x) \tag{5.15}$$

Now we use the following known results [39], After applying the Laplace transform on I-function used as a kernel function, with the help of known results in the form of H-function

$$\int_o^\infty e^{-us} H\left[\frac{u}{z} \left| \begin{matrix} \{(1-A_j, E_j)_{1,Q}\}: \{(A_j, E_j)_{Q+1,T}\} \\ \{(B_j, F_j)_{1,P}\}, \{(1-B_j, F_j)_{P+1,S}\} \end{matrix} \right.\right] du$$

$$= \frac{1}{s} H\left[\frac{1}{zs} \left| \begin{matrix} (1,1)\{(1-A_j, E_j)_{1,Q}\}: \{(A_j, E_j)_{Q+1,T}\} \\ \{(B_j, F_j)_{1,P}\}, \{(1-B_j, F_j)_{P+1,S}\} \end{matrix} \right.\right]$$

Then we get the following results in terms of I-function.

$$\int_o^\infty e^{-us} I\left[\frac{u}{z} \left| \begin{matrix} \{(1-A_j, E_j)_{1,Q}\}: \{(A_{ji}, E_{ji})_{Q+1,T_i}\} \\ \{(B_j, F_j)_{1,P}\}, \{(1-B_{ji}, F_{ji})_{P+1,Si}\} \end{matrix} \right.\right] du$$

$$= \frac{1}{s} I\left[\frac{1}{zs} \left| \begin{matrix} (1,1)\{(1-A_j, E_j)_{1,Q}\}: \{(A_{ji}, E_{ji})_{Q+1,T_i}\} \\ \{(B_j, F_j)_{1,P}\}, \{(1-B_{ji}, F_{ji})_{P+1,S_i}\} \end{matrix} \right.\right] \tag{5.16}$$

$\mathrm{Re}(\xi) > 0$, $\mathrm{Re}(1 + \min B_j/F_j) > 0$, ($j = 1, 2, \ldots.. P$).

The I-function converges absolutely in the ξ-plane if

i. > 0, $|\arg z| < \pi\frac{\lambda}{2}$

ii. $\lambda \geq 0$, $|\arg z| \geq \pi\frac{\lambda}{2}$, Re $(\mu+1) < 0$

i.

$$\lambda = \sum_{j=1}^{m} \alpha_j + \sum_{j=1}^{n} \beta_j - \max_{1 \leq i \leq r}\left[\sum_{j=n+1}^{pi} \alpha_{ji} + \sum_{j=m+1}^{qi} \beta_{ji}\right] \tag{5.17}$$

$$\mu = \sum_{j=1}^{m}(b_j - \beta_j\sigma) - \sum_{j=1}^{n}(a_j - \alpha_j\sigma) - \min_{1 \leq i \leq r}\left[\sum_{j=n+1}^{p_i}(a_j - \alpha_j\sigma) - \sum_{j=m+1}^{p_i}(b_j - \beta_j\sigma) + \frac{p_i}{2}, \frac{q_i}{2}\right] \tag{5.18}$$

If $\sigma = 0$, then the set of above expressions (5.17) and (5.18) reduce to

$$\lambda = \sum_{j=1}^{m} \alpha_j + \sum_{j=1}^{n} \beta_j - \max_{1 \leq i \leq r}\left[\sum_{j=n+1}^{p_i} \alpha_{ji} + \sum_{j=m+1}^{q_i} \beta_{ji}\right]$$

$$\mu = \sum_{j=1}^{m}(b_j) - \sum_{j=1}^{n}(a_j) - \min_{1 \leq i \leq r}\left[\sum_{j=n+1}^{p_i}(a_j) - \sum_{j=m+1}^{q_i}(b_j) + \frac{p_i}{2}, \frac{q_i}{2}\right]$$

But by using formulae [40]

$$I\left[x \left| \begin{matrix} \{(1-A_j, E_j)_{1,Q}\}, \{(A_{ji}, E_{ji})_{Q+1,T_i}\} \\ \{(B_j, F_j)_{1,P}\}, \{(1-B_{ji}, F_{ji})_{P+1,S_i}\} \end{matrix} \right.\right]$$
$$= I\left[\frac{1}{x} \left| \begin{matrix} \{(B_j, F_j)_{1,P})\}, \{(1-B_{ji}, F_{ji})_{P+1,S_i}\} \\ \{(1-A_j, E_j)_{1,Q}\}, \{(A_{ji}, E_{ji})_{Q+1,T_i}\} \end{matrix} \right.\right] \tag{5.19}$$

and

$$x^{\rho} I\left[x\left|\begin{matrix}\{(1-A_j,E_j)_{1,Q}\},\{(A_{ji},E_{ji})_{Q+1,T_i}\}\\ \{(B_j,F_j)_{1,P}\},\{(1-B_{ji},F_{ji})_{P+1,S_i}\}\end{matrix}\right.\right]$$

$$= I\left[x\left|\begin{matrix}\{(1-A_j-\rho E_j,E_j)_{1,Q}\}:\{(A_{ji},+\rho E_{ji},E_{ji})_{Q+1,Ti}\}\\ \{(B_j,+\rho F_j,F_j)_{1,P}\},\{(1-B_{ji}-\rho F_{ji},F_{ji})_{P+1,S_i}\}\end{matrix}\right.\right] \tag{5.20}$$

we can transform (5.16) into

$$\int_o^{\infty} e^{-us} I\left[\frac{u}{z}\left|\begin{matrix}\{(1-A_j,E_j)_{1,Q}\},\{(A_{ji},E_{ji})_{Q+1,T_i})\\ \{(B_j,F_j)_{1,P}\},\{1-B_{ji},F_{ji})_{P+1,S_i}\}\end{matrix}\right.\right] du$$

$$= zI\left[zs\left|\begin{matrix}\{(B_j+F_j,F_j)_{1,P})\},\{(1-B_{ji}-F_{ji},F_{ji})_{P+1,S_i}\}\\ (0,1)\{1-A_j-E_j,E_j)_{1,Q}\},\{(A_{ji}+E_{ji},E_{ji})_{Q+1,T_i}\}\end{matrix}\right.\right] \tag{5.21}$$

under the I-function conditions.

Applying Equation (5.21) in (5.15) we get

$$\int_o^{\infty} f(u)\int_o^{\infty} e^{-uy} I\left[xy\left|\begin{matrix}\{(1-A_j,E_j)_{1Q_1}\},\{(A_{ji},E_j)_{Q_1+1,T_1})\\ \{(B_j,F_j)_{1,P_1}\},\{1-B_{ji},F_{ji})_{P_1+1,S_1}\}\end{matrix}\right.\right] dy\,du = G_1(x) \tag{5.22}$$

$P_1 = q$, $Q_1 = p-1$, $T_1 = s-1$, $S_1 = t$

$[A_{ji} = 1 - b_{1+ji} - f_{1+ji}, E_j = f_{1+j}]$ (here $j = 1,2,\ldots,T_1$)

$[B_{ji} = 1 - a_{ji} - e_{ji}, F_{ji} = e_{ji}]$ (here $j = 1,2,\ldots,S_1$)

and $G_1(x) = \frac{1}{x} g_1\left(\frac{1}{x}\right)$

Reversing the order of integration in (5.22) we arrive at the following set of integral equations.

$$\int_o^{\infty} e^{-ux} f(u)\,du = h_1(x) \tag{5.23}$$

$$\int_o^{\infty} I\left[ux\left|\begin{matrix}\{(1-A_j,E_j)_{1,Q_l}\}:\{(A_{ji},E_{ji})_{Q_l+1,T_{il}}\}\\ \{(B_j,F_j)_{1,P_l}\},\{(1-B_{ji},F_{ji})_{P_l+1,S_{il}}\}\end{matrix}\right.\right] h_1(u)\,du = G_1(x) \tag{5.24}$$

Here, $h_1(x)$ is Laplace transform of $f(u)$, Therefore, the inversion formula will be.

$$f(u) = \lim_{T\to\infty} \frac{1}{2\pi i}\int_{\sigma_0-iT}^{\sigma_0+iT} e^{uz}\, h_l(z)\,dz \tag{5.25}$$

where $\sigma_o = \text{Re}(z)$

Hence, if the integral Equation (5.23) is assumed to have known inversion formulae, then $h_l(x)$ will be obtained in terms of prescribed function $G_1(x)$. Consequently, the unknown function $f(x)$ will be determined with the help of (5.25). Actually the selection of the form (5.12) implies that the resulting integral Equation (5.26) is standard one.

Accordingly, we have the following theorem:

Theorem 5.1

If $f(x)$ and $h_l(x)$ are solution of the integral equations.

$$g(x)=\int_o^{\infty} I\left[ux\left|\begin{matrix}\{(1-a_j,e_j)_{1,q}\},\{(a_{ji},e_{ji})_{q+1,t_i}\}\\ \{(b_1,1)\},\{(b_j,f_j)_{2,p}\},\{1-b_{ji},f_{ji})_{p+1,s_i}\end{matrix}\right.\right] f(u)\,du \tag{5.26}$$

$$G_l(x)=\int_0^{\infty} I\left[ux\left|\begin{matrix}\{(1-A_j,E_j)_{1,Q_l}\}:\{(A_{ji},E_{ji})_{Q_l+1,T_{il}}\}\\ \{(B_j,F_j)_{1,P_l}\},\{(1-B_{ji},F_{ji})_{P_l+1,S_{il}}\}\end{matrix}\right.\right] h_l(u)\,du \tag{5.27}$$

then,

$$f(u) = \lim_{T\to\infty} \frac{1}{2\pi i}\int_{\sigma_0-iT}^{\sigma_0+iT} e^{zx}\, h_l(z)\,dz \tag{5.28}$$

where

$$G_1(x)=\frac{1}{x}R\left[b_1,0;1\right] g\left(\frac{1}{x}\right) \tag{5.29}$$

We consider the integral equation

$$g(x)=\int_0^{\infty} I\left[ux\left|\begin{matrix}\{(1-a_j,e_j),_{1\ q}\};\{(a_{ji},e_{ji})_{\ q+1,t_i}\}\\ \{(b_j,f_j)_{1,p}\};\{1-b_{ji},f_{ji})_{p+1,s_i}\}\end{matrix}\right.\right] F(u)\,du \tag{5.30}$$

Here $g(x)$ is known and $f(x)$ is to be determined. The notation of the I-function, used as a kernel function, which is in different form of equation (1.22).

And Mellin transform of I-function is given by

$$\mathrm{M}\left[I\left[u \left| \begin{matrix} \{1-a_j, e_j)_{1,q}\}; \{(a_{ji}, e_{ji})_{q+1,t_i}\} \\ \{b_j, f_j)_{1,p}\}; \{1-b_{ji}, f_{ji})_{p+1,s_i}\} \end{matrix} \right. \right] \right]$$

$$= \frac{\prod_{j=1}^{p} \Gamma(b_j + f_j \xi) \prod_{j=1}^{q} \Gamma(1 - a_j - e_j \xi)}{\sum_{i=1}^{r} \left\{ \prod_{j=p+1}^{s_i} \Gamma(1 - b_{ji} - f_{ji}\xi) \prod_{j=q+1}^{t_i} \Gamma(a_{ji} + e_{ji}\xi) \right\}} \tag{5.31}$$

The Mellin transform of a function $f(x)$ is defined as

$$F(\xi) = M[f(u)] = \int_0^\infty u^{\xi-1} f(u)\, du \tag{5.32}$$

The Parseval's theorem of the transform is

$$\int_0^\infty g(ux) f(u)\, du = \lim_{T\to\infty} \frac{1}{2\pi i} \int_{\sigma_0 - iT}^{\sigma_0 + iT} x^{-\xi} g(\xi) f(1-\xi)\, d\xi \tag{5.33}$$

From Equation (5.31) and (5.33) we obtain

$$\frac{1}{2\pi i} \int_L \frac{\prod_{j=1}^{p} \Gamma b_j + f_j \xi) \prod_{j=1}^{q} \Gamma 1 - a_j - e_j \xi)}{\sum_{i=1}^{r} \left\{ \prod_{p+1}^{s_i} \Gamma(1 - b_j - f_j \xi) \prod_{q+1}^{t_i} \Gamma(a_j + e_j \xi) \right.} x^{-\xi} F(1-\xi)\, d\xi \tag{5.34}$$

$$- \; g(x)$$

where $F(\xi) = M[f(u)]$

Now we shall use following definition of Beta function

$$\int_x^\infty \left(v^{1/c} - x^{1/c}\right)^{b-\beta-1} v^{\frac{1}{c} - \frac{b}{c} - s - 1}\, dv = \frac{c\, \Gamma(b-\beta)\Gamma(\beta + cs)}{\Gamma(b+cs)}\; x^{-\frac{\beta}{c} - s} \tag{5.35}$$

Replace x with v in (5.35) and multiply by

$$\left(v^{\frac{1}{f_1}} - x^{\frac{1}{f_1}} \right)^{b_1 - \mu_1 - 1} v^{\frac{1}{f_1} - \frac{b_1}{f_1} - 1},$$

and integrate under the integral sign with respect to v from x to ∞ where $x > 1$ we obtain

$$\int_x^\infty \left(v^{\frac{1}{f_1}} - x^{\frac{1}{f_1}} \right)^{b_1-\mu_1-1} v^{\frac{1}{f_1}-\frac{b_1}{f_1}-1} dv$$

$$= \frac{f_1\, \Gamma(b_1-\mu_1)\, \Gamma(\mu_1+f_1\xi)}{\Gamma(b_1+f_1\xi)} x^{-\frac{\mu_1}{f_1}}$$

From Equation (5.34) we have

$$\frac{1}{2\pi i}\int_L \frac{\prod_{j=2}^{P}\Gamma(b_j+f_j\xi)\,\Gamma(b_1+f_1\xi)\,\Gamma(\mu_1+f_1\xi)\prod_{j=1}^{q}\Gamma(1-a_j-e_j\xi)}{\sum_{i=1}^{r}\left\{\prod_{j=p+1}^{s_i}\Gamma(1-b_{ji}-f_{ji}\xi)\prod_{j=q+1}^{t_i}\Gamma(a_{ji}+e_{ji}\xi)\right\}}$$

$$\times \frac{1}{\Gamma(b_1+f_1\xi)} x^{-\xi} f(1-\xi)\, d\xi \tag{5.36}$$

$$= \frac{1}{f_1\,\Gamma(b_1-\mu_1)} x^{\frac{\mu_1}{f_1}} \int_x^\infty (v^{\frac{1}{f_1}} - x^{\frac{1}{f_1}})^{b_1-\mu_1-1} v^{\frac{1}{f_1}-\frac{b_1}{f_1}-1} g(v)\, dv$$

Let us denote right-hand side of above equations by $g_1(x)$, Then it is evident that

$$g_1(x) = \Re\left[b_1-\mu_1, \frac{\mu_1}{f_1} : \frac{1}{f_1} \right] g(x) \tag{5.37}$$

where $\Re$ is second operator of fractional integration defined in (5.35) and (1.26) again, if we apply the operator, we get

$$\frac{1}{2\pi i}\int_L \frac{\prod_{j=3}^{p}\Gamma(b_j+f_j\xi)\,\Gamma(\mu_1+f_1\xi)\,\Gamma(\mu_2+f_2\xi)\prod_{j=1}^{q}\Gamma(1-a_j-e_j\xi)}{\sum_{i=1}^{r}\left[\prod_{p+1}^{s_i}\Gamma(1-b_{ji}-f_{ji}\xi)\prod_{q+1}^{t_i}\Gamma(a_{ji}+e_{ji}\xi)\right]} x^{-\xi} F(1-\xi) d\xi$$

$$= g_2(x) \tag{5.38}$$

where

$$g_2(x) = \Re\left[b_2-\mu_2, \frac{\mu_2}{f_2} : \frac{1}{f_2} \right] g_1(x) \tag{5.39}$$

In a similar manner, we have

$$\frac{1}{2\pi i}\int_L \frac{\prod_{j=3}^{P}\Gamma(b_j+f_j\xi)\,\Gamma(\mu_1+f_1\xi)\,\Gamma(\mu_2+f_2\xi)\prod_{j=1}^{q}\Gamma(1-a_j-e_j\xi)}{\sum_{i=1}^{r}\left[\prod_{p+1}^{s_i}\Gamma(1-b_{ji}-f_{ji}\xi)\prod_{q+1}^{t_i-1}\Gamma(a_{ji}+e_{ji}\xi)\right]\Gamma(v_L+e_L\xi)} \tag{5.40}$$

$$\times\, x^{-\xi}F(1-\xi)d\xi \;=\; g_3(x)$$

$$g_3(x)=\left[v_L-a_L,\;\frac{a_L}{e_L}\;:\;\frac{1}{e_L}\right]g_2(x)$$

Here, subscripts, $L = tr$.

Continuing in this manner, and making use of Perseval's theorem [38], we obtain from (5.36), (5.38), and (5.40) respectively, that

$$g_1(x)=\int_0^\infty I\left[ux\left|\begin{matrix}\{(1-a_j,e_j)_{1,q}\}:\{(a_{ji},e_{ji})_{q+1,t_i}\}(v_L,e_L)\\ \{(\mu_1,f_1)\,(b_j,f_j)_{2,p}\}\{(1-b_{ji},f_{ji})_{p+1,s_i}\}\end{matrix}\right.\right]f(u)\,du \tag{5.41}$$

$$g_2(x)=\int_0^\infty I\left[ux\left|\begin{matrix}\{(1-a_j,e_j)_{1q}\}:\{(a_{ji},e_{ji})_{q+1,t_i})\}(v_L,e_L)\\ \{(\mu_1,f_1),(\mu_2,f_2)\},\{b_j,f_j)_{3,P}\}\{(1-b_{ji},f_{ji})_{p+1,s_i}\}\end{matrix}\right.\right]f(u)\,du \tag{5.42}$$

$$g_3(x)=\int_0^\infty I\left[ux\left|\begin{matrix}\{(1-a_j,e_j)_{1,q}\},\{(a_{ji},e_{ji})_{q+1,t_i-1})(v_L,e_L)\\ \{(\mu_1,f_1)\{(\mu_2,f_2)\},\{b_j,f_j)_{3,P}\}\{(1-b_{ji},f_{ji})_{p+1,s_i}\}\end{matrix}\right.\right]f(u)du \tag{5.43}$$

where the functions $g_1(x)$, $g_2(x)$ and $g_3(x)$ are given in (5.37), (5.39), and (5.40)

Now, we split the equation either of the form (5.41), (5.42), and (5.43), or in any suitable form, into the pair of known integral equations. We have already indicated that the choice of the form depends on the type and number of parameters in (5.1). The criteria for the selection will be clearer when we arrive at the final solution.

First we assure that any one of f_j $(j = 1, 2 \ldots p)$ is equal to unity or reducible to unity, say $f_1 = 1$. In this case, we consider the integral Equation (5.41) may be written in the following form where $\mu_1 = 0$

$$\int_o^\infty I\left[ux\left|\begin{matrix}\{(1-a_j,e_j)_{1,q}\}:\{(a_{ji},e_{ji})_{q+1,t_i}\}\\\{(0,1)\},\{(b_j,f_j)_{2,p}\}\{1-b_{ji},f_{ji})_{p+1,s_i}\}\end{matrix}\right.\right]f(u)du=g_1(x) \tag{5.44}$$

Now we use the following known results [39]. After applying the Laplace transform on I-function used as a kernel function, with the help of known results in the form of H-function

$$\int_o^\infty e^{-us}H\left[\frac{u}{z}\left|\begin{matrix}\{(1-A_j,E'_j)_{1,Q}\}:\{(A_j,E_j)_{Q+1,T}\}\\\{(B_j,F_j)_{1,P}\},\{(1-B_j,F_j)_{P+1,S}\}\end{matrix}\right.\right]du$$

$$\int_o^\infty e^{-us}H\left[\frac{u}{z}\left|\begin{matrix}\{(1-A_j,E_j)_{1,Q}\}:\{(A_j,E_j)_{Q+1,T}\}\\\{(B_j,F_j)_{1,P}\},\{(1-B_j,F_j)_{P+1,S}\}\end{matrix}\right.\right]du$$

$$=\frac{1}{s}H\left[\frac{1}{zs}\left|\begin{matrix}(1,1)\{(1-A_j,E_j)_{1,Q}\}:\{(A_j,E_j)_{Q+1,T}\}\\\{(B_j,F_j)_{1,P}\},\{(1-B_j,F_j)_{P+1,S}\}\end{matrix}\right.\right]$$

Then we get the following results in terms of I-function.

$$\int_o^\infty e^{-us}I\left[\frac{u}{z}\left|\begin{matrix}\{(1-A_j,E_j)_{1,Q}\}:\{(A_{ji},E_{ji})_{Q+1,T_i}\}\\\{(B_j,F_j)_{1,P}\},\{(1-B_{ji},F_{ji})_{P+1,Si}\}\end{matrix}\right.\right]du$$

$$=\frac{1}{s}I\left[\frac{1}{zs}\left|\begin{matrix}(1,1)\{(1-A_j,E_j)_{1,Q}\}:\{(A_{ji},E_{ji})_{Q+1,T_i}\}\\\{(B_j,F_j)_{1,P}\},\{(1-B_{ji},F_{ji})_{P+1,S_i}\}\end{matrix}\right.\right] \tag{5.45}$$

Re $(\xi)>0$, Re$(1+\min B_j/F_j)>0$, $(j=1,2,\ldots..P)$,

The I-function converges absolutely in the ξ -plane if

i. $\lambda>0$, $|\arg z|<\pi\frac{\lambda}{2}$

ii. $\lambda\geq 0$, $|\arg z|\geq\pi\frac{\lambda}{2}$, Re $(\mu+1)<0$

$$\lambda=\sum_{j=1}^{m}\alpha_j+\sum_{j=1}^{n}\beta_j-\max_{1\leq i\leq r}\left[\sum_{j=n+1}^{pi}\alpha_{ji}+\sum_{j=m+1}^{qi}\beta_{ji}\right] \tag{5.46}$$

$$\mu = \sum_{j=1}^{m}(b_j - \beta_j\sigma) - \sum_{j=1}^{n}(a_j - \alpha_j\sigma) - \min_{1 \le i \le r}\left[\sum_{j=n+1}^{p_i}(a_j - \alpha_j\sigma) - \sum_{j=m+1}^{p_i}(b_j - \beta_j\sigma) + \frac{p_i}{2}, \frac{q_i}{2}\right] \tag{5.47}$$

If $\sigma = 0$, then the set of above expressions (5.46) and (5.47) reduce to

$$\lambda = \sum_{j=1}^{m}\alpha_j + \sum_{j=1}^{n}\beta_j - \max_{1 \le i \le r}\left[\sum_{j=n+1}^{p_i}\alpha_{ji} + \sum_{j=m+1}^{q_i}\beta_{ji}\right]$$

$$\mu = \sum_{j=1}^{m}(b_j) - \sum_{j=1}^{n}(a_j) - \min_{1 \le i \le r}\left[\sum_{j=n+1}^{p_i}(a_j) - \sum_{j=m+1}^{q_i}(b_j) + \frac{p_i}{2}, \frac{q_i}{2}\right]$$

But by using the formulae [40]

$$I\left[x \left| \begin{matrix} \{(1-A_j, E_j)_{1,Q}\}, \{(A_{ji}, E_{ji})_{Q+1,T_i}\} \\ \{(B_j, F_j)_{1,P}\}, \{(1-B_{ji}, F_{ji})_{P+1,S_i}\} \end{matrix} \right.\right] = I\left[\frac{1}{x} \left| \begin{matrix} \{(B_j, F_j)_{1,P})\}, \{(1-B_{ji}, F_{ji})_{P+1,S_i}\} \\ \{(1-A_j, E_j)_{1,Q}\}, \{(A_{ji}, E_{ji})_{Q+1,T_i}\} \end{matrix} \right.\right] \tag{5.48}$$

and

$$x^{\rho} I\left[x \left| \begin{matrix} \{(1-A_j, E_j)_{1,Q}\}, \{(A_{ji}, E_{ji})_{Q+1,T_i}\} \\ \{(B_j, F_j)_{1,P}\}, \{(1-B_{ji}, F_{ji})_{P+1,S_i}\} \end{matrix} \right.\right] = I\left[x \left| \begin{matrix} \{(1-A_j-\rho E_j, E_j)_{1,Q}\}: \{(A_{ji}, +\rho E_{ji}, E_{ji})_{Q+1,T_i}\} \\ \{(B_j, +\rho F_j, F_j)_{1,P}\}, \{(1-B_{ji}-\rho F_{ji}, F_{ji})_{P+1,S_i}\} \end{matrix} \right.\right] \tag{5.49}$$

we can transform (5.45) into

$$\int_o^\infty e^{-us}\, I\left[\frac{u}{z}\left|\begin{matrix}\{(1-A_j,E_j)_{1,Q}\},\{(A_{ji},E_{ji})_{Q+1,T_i})\\ \{(B_j,F_j)_{1,P}\},\{1-B_{ji},F_{ji})_{P+1,S_i}\}\end{matrix}\right.\right]du$$

$$= z I\left[zs\left|\begin{matrix}\{(B_j+F_j,F_j)_{1,P})\},\{(1-B_{ji}-F_{ji},F_{ji})_{P+1,S_i}\}\\ (0,1)\{1-A_j-E_j,E_j)_{1,Q}\},\{(A_{ji}+E_{ji},E_{ji})_{Q+1,T_i}\}\end{matrix}\right.\right] \tag{5.50}$$

under the I-function conditions.

Applying Equation (5.50) in (5.44) we get

$$\int_o^\infty f(u)\int_o^\infty e^{-uy}\, I\left[xy\left|\begin{matrix}\{(1-A_j,E_j)_{1Q_1}\},\{(A_{ji},E_j)_{Q_1+1,T_1})\\ \{(B_j,F_j)_{1,P_1}\},\{1-B_{ji},F_{ji})_{P_1+1,S_1}\}\end{matrix}\right.\right]dy\,du = G_1(x) \tag{5.51}$$

$P_1 = q$, $Q_1 = p -=1$, $T_1 = s-1$, $S_1 = t$

$[A_{ji} = 1 - b_{1+ji} - f_{1+ji}, E_j = f_{1+j}]$ (here $j = 1,2,\ldots, T_1$)

$[B_{ji} = 1 - a_{ji} - e_{ji}, F_{ji} = e_{ji}]$ (here $j = 1,2,\ldots, S_1$)

and $G_1(x) = \frac{1}{x} g_1\left(\frac{1}{x}\right)$

Reversing the order of integration in (5.51) we arrive at the following set of integral equations.

$$\int_0^\infty e^{-ux} f(u)\,du = h_1(x) \tag{5.52}$$

$$\int_0^\infty I\left[ux\left|\begin{matrix}\{(1-A_j,E_j)_{1,Q_l}\}:\{(A_{ji},E_{ji})_{Q_l+1,T_{il}}\}\\ \{(B_j\ ,F_j)_{1,P_l}\},\{(1-B_{ji},F_{ji})_{P_l+1,S_{il}}\}\end{matrix}\right.\right]h_1(u)\,du = G_1(x) \tag{5.53}$$

Here, $h_1(x)$ is Laplace transform of $f(u)$, Therefore the inversion formula will be.

$$f(u) = \lim_{T\to\infty}\frac{1}{2\pi i}\int_{\sigma_0-iT}^{\sigma_0+iT} e^{uz}\, h_l(z)\,dz \tag{5.54}$$

where $\sigma_o = \text{Re}(z)$

Hence, if the integral equation (5.52) is assumed to have known inversion formulae, then $h_l(x)$ will be obtained in terms of prescribed function $G_1(x)$. Consequently, the unknown function $f(x)$ will be determined with the help of (5.54). Actually the selection of the form (5.41) implies that the resulting integral equation (5.55) is standard one.

Accordingly, we have the following theorem:

Theorem 5.2

If $f(x)$ and $h_l(x)$ are solution of the integral equations.

$$g(x)=\int_o^{\infty} I\left[ux\left|\begin{matrix}\{(1-a_j,e_j)_{1,q}\},\{(a_{ji},e_{ji})_{q+1,t_i}\}\\ \{(b_1,1)\},\{(b_j,f_j)_{2,p}\},\{1-b_{ji},f_{ji})_{p+1,s_i}\end{matrix}\right.\right] f(u)\,du \qquad (5.55)$$

$$G_l(x)=\int_0^{\infty} I\left[ux\left|\begin{matrix}\{(1-A_j,E_j)_{1,Q_l}\}:\{(A_{ji},E_{ji})_{Q_l+1,T_{il}}\}\\ \{(B_j,F_j)_{1,P_l}\},\{(1-B_{ji},F_{ji})_{P_l+1,S_{il}}\}\end{matrix}\right.\right] h_l(u)du \qquad (5.56)$$

then,

$$f(u) = \lim_{T\to\infty} \frac{1}{2\pi i}\int_{\sigma_0-iT}^{\sigma_0+iT} e^{zx}\, h_l(z)\,dz \qquad (5.57)$$

where

$$G_1(x)=\frac{1}{x}R\left[b_1,0;1\right] g\left(\frac{1}{x}\right) \qquad (5.58)$$

Theorem 5.3

If $f(x)$ and $h(x)$ are solution of the integral equations.

$$g_3(x)=\int_0^{\infty} I\left[ux\left|\begin{matrix}\{(1-a_j,e_j);\{a_{ji},e_{ji})_{q+1,s_i})(a_{si},1)\}\\ (\mu_1,f_1)(\mu_2,f_2);\{(b_j,f_{ji})_{3,p}(1-b_{ji},f_{ji})_{p+1,t_i}\}\end{matrix}\right.\right] f(u)\,du$$

then,

$$g_3(x)=\int_0^{\infty} I\left[ux\left|\begin{matrix}\{(1-a_j,e_j);\{a_{ji},e_{ji})_{q+1,s_i})(a_{si},1)\}\\ (b_1,1)(b_2,1);\{(b_j,f_{ji})_{3,p}(1-b_{ji},f_{ji})_{p+1,t_i}\}\end{matrix}\right.\right] f(u)\,du. \qquad (5.59)$$

and use $\mu_1 = (-k-m)$, $\mu_2 = (-k+m)$, $v_L = (-2k)$

$$g_3(x) = \int_0^\infty I\left[ux\left|\begin{matrix}\{(1-a_j,e_j)_{1,q};\ \{a_{ji},e_{ji})_{q+1,s_i})(-2k,1)\}\\ (-k-m,1)(-k+m,1);\{(b_j,f_j)_{3,p}(1-b_{ji},\ f_{ji})_{p+1,t_i}\}\end{matrix}\right.\right] \tag{5.60}$$

$\times f(u)\, du$

5.2 I-FUNCTION AND WHITTAKER FUNCTION-BASED SOLUTION

We shall use the following known result [39]

$$\int_0^\infty (us)^{(-k-\frac{1}{2})} e^{(-\frac{1}{2}us)} W_{k+\frac{1}{2},m}(us) \times I\left[\frac{u}{z}\left|\begin{matrix}\{(1-A_j,E_j)_{1,Q}\};\{A_{ji},E_{ji})_{Q+1,S_i}\}\\ (B_j,F_j)_{,1,P}\};\{1-B_{ji},F_{ji})_{P+1,T_i}\}\end{matrix}\right.\right] du \tag{5.61}$$

$$= \frac{1}{s} I\left[\frac{1}{zs}\left|\begin{matrix}(1-k-m,1)\ (1-k+m,1),\{(1-A_j,E_j)_{1,Q}\},\{(A_{ji},E_{ji})_{Q+1,T_i}\}\\ \{(B_j,F_j)_{1,P}\},\{(1-B_{ji}+F_{ji})_{P+1,S_i,}\},(1-2k,1)\end{matrix}\right.\right] \tag{5.62}$$

provided Re(s) > 0, Re($-k \pm m + 1 + \min B_j/F_j$) > 0, ($j$ = 1,2,…,P) and following set of conditions are assumed to be satisfied.

i. $\lambda > 0, |\arg 1/z| < (1/2)\lambda\pi$
ii. $\lambda \geq 0, |\arg 1/z| < (1/2)\lambda\pi$, Re $(\mu + 1) < 0$,

where λ and μ are given in Equation (2.19) and (2.20)

With the help of (5.49), this integral may be written as

$$\int_0^\infty (us)^{-k-\frac{1}{2}} e^{-\frac{1}{2}us} W_{k+\frac{1}{2},m}(us) \times I\left[\frac{u}{z}\left|\begin{matrix}\{(1-A_j,E_j)_{1,Q}\};\ \{A_{ji},E_{ji})_{Q+1,S_i}\}\\ (B_j,F_j)_{,1,P}\};\ \{1-B_{ji},F_{ji})_{P+1,T_i}\}\end{matrix}\right.\right] du \tag{5.63}$$

$$= zI\left[zs\left|\begin{matrix}\{(B_j+F_j,F_j)_{1,P}\},\{(1-B_{ji}-F_{ji},F_{ji})_{P+1,T_i,}\},(2k,1)\\ (-k-m,1)\ (-k+m,1),\{(1-A_j-E_j,E_j)_{1,Q}\},\{(A_{ji}+E_{ji},E_{ji})_{Q+1,S_i}\end{matrix}\right.\right] \tag{5.64}$$

Making use of the result (5.60), we get

$$\int_0^\infty f(u) \int_0^\infty (uy)^{-k-\frac{1}{2}} e^{-\frac{1}{2}uy} W_{k+\frac{1}{2},m}(uy)\, h_3(u)\, du$$

$$\times\ I\left[xy \middle| \begin{matrix} \{(1-A_j, E_j)_{1,Q_3}\}; \{A_{ji}, E_{ji}): _{Q_3+1}, {}_{S_3}\} \\ \{(B_j, F_j)_{1,P_3}\}; \{1-B_{ji}, F_{ji}): _{P_3+1}, {}_{T_3}\} \end{matrix}\right] dy \tag{5.65}$$

where $P_3 = q$, $T_3 = s - 1$, $Q_3 = p - 2$, $S_3 = t - 2$;
$[A_{ji} = 1 - b_{2+j} - f_{2+j}, E_{ji} = f_{2+j},], (j = 1, 2,\ldots, S_{3i})$
$[B_{ji} = 1 - a_j - e_j, F_{ji} = e_j,],(j = 1, 2,\ldots, T_{3i})$
and $G_3(x) = (1/x)\, g_3\,(1/x)$.

Finally, reversing the order of the integration (5.65) and separating the u and y we get

$$\int_0^\infty (ux)^{-k-\frac{1}{2}} e^{-\frac{1}{2}ux} W_{k+\frac{1}{2},m}(ux)\, f(u)\, du = h_3(x)$$

$$\int_0^\infty I\left[ux \middle| \begin{matrix} \{(1-A_j, E_j)_{1,Q_3}\}; \{A_{ji}, E_{ji}): _{Q_3+1}, {}_{S_3}\} \\ \{(B_j, F_j)_{1,P_3}\}; \{1-B_{ji}, F_{ji}): _{P_3+1}, {}_{T_3}\} \end{matrix}\right] h(u)\, du \tag{5.66}$$

$$= G_3(x)$$

$h_3(x)$ is Meijer's Whittaker function transform of $f(u)$, we use the formula of Whittaker function, which is defined below.

$$h(u) = \lim_{T\to\infty} \frac{1}{2\pi i} \int_{c-iT}^{c+iT} (zu)^{\frac{1}{2}}\, I_v^{\frac{1}{2}(zu)}(zu)\, M_{k-\frac{1}{2},m}(zu)\, f(z)\, dz$$

$$f(x) = \lim_{T\to\infty} \frac{1}{2\pi i} \int_{c-iT}^{c+iT} (ux)^{\frac{1}{2}}\, e^{\frac{1}{2}ux}\, W_{k+\frac{1}{2},m}(ux)\, h(u)\, du \tag{5.67}$$

$$f(x) = \lim_{T\to\infty} \frac{\Gamma(1-k+m)}{\Gamma(1+2m)} \frac{1}{2\pi i} \int_{\sigma_o-iT}^{\sigma_o+it} (zx)^{1/2}\, e^{\frac{zx}{2}} M_{k-\frac{1}{2},m}(zx)\, h(z)\, dz \tag{5.68}$$

Hence, we have the following theorem;

b. *if*, $\sum_{Ql}^{Pl} B_j - \sum A_{ji}^{Sil} < \sum_{Pl+l}^{Til}(1-B_{ji}) - \sum^{Ql}(1-A_j)$

$\zeta_j\,(j=Q_{l-n+1},Q_{l-n+2},\ldots\ldots,Q_l)$ are constants, as

$$\mathrm{Re}(\zeta_j - E_j/2) > E_j/2, \quad (j=Q_{l-n+1},Q_{l-n+2},\ldots\ldots,Q_l)$$

$$\sum^{Pl} B_j - \sum_{Ql+1}^{Sil} A_{ji} = \sum_{Pl+1}^{Til}(1-B_{ji}) - \sum^{Ql-n}(1-A_j) \sum_{Ql-n+1}^{Q_l}(\zeta_j)$$

then $h_l(x)$ is given by

$$\frac{1}{2}[h_l(x+0)+h_l(x-0)]\int_0^\infty \Im_{Ql-n+1}\Big[\Im_{Ql-n+2}\ldots\ldots\Im_{Ql}\{G_l(u)\ldots\}\Big]$$

$$I\left(ux\left|\begin{matrix}\{A_{ji}+E_{ji},E_{ji})_{Q\,l+1,\;Sil}\};\{(1-A_j-E_j,E_j)_{1,Ql-n}\left(\zeta_j-E_j,E_j\right)_{Ql-n+1\cdot Ql}\}\\ \{(1-B_{ji}-F_{ji},F_{ji})P\,l+1,T_i\};\{(B_j+F_j,F_j)_{1\cdot P_l})\}\end{matrix}\right.\right)du \tag{5.75}$$

where

$$\Im_j w(x) = \Im\,[1-A_j-\zeta_j,\zeta\,E_j^{-1}-1,\;E_j^{-1}]w(x) \tag{5.76}$$

c. If,

$$\sum^{Pl} B_j - \sum_{Ql+1}^{Sil} A_{ji} < \sum_{Pl+1}^{Til}(1-B_{ji}) - \sum^{Ql-n}(1-A_j)$$

$$E_{S_{li}} = S_{li},\;\; A_{S_{li}} = A+n,\; \mathrm{Re}\left(A+\frac{E_{S_{li}}}{2D}\right) > \frac{E_{S_{li}}}{2D}$$

Here n is a positive integer such that

$$\sum^{Pl} B_j - \sum_{Ql+1}^{Sil-1} A_{ji} - A = \sum_{Pl+1}^{Til}(1-B_{ji}) - \sum^{Ql}(1-A_j)$$

$$h_l(x) = \lim_{T\to\infty}\frac{\Gamma(1-B_1-A_1)}{\Gamma(1+B_1-A-n)}\,\frac{1}{2\pi i}\int_{\sigma_0-iT}^{\sigma_0+iT} G(1-\xi)$$

$$\times\left[\sum_{k=0}^{n}{}^{n}c_k\,(-1)^k\,Q_k(\xi)\right]^{-1} d\xi$$

$$Q_k(\xi) = \frac{\prod_{2}^{P_l}\Gamma(B_j+F_j-F_j\xi)\prod^{Q_l}\Gamma(1-A_j-E_j+E_j\xi)}{\prod_{P_l+1}^{S_{li}}\Gamma(1-B_j-F_j+F_j\xi)\prod_{Q_l+1}^{T_{li}}\Gamma(A_j+E_j-E_j\xi)} \tag{5.77}$$

$$\times \frac{\Gamma(B_j+F_j+K-F_j\xi)}{\Gamma(A_j+F_j+K-F_j\xi)}$$

$$G(1-\xi) = M\left[G_l(u)\right] \tag{5.78}$$

Then the kernel (5.73) reduces to the symmetric Fourier kernel [15] so that

$$\tfrac{1}{2}\{h_l(x+0)+h_l(x-0)\}$$

$$= \int_0^\infty I\left[ux\left|\begin{matrix}\{(1-A_j,E_j)_{1,Ql}\};\ \{1-A_{ji}-E_{ji},\ E_{ji})_{1,Ql}\}\\ \{(B_j,F_j)_{1,P_i}\};\ \{B_{ji}+F_{ji},\ F_{ji})_{1,P_l}\}\end{matrix}\right.\right]G(u)\,du \tag{5.79}$$

$$(l=1,2,3)$$

CASE II

Since we know that by the special case of I-function

$$I\left(x\left|\begin{matrix}\{(1-A_j,E_j,)_{1,P}\};\ \{(A_{ji}\ F_{ji})_{P_{l+1},T_{il}}\}\\ \{(B_j,E_j)_{1,Q}\};\ \{(1-B_{ji},E_{ji})_{Q+1,S_{il}}\}\end{matrix}\right.\right)$$

$$= \sigma I\left(x^{\sigma}\left|\begin{matrix}\{(1-A_j,\ \sigma E_j)_{1,Q}\};\ \{(A_{ji},\ \sigma E_{ji})_{Q+1,Si}\}\\ \{(B_j,\ \sigma F_j)_{1,P}\};\ \{(1-B_{ji},\sigma F_{ji})_{P+1,Ti})\}\end{matrix}\right.\right) \tag{5.80}$$

It was described in the form of H-function as,

$$II\left(x\left|\begin{matrix}\{(1-A_j,E_j,)_{1,P}\};\ \{(A_{ji}\ F_{ji})_{P_{l+1},T_{il}}\}\\ \{(B_j,E_j)_{1,Q}\};\ \{(1-B_{ji},E_{ji})_{Q+1,S_{il}}\}\end{matrix}\right.\right)$$

$$= \sigma H\left(x^{\sigma}\left|\begin{matrix}\{(1-A_j,\ \sigma E_j)_{1,Q}\};\ \{(A_{ji},\ \sigma E_{ji})_{Q+1,\ Si}\}\\ \{(B_j,\ \sigma F_j)_{1,P}\};\ \{(1-B_{ji},\sigma F_{ji})_{P+1,Ti})\}\end{matrix}\right.\right) \tag{5.81}$$

and the second result of (5.80), then Equation (5.73) transforms to (if $r = 1$ and $E_{ji} = E_j$, $F_{ji} = F_j$),

$$G_l\left(\frac{x}{2}\right) = z^{1-2B_1}\int_0^\infty (ux)^{2B_1}\, I\left[\frac{u^2x^2}{4}\left|\begin{matrix}(\ldots\ldots\ldots\ldots\ldots)\\ \{(0,1);\ (\tfrac{1}{2}+2B_1,\ 2F_2)\}\end{matrix}\right.\right]h_l(u)\,du \tag{5.82}$$

here, $(l = 1,\ 2,\ 3)$

Again with the help of relation.

$$H\left[x\left|\begin{matrix}(\ldots\ldots\ldots\ldots\ldots\ldots)\\ \{(0,1)\,;\,(1+\lambda,\mu)\end{matrix}\right.\right]=J_{\lambda}^{\mu}(x)$$

where $J_{\lambda}^{\mu}(x)$ is Bessel–Maitland function, defined in (3.25).

Now Equation (5.80) reduces to this form

$$G_l\left(\frac{x}{2}\right)=z^{1-2B_1}\int_0^{\infty}(ux)^{2B_1}\;J_{2B_1-\frac{1}{2}}^{2F_2}\left(\frac{u^2x^2}{4}\right)h_l(u)\,du \tag{5.83}$$

$(l=1,2,3)$

Now, using the inversion formula of generalized Hankel transform [41].

$$h_l(x)=\frac{1}{2F_2}(z)^{(2B_2+\frac{1}{2})(1-\frac{1}{F_2})}\int_0^{\infty}(ux)^{(2B_1+\frac{1}{2})/(F_2-2B_1-1)}$$
$$\times J_{(\frac{1}{2}+2B_1-2F_2)/2F_2}^{\frac{1}{2F_2}}\left(\frac{u^2x^2}{4}\right)^{\frac{1}{2F_2}}G_l\left(\frac{u}{2}\right)du \tag{5.84}$$

PARTICULAR CASE I

We consider the case when $e_j=f_j=e_{ji}=f_{ji}=1$, and $r=1$, so that

$$g(x)=\int_0^{\infty}G_{s,t}^{p,q}\left(ux\left|\begin{matrix}(a_j)_{1,s}\\(b_j)_{1,t}\end{matrix}\right.\right)f(u)\,du \tag{5.85}$$

$$G_l(x)=\int_0^{\infty}G_{S_l,T_l}^{P_l,Q_l}\left(ux\left|\begin{matrix}(A_j)_{1,S_l}\\(B_j)_{1,T_l}\end{matrix}\right.\right)h_l(u)\,du \tag{5.86}$$

Then

$$f(x)=\lim_{T\to\infty}\frac{1}{2\pi i}\int_{\sigma_o-iT}^{\sigma_o+it}e^{(zx)}h_l(z)\,dz \tag{5.87}$$

$$f(x)=\lim_{T\to\infty}\frac{\Gamma(1-k+m)}{\Gamma(1+2m)}\frac{1}{2\pi i}\int_{\sigma_o-iT}^{\sigma_o+it}(zx)^{1/2}\,e^{\frac{zx}{2}}M_{k-\frac{1}{2},m}(zx)\,h_3(z)\,dz \tag{5.88}$$

and

$$G_l(x)=\frac{1}{x}\,\Re\,[b_1,0;1]\,g\left(\frac{1}{x}\right)$$

$$G_3(x)=\frac{1}{x}\,\Re\,[a_{si}+2k,-2k{:}1]\;\Re\,[b_2+k-m,-k+m:1] \tag{5.89}$$

$$\times\,\Re\,[b_1+k+m,-k-m:1]\;g\left(\frac{1}{x}\right)$$

$P_1=q$, $Q=p-1$, $S_1=t-=1$, $T_1=s$, $P_3=q$, $T_3=s-1$
$Q_3=p-2$, $S_3=t-2$
$A_{ji}=-b1_{+j}\;(j=1,2\ldots S_1)$,
$B_{ji}=-aj\;(j=1,2,\ldots\ldots,T_1)$, $A_{ji}=-ib2_{+ji}\;(j=1,2\ldots\ldots S_3)$
$(i=1,2\ldots\ldots\ldots r=1)$
$\Re$ is the operator of fractional integration are defined in (1.43) and
$(S_1+T_1)<2\,(P_1+Q_1)$
$[P_l+Q_l-\frac{1}{2}(S_l+T_l)\,]\,\pi<(\arg x)\;(l=1,3)$

Similarly, we take $e_j=f_j=e_{ji}=f_{ji}=1/2$, $r=1$,

so that, if

$$g(x)=z\int_0^\infty G_{s,t}^{p,q}\left(u^2x^2\middle|\begin{matrix}(a_j)_{1,s}\\(b_j)_{1,t}\end{matrix}\right)f(u)du \tag{5.90}$$

and

$$G_2(x)=\int_0^\infty G_{s2,t2}^{p2,q2}\left(u^2x^2\middle|\begin{matrix}(A_j)_{1,s2}\\(B_j)_{1,T2}\end{matrix}\right)h_2(u)du \tag{5.91}$$

$$f(x)=\lim_{T\to\infty}\frac{1}{2\pi i}\int_{\sigma_0-iT}^{\sigma_0-iT}(2zx)^{\frac{1}{2}}\,I_V(2zx)\,h_2(z)dz \tag{5.92}$$

$$G_2(x)=\frac{1}{2\sqrt{2}x}\,\Re\left[b_2-\frac{1}{4}+\frac{1}{2}v,\frac{1}{2}-v{:}2\right]$$

$$\times\,\Re\left[b_1-\frac{1}{4}+\frac{1}{2}v,\frac{1}{2}+v{:}2\right]g\left(\frac{1}{x}\right) \tag{5.93}$$

Again, if in Equation (5.90) $p = 2, q = 1, t = 2, s = 2, a_2 = -(1/2)\mu, b_2 = (1/2)\mu$,

$$G_{0,2}^{1,0}\left(\frac{x^2}{2}\middle|\frac{1}{2}\mu,-\frac{1}{2}\mu\right)=J_\mu(x) \tag{5.94}$$

Then the solution of the integral equation

$$g(x)=\int_0^\infty G_{2,2}^{2,1}\left(u^2x^2\middle|\begin{matrix}-\frac{1}{2}\mu,\frac{1}{2}\mu\\ b_1,b_2\end{matrix}\right)f(u)\,du=J_\mu(x) \tag{5.95}$$

which is given by

$$f(x)=\lim_{T\to\infty}\frac{1}{2\pi i}\int_{\sigma_0-iT}^{\sigma_0-iT}(2zx)^{\frac{1}{2}}\,I_v(2zx)\,h_2(z)\,dz \tag{5.96}$$

$$h_2(z)=\int_0^\infty\frac{1}{2\sqrt{2}\,x}\,\Re\left[b_2-\frac{1}{4}+\frac{1}{2}v,\frac{1}{2}-v:2\right]\Re\left[b_2-\frac{1}{4}+\frac{1}{2}v,\frac{1}{2}-v:2\right] \tag{5.97}$$

Then we have made use, in the above deduction of Hankel inversion formula.

If we take $b_1 = (1/2)\nu$, $b_2 = (-1/2)\mu$, in Equation (5.95) then, $G_2(x) = [1/2\surd(2)\,x]\,g(1/x)$ above set consists of Equations (5.95), (5.96), and (5.97), transform to the following set of equations.

$$h_2(x)=\int_0^\infty(ux)^{\frac{1}{2}}\,K_v(ux)\,f(u)\,du \tag{5.98}$$

$$G(x)=\lim_{T\to\infty}\int_0^\infty(ux)^{\frac{1}{2}}\,J_\mu(ux)\,h_2(u)\,du \tag{5.99}$$

$$\frac{1}{x}G\left(\frac{1}{x}\right)=\int_0^\infty G_{2,2}^{2,1}\left(u^2x^2\middle|\begin{matrix}-\frac{1}{2}v,\frac{1}{2}v\\ \frac{1}{2}\mu,-\frac{1}{2}\mu\end{matrix}\right)f(u)\,du \tag{5.100}$$

This solution may also be obtained by using the Goyal [42] theorem.

5.3 DUAL INTEGRAL EQUATIONS

The G- and H-functions have been extensively used as the kernels in the theory of dual integral equations. In fact, the use of these functions in dual integral equations is based on the results and inversion formulas of ordinary integral transform theory. The earlier dual integral equations employed in various applications were given by Sneddon [43].

An elegant generalization of the dual integral equation is given by Fox [24] is

$$\int_0^\infty H\left[ux \left| \begin{matrix} \alpha_j, a_j \\ \beta_j, b_j \end{matrix} \right. :n\right] h(u)du = f(x), \quad 0 < x < 1 \tag{5.101}$$

$$\int_0^\infty H\left[ux \left| \begin{matrix} \lambda_j, a_j \\ \mu_j, b_j \end{matrix} \right. :n\right] h(u)du = g(x), \quad x > 1 \tag{5.102}$$

The H-function used here as kernel is defined as

$$H\left[x \left| \begin{matrix} \alpha_j, a_j \\ \beta_j, b_j \end{matrix} \right. :n\right] = \frac{1}{2\pi i}\int_L \prod_{j=1}^{n} \frac{\Gamma(\alpha_j + a_j\xi)}{\Gamma(\beta_j - b_j\xi)} x^{-\xi}\, d\xi \tag{5.103}$$

where the contour L is of Barnes type.

In particular

$$H\left[x \left| \begin{matrix} (\alpha_j, \frac{1}{2}) \\ (\beta_j, \frac{1}{2}) \end{matrix} \right. :1\right] = 2x^{\alpha+\beta+1} J_{\alpha+\beta+1}(x) \tag{5.104}$$

The set (5.101) and (5.102) have been further generalized by Saxena [25].

The following dual integral equations involving two H-functions as kernels have also been studied and solved by Saxena [31]

$$\int_0^\infty [H_1(ux) + H_2(ux)]a(u)\, du = E(x), \quad 0 < x < 1 \tag{5.105}$$

$$\int_0^\infty [L_1(ux) + L_2(ux)]a(u)\, du = G(x), \quad 0 < x < 1 \tag{5.106}$$

The functions $E(x)$ and $G(x)$ are prescribed and $a(x)$ is to be determined. $H_1(x)$, $H_2(x)$, $L_1(x)$, $L_2(x)$ are abbreviated forms of the H-functions:

$$H_1(ux) \equiv H_{p,q}^{m,n}\left[ux \left| \begin{matrix} \{(a_j, \alpha_j)_{1,m}\}; \{(b_j, \beta_j)_{1,p}\} \\ \{(c_j, \gamma_j)_{1,n}\}\ ; \{(d_j, \delta_j)_{1,q}\} \end{matrix} \right.\right] \tag{5.107}$$

$$H_2(ux) \equiv H_{p,q}^{m,n}\left[ux \left| \begin{matrix} \{(a_j, \alpha_j)_{1,m}\}; \{(f_j, \beta_j)_{1,p}\} \\ \{(g_j, \gamma_j)_{1,n}\}\ ; \{(d_j, \delta_j)_{1,q}\} \end{matrix} \right.\right] \tag{5.108}$$

$$L_1(ux) \equiv H^{m,n}_{p,q}\left[ux \left| \begin{matrix} \{(e_j,\alpha_j)_{1,m}\};\{(h_j,\beta_j)_{1,p}\} \\ \{(i_j,\gamma_j)_{1,n}\}\ ;\{(k_j,\delta_j)_{1,q}\} \end{matrix} \right.\right] \tag{5.109}$$

$$L_2(ux) \equiv H^{m,n}_{p,q}\left[ux \left| \begin{matrix} \{(l_j,\alpha_j)_{1,m}\};\{(h_j,\beta_j)_{1,p}\} \\ \{(i_j,\gamma_j)_{1,n}\}\ ;\{(w_j,\delta_j)_{1,q}\} \end{matrix} \right.\right] \tag{5.110}$$

5.3.1 Dual Integral Equations Involving Sum of Two I-Functions as Kernels

In this chapter, we will deal with dual integral equations, having linear sum of I-functions as its kernels. First we split these equations into four different pairs of ordinary dual integral equations. Then, using the property of fractional integral operators and Mellin transform, all of these pairs have been reduced to a single integral equation, having the sum of I-functions as a kernel. Finally, we obtain inversion of this single integral equation and hence, we arrive at the desired solution of dual integral equations given by Jat and Saxena [44].

Here we consider the pair

$$\int_0^\infty [I_1(ux)+I_2(ux)]a(u)\,du = E(x), \quad 0<x<1 \tag{5.111}$$

$$\int_0^\infty [I_3(ux)+I_4(ux)]a(u)\,du = G(x), \quad x>1 \tag{5.112}$$

where I_i $(i = 1,2,3,4)$ are non-identical and abbreviated forms of the functions

$$I_1(ux) \equiv I^{m,n}_{p_{ij},q_{ij}}\left[ux \left| \begin{matrix} \{(a_j,\alpha_j)_{1,m}\};\{(b_{ji},\beta_{ji})_{1,p_i}\} \\ \{(c_j,\gamma_j)_{1,n}\}\ ;\{(d_{ji},\delta_{ji})_{1,q_i}\} \end{matrix} \right.\right] \tag{5.113}$$

$$I_2(ux) \equiv I^{m,n}_{p_{ij},q_{ij}}\left[ux \left| \begin{matrix} \{(e_j,\alpha_j)_{1,m}\};\{(b_{ji},\beta_{ji})_{1,p_i}\} \\ \{(f_j,\gamma_j)_{1,n}\}\ ;\{(d_{ji},\delta_{ji})_{1,q_i}\} \end{matrix} \right.\right] \tag{5.114}$$

$$I_3(ux) \equiv I^{m,n}_{p_{ij},q_{ij}}\left[ux \left| \begin{matrix} \{(g_j,\alpha_j)_{1,m}\};\{(b_{ji},\beta_{ji})_{1,p_i}\} \\ \{(h_j,\gamma_j)_{1,n}\}\ ;\{(d_{ji},\delta_{ji})_{1,q_i}\} \end{matrix} \right.\right] \tag{5.115}$$

$$I_4(ux) \equiv I^{m,n}_{p_{ij},q_{ij}}\left[ux \left| \begin{matrix} \{(k_j,\alpha_j)_{1,m}\};\{(b_{ji},\beta_{ji})_{1,p_i}\} \\ \{(l_j,\gamma_j)_{1,n}\}\ ;\{(d_{ji},\delta_{ji})_{1,q_i}\} \end{matrix} \right.\right] \tag{5.116}$$

respectively.

5.3.2 Solution

To solve the pair of dual integral Equations (5.111) and (5.112), we take

$$E_1(x) = \int_0^\infty I_1(ux)a(u)\,du \quad 0 < x < 1 \tag{5.117}$$

$$E_2(x) = \int_0^\infty I_2(ux)a(u)\,du \quad 0 < x < 1 \tag{5.118}$$

$$G_1(x) = \int_0^\infty I_3(ux)a(u)\,du \quad x > 1 \tag{5.119}$$

$$G_2(x) = \int_0^\infty I_4(ux)a(u)\,du \quad x > 1 \tag{5.120}$$

where

$$E(x) = E_1(x) + E_2(x) \tag{5.121}$$

$$G(x) = G_1(x) + G_2(x) \tag{5.122}$$

In terms of these functions, the dual integral Equations (5.111) and (5.112) can be written in the form of the following four sets.

$$E_1(x) = \int_0^\infty I_1(ux)a(u)du, \quad 0 < x < 1 \tag{5.123}$$

$$G_1(x) = \int_0^\infty I_3(ux)a(u)du, \quad x > 1 \tag{5.124}$$

$$E_2(x) = \int_0^\infty I_2(ux)a(u)du, \quad 0 < x < 1 \tag{5.125}$$

$$G_1(x) = \int_0^\infty I_3(ux)a(u)du, \quad x > 1 \tag{5.126}$$

$$E_1(x) = \int_0^\infty I_1(ux)a(u)du, \quad 0 < x < 1 \tag{5.127}$$

$$G_2(x)=\int_0^\infty I_4(ux)a(u)du, \quad x>1 \tag{5.128}$$

$$E_2(x)=\int_0^\infty I_2(ux)a(u)du, \quad 0<x<1 \tag{5.129}$$

$$G_2(x)=\int_0^\infty I_4(ux)a(u)du, \quad x>1 \tag{5.130}$$

The Mellin transform of I-function is given below

$$M\left[I\left[u\left|\begin{matrix}\{1-a_j,e_j)_{1,q}\};\ \{(a_{ji},e_{ji})_{q+1,t_i}\}\\ \{b_j,f_j)_{1,p}\};\ \{1-b_{ji},f_{ji})_{p+1,s_i}\}\end{matrix}\right.\right]\right]$$

$$=\frac{\prod_{j=1}^{p}\Gamma(b_j+f_j\xi)\prod_{j=1}^{q}\Gamma(1-a_j-e_j\xi)}{\sum_{i=1}^{r}\left\{\prod_{j=p+1}^{s_i}\Gamma(1-b_{ji}-f_{ji}\xi)\prod_{j=q+1}^{t_i}\Gamma(a_{ji}+e_{ji}\xi)\right\}} \tag{5.131}$$

The Mellin transforms of the functions $f(x)$ and $g(x)$ are defined as

$$F(\xi)=M[f(u)]=\int_0^\infty u^{\xi-1}f(u)du$$

$$G(\xi)=M[g(u)]=\int_0^\infty u^{\xi-1}g(u)du \tag{5.132}$$

The Parseval's theorem of the transforms is given as

$$\int_0^\infty g(ux)f(u)du=\lim_{T\to\infty}\frac{1}{2\pi i}\int_{\sigma_0-iT}^{\sigma_0+iT}x^{-\xi}G(\xi)F(1-\xi)d\xi \tag{5.133}$$

Using this theorem, the pair (5.134) and (5.135) is reduced to

$$\frac{1}{2\pi i}\int_L\frac{\prod_{j=1}^{n}\Gamma(c_j+\gamma_j\xi)\prod_{j=1}^{m}\Gamma(a_j-\alpha_j\xi)}{\sum_{i=1}^{r}\left[\prod_{j=1}^{q_i}\Gamma(d_{ji}-\delta_{ji}\xi)\prod_{j=1}^{p_i}\Gamma(b_{ji}+\beta_{ji}\xi)\right]}x^{-\xi}A(1-\xi)d\xi \tag{5.134}$$

$$=E_1(x),\quad 0<x<1$$

$$\frac{1}{2\pi i}\int_L \frac{\prod_{j=1}^{n}\Gamma\left(h_j+\gamma_j\xi\right)\prod_{j=1}^{m}\Gamma\left(g_j-\alpha_j\xi\right)}{\sum_{i=1}^{r}\left[\prod_{j=1}^{q_i}\Gamma\left(d_{ji}-\delta_{ji}\xi\right)\prod_{j=1}^{p_i}\Gamma\left(b_{ji}+\beta_{ji}\xi\right)\right]}x^{-\xi}A(1-\xi)\,d\xi \tag{5.135}$$

$$= G_1(x), \quad x>1$$

$$A(\xi) = M[a(u)]$$

In (5.134) replace x by v, multiply both the sides by

$$\left(x^{\frac{1}{\alpha_p}} - v^{\frac{1}{\alpha_p}}\right)^{a_p-g_p-1} v^{\frac{g_p}{\alpha_p}-1}$$

and integrate, through the integral sign with respect to v from 0 to x where $0 < x < 1$. With the use of

$$\int_0^x \left(v^{1/c}-x^{1/c}\right)^{d-e-1} v^{\frac{e}{c}-s-1}\,dv = c\Gamma(d-e)\frac{\Gamma(e-cs)}{\Gamma(d-cs)}x^{\frac{d}{c}-\frac{1}{c}-s} \tag{5.136}$$

we obtain

$$\frac{1}{2\pi i}\int_L \frac{\prod_{j=1}^{n}\Gamma\left(c_j+\gamma_j\xi\right)\prod_{j=1}^{m-1}\Gamma\left(a_j-\alpha_j\xi\right)\Gamma(g_m-\alpha_m\xi)}{\sum_{i=1}^{r}\left[\prod_{j=1}^{q_i}\Gamma\left(d_{ji}-\delta_{ji}\xi\right)\prod_{j-1}^{p_i}\Gamma\left(b_{ji}+\beta_{ji}\xi\right)\right]}x^{-\xi}A(1-\xi)\,d\xi$$

$$= \frac{1}{\alpha_m\Gamma(a_m-g_m)}\int_0^x (x^{\frac{1}{\alpha_m}} - v^{\frac{1}{\alpha_m}})^{a_m-g_m-1} v^{\frac{g_m}{\alpha_m}-1} E_1(v)\,dv \tag{5.137}$$

$$= \Im_m\{E_1(x)\}$$

where

$$\Im[\gamma,\varepsilon:m]w(x) = \frac{m}{\Gamma\gamma}x^{-\varepsilon-\gamma m+m-1}\int_0^x (x^m - v^m)^{\gamma-1} v^{\varepsilon} w(v)\,dv \tag{5.138}$$

In terms of the definition of fractional integral operator (5.138), we write

$$\Im_j[w(x)] = \Im[a_j - g_j, \tfrac{g_j}{\alpha_j} - 1 : \tfrac{1}{\alpha_j}]w(x) \tag{5.139}$$

If at the same time step by step, we apply the operator to the right-hand side of (5.137) for

$j = m-1, m-2, \ldots\ldots 2, 1$, we obtain

$$\frac{1}{2\pi i}\int_L \frac{\prod_{j=1}^{n}\Gamma(c_j+\gamma_j\xi)\prod_{j=1}^{m}\Gamma(g_j-\alpha_j\xi)}{\sum_{i=1}^{r}\left[\prod_{j=1}^{q_i}\Gamma(d_{ji}-\delta_{ji}\xi)\prod_{j=1}^{p_i}\Gamma(b_{ji}+\beta_{ji}\xi)\right]} x^{-\xi} A(1-\xi)\,d\xi \tag{5.140}$$

$$=\Im_c[\Im_2....\Im_m\{E_1(x)\}],0<x<1$$

Next we consider (5.135), replace x with v, multiply by

$$\left(x^{\frac{1}{\gamma n}}-v^{\frac{1}{\gamma n}}\right)^{hn-cn-1} v^{\frac{hn}{\gamma n}-1} \tag{5.141}$$

and integrate with respect to v using the definition of fractional integral operator.

$$\int_x^{\infty}\left(v^{1/c}-x^{1/c}\right)^{d-e-1} v^{\frac{1}{c}-\frac{d}{c}-s-1}\,dv = c\Gamma(d-e)\frac{\Gamma(e+cs)}{\Gamma(d+cs)}\,x^{-\frac{e}{c}-s} \tag{5.142}$$

The result is

$$\frac{1}{2\pi i}\int_L \frac{\prod_{j=1}^{n-1}\Gamma(h_j+\gamma_j\xi)\Gamma(c_n+\gamma_n\xi)\prod_{j=1}^{m}\Gamma(g_j-\alpha_j\xi)}{\sum_{i=1}^{r}\left[\prod_{j=1}^{q_i}\Gamma(d_{ji}-\delta_{ji}\xi)\prod_{j=1}^{p_i}\Gamma(b_{ji}+\beta_{ji}\xi)\right]} x^{-\xi} A(1-\xi)\,d\xi$$

$$=\frac{1}{\gamma_n\Gamma(h_n-c_n)}\int_x^{\infty}(v^{\frac{1}{\gamma n}}-x^{\frac{1}{\gamma n}})^{hn-cn-1}\,v^{\frac{1}{\gamma n}-\frac{hn}{\gamma n}-1}\,G_1(v)dv, \qquad x>1 \tag{5.143}$$

$$=\Re_n\{G_1(x)\}$$

Here we use the second operator of the fractional integral, defined in (5.142). In the term of this operator the right-hand side of (5.143) may be represented as

$$\Re_j[w(x)]=\Re\left[h_j-c_j,\frac{c_j}{\gamma_j}-1:\frac{1}{\gamma_j}\right]w(x) \tag{5.144}$$

so that

$$\Re[\gamma,\varepsilon:m]w(x)=\frac{m}{\Gamma\gamma}x^{\varepsilon}\int_x^{\infty}(v^m-x^m)^{\gamma-1}v^{-\varepsilon-\gamma m+m-1}\,w(v)dv$$

We apply this operator term by term for $j = n - 1, n - 2 \ldots 2, 1$, to arrive at

$$\frac{1}{2\pi i}\int_L \frac{\prod_{j=1}^{n}\Gamma(c_j+\gamma_j\xi)\prod_{j=1}^{m}\Gamma(g_j-\alpha_j\xi)}{\sum_{i=1}^{r}\left[\prod_{j=1}^{q_i}\Gamma(d_{ji}-\delta_{ji}\xi)\prod_{j=1}^{p_i}\Gamma(b_{ji}+\beta_{ji}\xi)\right]} x^{-\xi} A(1-\xi)\,d\xi \tag{5.146}$$

$$= \Re_1[\Re_2 \ldots .\Re_n\{G_1(x)\}],\ x > 1$$

Now we consider (5.118) and (5.119), which can be written as

$$\frac{1}{2\pi i}\int_L \frac{\prod_{j=1}^{n}\Gamma(f_j+\gamma_j\xi)\prod_{j=1}^{m}\Gamma(e_j-\alpha_j\xi)}{\sum_{i=1}^{r}\left[\prod_{j=1}^{q_i}\Gamma(d_{ji}-\delta_{ji}\xi)\prod_{j=1}^{p_i}\Gamma(b_{ji}+\beta_{ji}\xi)\right]} x^{-\xi} A(1-\xi)\,d\xi \tag{5.147}$$

$$= E_2(x), \quad 0 < x < 1$$

$$\frac{1}{2\pi i}\int_L \frac{\prod_{j=1}^{n}\Gamma(h_j+\gamma_j\xi)\prod_{j=1}^{m}\Gamma(g_j-\alpha_j\xi)}{\sum_{i=1}^{r}\left[\prod_{j=1}^{q_i}\Gamma(d_{ji}-\delta_{ji}\xi)\prod_{j=1}^{p_i}\Gamma(b_{ji}+\beta_{ji}\xi)\right]} x^{-\xi} A(1-\xi)\,d\xi \tag{5.148}$$

$$= G_1(x), \quad x > 1$$

If at the same time we apply the first operator successively to (5.147) for $j = m, m - 1, m - 2, \ldots, 2, 1$, then

$$\frac{1}{2\pi i}\int_L \frac{\prod_{j=1}^{n}\Gamma(f_j+\gamma_j\xi)\prod_{j=1}^{m}\Gamma(g_j-\alpha_j\xi)}{\sum_{i=1}^{r}\left[\prod_{j=1}^{q_i}\Gamma(d_{ji}-\delta_{ji}\xi)\prod_{j=1}^{p_i}\Gamma(b_{ji}+\beta_{ji}\xi)\right]} x^{-\xi} A(1-\xi)\,d\xi \tag{5.149}$$

$$= \Im_1[\Im_2 \ldots .\Im_m\{E_2(x)\}],\ 0 < x < 1$$

Similarly from Equation (5.148) using the definition of second fractional operator we get

$$\frac{1}{2\pi i}\int_L \frac{\prod_{j=1}^{n}\Gamma(f_j+\gamma_j\xi)\prod_{j=1}^{m}\Gamma(g_j-\alpha_j\xi)}{\sum_{i=1}^{r}\left[\prod_{j=1}^{q_i}\Gamma(d_{ji}-\delta_{ji}\xi)\prod_{j=1}^{p_i}\Gamma(b_{ji}+\beta_{ji}\xi)\right]} x^{-\xi} A(1-\xi)\,d\xi \tag{5.150}$$

$$= \Re_1[\Re_2 \ldots .\Re_n\{G_1(x)\}],\ x > 1$$

Adding the relation (5.140) and (5.149), we obtain

$$\frac{1}{2\pi i}\int_L \frac{\prod_{j=1}^{n}\Gamma\left(c_j+\gamma_j\xi\right)\prod_{j=1}^{m}\Gamma\left(g_j-\alpha_j\xi\right)}{\sum_{i=1}^{r}\left[\prod_{j=1}^{q_i}\Gamma\left(d_{ji}-\delta_{ji}\xi\right)\prod_{j=1}^{p_i}\Gamma\left(b_{ji}+\beta_{ji}\xi\right)\right]}x^{-\xi}A(1-\xi)d\xi$$

$$+\frac{1}{2\pi i}\int_L \frac{\prod_{j=1}^{n}\Gamma\left(f_j+\gamma_j\xi\right)\prod_{j=1}^{m}\Gamma\left(g_j-\alpha_j\xi\right)}{\sum_{i=1}^{r}\left[\prod_{j=1}^{q_i}\Gamma\left(d_{ji}-\delta_{ji}\xi\right)\prod_{j=1}^{p_i}\Gamma\left(b_{ji}+\beta_{ji}\xi\right)\right]}x^{-\xi}A(1-\xi)d\xi \tag{5.151}$$

$$=\Im_1[\Im_2....\Im_m\{E_1(x)\}]+\Im_1[\Im_2....\Im_m\{E_2(x)\}],\ 0<x<1$$

Adding Equations (5.146) and (5.150), we get

$$\frac{1}{2\pi i}\int_L \frac{\prod_{j=1}^{n}\Gamma\left(c_j+\gamma_j\xi\right)\prod_{j=1}^{m}\Gamma\left(g_j-\alpha_j\xi\right)}{\sum_{i=1}^{r}\left[\prod_{j=1}^{q_i}\Gamma\left(d_{ji}-\delta_{ji}\xi\right)\prod_{j=1}^{p_i}\Gamma\left(b_{ji}+\alpha_{ji}\xi\right)\right]}x^{-\xi}A(1-\xi)d\xi$$

$$+\frac{1}{2\pi i}\int_L \frac{\prod_{j=1}^{n}\Gamma\left(f_j+\gamma_j\xi\right)\prod_{j=1}^{m}\Gamma\left(g_j-\alpha_j\xi\right)}{\sum_{i=1}^{r}\left[\prod_{j=1}^{q_i}\Gamma\left(d_{ji}-\delta_{ji}\xi\right)\prod_{j=1}^{p_i}\Gamma\left(b_{ji}+\beta_{ji}\xi\right)\right]}x^{-\xi}A(1-\xi)d\xi \tag{5.152}$$

$$=\Re_1[\Re_2....\Re n\{G_1(x)\}]+\Re_1[\Re_2....\Re n\{G_1(x)\}],\quad x>1$$

Here, we observe that the left-hand side of (5.151) and (5.152) are identical. Hence combining both results, we obtain

$$\frac{1}{2\pi i}\int_L \frac{\prod_{j=1}^{n}\Gamma\left(c_j+\gamma_j\xi\right)\prod_{j=1}^{m}\Gamma\left(g_j-\alpha_j\xi\right)}{\sum_{i=1}^{r}\left[\prod_{j=1}^{q_i}\Gamma\left(d_{ji}-\delta_{ji}\xi\right)\prod_{j=1}^{p_i}\Gamma\left(b_{ji}+\beta_{ji}\xi\right)\right]}x^{-\xi}A(1-\xi)d\xi$$

$$+\frac{1}{2\pi i}\int_L \frac{\prod_{j=1}^{n}\Gamma\left(f_j+\gamma_j\xi\right)\prod_{j=1}^{m}\Gamma\left(g_j-\alpha_j\xi\right)}{\sum_{i=1}^{r}\left[\prod_{j=1}^{q_i}\Gamma\left(d_{ji}-\delta_{ji}\xi\right)\prod_{j=1}^{p_i}\Gamma\left(b_{ji}+\beta_{ji}\xi\right)\right]}x^{-\xi}A(1-\xi)d\xi=T_1(x) \tag{5.153}$$

Here

$$T_1(x)=\Im_1[\Im_2....\Im_m\{E_1(x)\}]+\Im_1[\Im_2....\Im_m\{E_2(x)\}] + \Re_1[\Re_2...\Re_n\{G_1(x)\}]+\Re_1[\Re_2...\Re_n\{G_1(x)\}] \tag{5.154}$$

Again using Parseval's Theorem in (1.37) and (1.38), we find that it takes the form of

$$\int_0^\infty \left\{ I\left[ux \left| \begin{matrix} \{(g_j,\alpha_j)_{1,m}\};\{(b_{ji},\beta_{ji})_{1,p_i}\} \\ \{(c_j,\gamma_j)_{1,n}\} \ ;\{(d_{ji},\delta_{ji})_{1,q_i}\} \end{matrix} \right. \right] + I\left[ux \left| \begin{matrix} \{(g_j,\alpha_j)_{1,m}\};\{(b_{ji},\beta_{ji})_{1,p_i}\} \\ \{(f_j,\gamma_j)_{1,n}\};\{(d_{ji},\delta_{ji})_{1,q_i}\} \end{matrix} \right. \right] \right\} a(u)\,du = T_1(x),\ x>0 \tag{5.155}$$

Next, Equations (5.117) and (5.120) are transformed to

$$\int_0^\infty I\left[ux \left| \begin{matrix} \{(a_j,\alpha_j)_{1,m}\};\{(b_{ji},\beta_{ji})_{1,p_i}\} \\ \{(c_j,\gamma_j)_{1,n}\} \ ;\{(d_{ji},\delta_{ji})_{1,q_i}\} \end{matrix} \right. \right] a(u)\,du = E_1(x) \tag{5.156}$$

$$\int_0^\infty I\left[ux \left| \begin{matrix} \{(k_j,\alpha_j)_{1,m}\};\{(b_{ji},\beta_{ji})_{1,p_i}\} \\ \{(l_j,\gamma_j)_{1,n}\} \ ;\{(d_{ji},\delta_{ji})_{1,q_i}\} \end{matrix} \right. \right] a(u)\ du = G_2(x) \tag{5.157}$$

In the above equation, replace x by v, multiply with

$$\left(x^{\frac{1}{\alpha m}} - v^{\frac{1}{\alpha m}} \right)^{am-km-1} v^{\frac{km}{\alpha m}-1}$$

and integrate, using (5.136) through the integral sign with respect to v from 0 to x, where $0 < x < 1$. Consequently, we arrive at

$$\frac{1}{2\pi i}\int_L \frac{\prod_{j=1}^{n}\Gamma(c_j+\gamma_j\xi)\prod_{j=1}^{m-1}\Gamma(a_j-\alpha_j\xi)\Gamma(k_m-\alpha_m\xi)}{\sum_{i=1}^{r}\left[\prod_{j=1}^{q_i}\Gamma(d_{ji}-\delta_{ji}\xi)\prod_{j=1}^{p_i}\Gamma(b_{ji}+\beta_{ji}\xi)\right]} x^{-\xi} A(1-\xi)\,d\xi \tag{5.158}$$

$$= \frac{1}{\alpha_m\Gamma(a_m-k_m)}\int_0^x (x^{\frac{1}{\alpha_m}} - v^{\frac{1}{\alpha_m}})^{a_m-l_m-1} v^{\frac{k_m}{\alpha_m}-1} E_1(v)\,dv$$

Representing

$$\Im_j[w(x)] = \Im_j^2\left[a_j-k_j, \frac{k_j}{\alpha_j}-1 : \frac{1}{\alpha_j} \right] w(x) \tag{5.159}$$

and using the second fractional integral operator, we arrive at the solution

$$\frac{1}{2\pi i}\int_L \frac{\prod_{j=1}^{n}\Gamma\left(c_j+\gamma_j\xi\right)\prod_{j=1}^{m}\Gamma\left(k_j-\alpha_j\xi\right)}{\sum_{i=1}^{r}\left[\prod_{j=1}^{q_i}\Gamma\left(d_{ji}-\delta_{ji}\xi\right)\prod_{j=1}^{p_i}\Gamma\left(b_{ji}+\beta_{ji}\xi\right)\right]} x^{-\xi} A(1-\xi)\,d\xi \tag{5.160}$$

$$= \Im^2{}_1[\Im^2{}_2 \ldots \Im^2{}_m\{E_1(x)\}], 0<x<1$$

When we use the second fractional integral operator, we arrive at

$$\frac{1}{2\pi i}\int_L \frac{\prod_{j=1}^{n}\Gamma\left(c_j+\gamma_j\xi\right)\prod_{j=1}^{m}\Gamma\left(k_j-\alpha_j\xi\right)}{\sum_{i=1}^{r}\left[\prod_{j=1}^{q_i}\Gamma\left(d_{ji}-\delta_{ji}\xi\right)\prod_{j=1}^{p_i}\Gamma\left(b_{ji}+\beta_{ji}\xi\right)\right]} x^{-\xi} A(1-\xi)\,d\xi \tag{5.161}$$

$$= \Re^2{}_1[\Re^2{}_2 \ldots \Re^2{}_m\{G_2(x)\}],\ x>1$$

Next we consider the pair of Equations (5.118) and (5.120), which can be written as

$$\frac{1}{2\pi i}\int_L \frac{\prod_{j=1}^{n}\Gamma\left(f_j+\gamma_j\xi\right)\prod_{j=1}^{m}\Gamma\left(k_j-\alpha_j\xi\right)}{\sum_{i=1}^{r}\left[\prod_{j=1}^{q_i}\Gamma\left(d_{ji}-\delta_{ji}\xi\right)\prod_{j=1}^{p_i}\Gamma\left(b_{ji}+\beta_{ji}\xi\right)\right]} x^{-\xi} A(1-\xi)\,d\xi \tag{5.162}$$

$$= \Im^2{}_1[\Im^2{}_2 \ldots \Im^2{}_m\{E_2(x)\}], 0<x<1$$

and

$$\frac{1}{2\pi i}\int_L \frac{\prod_{j=1}^{n}\Gamma\left(f_j+\gamma_j\xi\right)\prod_{j=1}^{m}\Gamma\left(k_j-\alpha_j\xi\right)}{\sum_{i=1}^{r}\left[\prod_{j=1}^{q_i}\Gamma\left(d_{ji}-\delta_{ji}\xi\right)\prod_{j=1}^{p_i}\Gamma\left(b_{ji}+\beta_{ji}\xi\right)\right]} x^{-\xi} A(1-\xi)\,d\xi \tag{5.163}$$

$$= \Re^2{}_1[\Re^2{}_2 \ldots \Re^2{}_m\{G_2(x)\}],\quad x>1$$

On adding Equations (5.139) and (5.141) we get

$$\frac{1}{2\pi i}\int_L \frac{\prod_{j=1}^{n}\Gamma(c_j+\gamma_j\xi)\prod_{j=1}^{m}\Gamma(k_j-\alpha_j\xi)}{\sum_{i=1}^{r}\left[\prod_{j=1}^{q_i}\Gamma(d_{ji}-\delta_{ji}\xi)\prod_{j=1}^{p_i}\Gamma(b_{ji}+\beta_{ji}\xi)\right]} x^{-\xi} A(1-\xi)\,d\xi$$

$$+\frac{1}{2\pi i}\int_L \frac{\prod_{j=1}^{n}\Gamma(f_j+\gamma_j\xi)\prod_{j=1}^{m}\Gamma(k_j-\alpha_j\xi)}{\sum_{i=1}^{r}\left[\prod_{j=1}^{q_i}\Gamma(d_{ji}-\delta_{ji}\xi)\prod_{j=1}^{p_i}\Gamma(b_{ji}+\beta_{ji}\xi)\right]} x^{-\xi} A(1-\xi)\,d\xi \tag{5.164}$$

$$= \Im_1[\Im_2....\Im_m\{E_1(x)\}] + \Im_1[\Im_2....\Im_m\{E_2(x)\}], \quad 0<x<1$$

Adding relations (5.140) and (5.142), we arrive at

$$\frac{1}{2\pi i}\int_L \frac{\prod_{j=1}^{n}\Gamma(c_j+\gamma_j\xi)\prod_{j=1}^{m}\Gamma(k_j-\alpha_j\xi)}{\sum_{i=1}^{r}\left[\prod_{j=1}^{q_i}\Gamma(d_{ji}-\delta_{ji}\xi)\prod_{j=1}^{p_i}\Gamma(b_{ji}+\beta_{ji}\xi)\right]} x^{-\xi} A(1-\xi)\,d\xi$$

$$+\frac{1}{2\pi i}\int_L \frac{\prod_{j=1}^{n}\Gamma(f_j+\gamma_j\xi)\prod_{j=1}^{m}\Gamma(k_j-\alpha_j\xi)}{\sum_{i=1}^{r}\left[\prod_{j=1}^{q_i}\Gamma(d_{ji}-\delta_{ji}\xi)\prod_{j=1}^{p_i}\Gamma(b_{ji}+\beta_{ji}\xi)\right]} x^{-\xi} A(1-\xi)\,d\xi \tag{5.165}$$

$$= \Re_1[\Re_2....\Re_m\{G_2(x)\}] + \Re_1[\Re_2....\Re_m\{G_2(x)\}], \quad x>1$$

Here we observe that the left-hand side of Equations (5.164) and (5.165) are identical, hence combining both of the results we obtain

$$\frac{1}{2\pi i}\int_L \frac{\prod_{j=1}^{n}\Gamma(c_j+\gamma_j\xi)\prod_{j=1}^{m}\Gamma(k_j-\alpha_j\xi)}{\sum_{i=1}^{r}\left[\prod_{j=1}^{q_i}\Gamma(d_{ji}-\delta_{ji}\xi)\prod_{j=1}^{p_i}\Gamma(b_{ji}+\beta_{ji}\xi)\right]} x^{-\xi} A(1-\xi)\,d\xi$$

$$+\frac{1}{2\pi i}\int_L \frac{\prod_{j=1}^{n}\Gamma(f_j+\gamma_j\xi)\prod_{j=1}^{m}\Gamma(k_j-\alpha_j\xi)}{\sum_{i=1}^{r}\left[\prod_{j=1}^{q_i}\Gamma(d_{ji}-\delta_{ji}\xi)\prod_{j=1}^{p_i}\Gamma(b_{ji}+\beta_{ji}\xi)\right]} x^{-\xi} A(1-\xi)\,d\xi = T_2(x) \tag{5.166}$$

Here

$$\begin{aligned} T_2(x) &= \Im_1[\Im_2....\Im_m\{E_1(x)\}] + \Im_1[\Im_2....\Im_m\{E_2(x)\}] \\ &+ \Re_1[\Re_2...\Re_n\{G_2(x)\}] + \Re_1[\Re_2...\Re_n\{G_2(x)\}] \end{aligned} \tag{5.167}$$

Now we combine Equations (5.154) and (5.167) to obtain

$$\frac{1}{2\pi i}\int_L \frac{\prod_{j=1}^{n}\Gamma(c_j+\gamma_j\xi)\prod_{j=1}^{m}\Gamma(g_j-\alpha_j\xi)}{\sum_{i=1}^{r}\left[\prod_{j=1}^{q_i}\Gamma(d_{ji}-\delta_{ji}\xi)\prod_{j=1}^{p_i}\Gamma(b_{ji}+\beta_{ji}\xi)\right]}x^{-\xi}A(1-\xi)d\xi$$

$$+\frac{1}{2\pi i}\int_L \frac{\prod_{j=1}^{n}\Gamma(f_j+\gamma_j\xi)\prod_{j=1}^{m}\Gamma(g_j-\alpha_j\xi)}{\sum_{i=1}^{r}\left[\prod_{j=1}^{q_i}\Gamma(d_{ji}-\delta_{ji}\xi)\prod_{j=1}^{p_i}\Gamma(b_{ji}+\beta_{ji}\xi)\right]}x^{-\xi}A(1-\xi)d\xi$$

$$+\frac{1}{2\pi i}\int_L \frac{\prod_{j=1}^{n}\Gamma(c_j+\gamma_j\xi)\prod_{j=1}^{m}\Gamma(k_j-\alpha_j\xi)}{\sum_{i=1}^{r}\left[\prod_{j=1}^{q_i}\Gamma(d_{ji}-\delta_{ji}\xi)\prod_{j=1}^{p_i}\Gamma(b_{ji}+\beta_{ji}\xi)\right]}x^{-\xi}A(1-\xi)d\xi \tag{5.168}$$

$$+\frac{1}{2\pi i}\int_L \frac{\prod_{j=1}^{n}\Gamma(f_j+\gamma_j\xi)\prod_{j=1}^{m}\Gamma(k_j-\alpha_j\xi)}{\sum_{i=1}^{r}\left[\prod_{j=1}^{q_i}\Gamma(d_{ji}-\delta_{ji}\xi)\prod_{j=1}^{p_i}\Gamma(b_{ji}+\beta_{ji}\xi)\right]}x^{-\xi}A(1-\xi)d\xi = T_1(x)+T_2(x)$$

In this way, our problem has been reduced to find the solution of a single integral equation.

$$\int_0^\infty [I_1'(ux)+I_2'(ux)+I_3'(ux)+I_4'(ux)]a(u)du = T(x) \tag{5.169}$$

where

$$I_1'(ux)=I\left[ux\left|\begin{matrix}\{(g_j,\alpha_j)_{1,m}\};\{(b_{ji},\beta_{ji})_{1,p_{ji}}\}\\ \{(c_j,\gamma_j)_{1,n}\}\ ;\{(d_{ji},\delta_{ji})_{1,q_{ji}}\}\end{matrix}\right.\right]$$

$$I_2'(ux)=I\left[ux\left|\begin{matrix}\{(g_j,\alpha_j)_{1,m}\};\{(b_{ji},\beta_{ji})_{1,p_{ji}}\}\\ \{(f_j,\gamma_j)_{1,n}\}\ ;\{(d_{ji},\delta_{ji})_{1,q_{ji}}\}\end{matrix}\right.\right]$$

$$I_3'(ux)=I\left[ux\left|\begin{matrix}\{(k_j,\alpha_j)_{1,m}\};\{(b_{ji},\beta_{ji})_{1,p_{ji}}\}\\ \{(c_j,\gamma_j)_{1,n}\}\ ;\{(d_{ji},\delta_{ji})_{1,q_{ji}}\}\end{matrix}\right.\right]$$

$$I_4'(ux)=I\left[ux\left|\begin{matrix}\{(k_j,\alpha_j)_{1,m}\};\{(b_{ji},\beta_{ji})_{1,p_{ji}}\}\\ \{(f_j,\gamma_j)_{1,n}\}\ ;\{(d_{ji},\delta_{ji})_{1,q_{ji}}\}\end{matrix}\right.\right]$$

and

$$T(x) = T_1(x) + T_2(x) \tag{5.170}$$

To solve this equation, we make use of the following theorems given by Saxena [45].

Theorem 5.5

If $h(x)$ is solution of the integral equation

$$f(x) = \int_0^{\infty} H^l(ux)h(u)\,du \tag{5.171}$$

Then, $H^l(x)$ and K^l (x) are unsymmetric Fourier kernels and its respective Mellin transforms.

Mellin transforms $\overline{H}(\xi)$ and $\overline{K}(\xi)$ satisfy the condition

$$\overline{H}(\xi)\,\overline{K}(1-\xi)=1$$

The existing conditions of the kernels are understood and described here.

Theorem 5.6

If $K(x)$ is a kernel function expressed as

$$K(x) = \sum_{l=1}^{n} K_l(x) \tag{5.172}$$

n is finite then, there exists a function $L_1(x)$ such that $K_l(x)$ and $L_1(x)$ are unsymmetric Fourier kernels, of which the Mellin transforms of $K_r(x)$ and $L_r(x)$, respectively, satisfy the condition.

i. $K_l(\xi)L_l(1-\xi)$ =1

ii. $f(u)$ and $h(u)$ are function of bounded variation near $u = x$, $f(x)$ $L(0, \infty)$ and

then solution of the integral equation

$$f(x)=\int_0^{\infty} K(ux)h(u)\,du \tag{5.173}$$

is given by

$$\frac{1}{2}\{h(x+0)+h(x-0)\}=\frac{1}{2\pi i}\lim_{T\to\infty}\int_{\frac{1}{2}-iT}^{\frac{1}{2}+iT}\varphi(\xi)F(1-\xi)x^{-\xi}\,d\xi \tag{5.174}$$

where

$$\varphi(\xi)=\left[\sum_{r=1}^{n}\frac{1}{L_s(\xi)}\right]^{-1} \tag{5.175}$$

SOLUTION

We take in Equation (5.169)

$$K(x)=\sum_{r=1}^{4}I'_r(x) \tag{5.176}$$

and applying the above theorem, we can establish the following so that,

$$\overline{K}_1(\xi)=M[I'_1]=\frac{\prod_{j=1}^{n}\Gamma(c_j+\gamma_j\xi)\prod_{j=1}^{m}(g_j-\alpha_j\xi)}{\sum_{i=1}^{r}\left[\prod_{j=1}^{q_i}\{D_j(\xi)\}\prod_{j=1}^{p_i}\{B_j(\xi)\}\right]} \tag{5.177}$$

$$\overline{K_2}(\xi)=M[I'_2]=\frac{\prod_{j=1}^{n}\Gamma(f_j+\gamma_j\xi)\prod_{j=1}^{m}(g_j-\alpha_j\xi)}{\sum_{i=1}^{r}\left[\prod_{j=1}^{q_i}\{D_j(\xi)\}\prod_{j=1}^{p_i}\{B_j(\xi)\}\right]} \tag{5.178}$$

$$\overline{K_3}(\xi)=M[I'_3]=\frac{\prod_{j=1}^{n}\Gamma(c_j+\gamma_j\xi)\prod_{j=1}^{m}(k_j-\alpha_j\xi)}{\sum_{i=1}^{r}\left[\prod_{j=1}^{q_i}\{D_j(\xi)\}\prod_{j=1}^{p_i}\{B_j(\xi)\}\right]} \tag{5.179}$$

$$\overline{K_4}(\xi)=M[I'_4]=\frac{\prod_{j=1}^{n}\Gamma(f_j+\gamma_j\xi)\prod_{j=1}^{m}(k_j-\alpha_j\xi)}{\sum_{i=1}^{r}\left[\prod_{j=1}^{q_i}\{D_j(\xi)\}\prod_{j=1}^{p_i}\{B_j(\xi)\}\right]} \tag{5.180}$$

$$M[I_1'(ux)+I_2'(ux)+I_3'(ux)+I_4'(ux)]A(1-\xi)=M[T(x)]$$

$$a(x)=\frac{1}{2\pi i}\int \varphi(\xi)\overline{T}(1-\xi)\,x^{-\xi}\,d\xi \tag{5.181}$$

where

or

$$a(x)=\int_0^\infty\left[\sum_{r=1}^4 I''(ux)\right]T(u)du \tag{5.182}$$

$$L_1(\xi)=\frac{\sum_{i=1}^r\left[\prod_{j=1}^{q_i}\{D_j(\xi)\}\prod_{j=1}^{p_i}\{B_j(\xi)\}\right]}{\prod_{j=1}^n\Gamma(c_j+\gamma_j\xi)\prod_{j=1}^m(g_j-\alpha_j\xi)}$$

$$L_2(\xi)=\frac{\sum_{i=1}^r\left[\prod_{j=1}^{q_i}\{D_j(\xi)\}\prod_{j=1}^{p_i}\{B_j(\xi)\}\right]}{\prod_{j=1}^n\Gamma(f_j+\gamma_j\xi)\prod_{j=1}^m(g_j-\alpha_j\xi)}$$

$$L_3(\xi)=\frac{\sum_{i=1}^r\left[\prod_{j=1}^{q_i}\{D_j(\xi)\}\prod_{j=1}^{p_i}\{B_j(\xi)\}\right]}{\prod_{j-1}^n\Gamma(c_j+\gamma_j\xi)\prod_{j=1}^m(k_i-\alpha_i\xi)}$$

$$L_4(\xi)=\frac{\sum_{i=1}^r\left[\prod_{j=1}^{q_i}\{D_j(\xi)\}\prod_{j=1}^{p_i}\{B_j(\xi)\}\right]}{\prod_{j=1}^n\Gamma(f_j+\gamma_j\xi)\prod_{j=1}^m(k_j-\alpha_j\xi)}$$

Using Equation (5.181) and putting all values, we get

$$\varphi(\xi)=\left[\frac{1}{L_1(\xi)}+\frac{1}{L_2(\xi)}+\frac{1}{L_3(\xi)}+\frac{1}{L_4(\xi)}\right]^{-1} \tag{5.183}$$

Therefore, with the help of known results, we find that their respective reciprocal kernels are

$$I_1''(ux) = I\left[ux\left|\begin{matrix}\{(g_j-\alpha_j,\alpha_j)_{1,m}\};\{(b_{ji}+\beta_{ji},\beta_{ji})_{1,p_{ji}}\}\\ \{(c_j+\gamma_j,\gamma_j)_{1,n}\}\;;\{(d_{ji}-\delta_{ji},\delta_{ji})_{1,q_{ji}}\}\end{matrix}\right.\right] \tag{5.184}$$

$$I_2''(ux) = I\left[ux\left|\begin{matrix}\{(g_j-\alpha_j,\alpha_j)_{1,m}\};\{(b_{ji}+\beta_{ji},\beta_{ji})_{1,p_{ji}}\}\\ \{(f_j+\gamma_j,\gamma_j)_{1,n}\}\;;\{(d_{ji}-\delta_{ji},\delta_{ji})_{1,q_{ji}}\}\end{matrix}\right.\right] \tag{5.185}$$

$$I_3''(ux) = I\left[ux\left|\begin{matrix}\{(k_j-\alpha_j,\alpha_j)_{1,m}\};\{(b_{ji}+\beta_{ji},\beta_{ji})_{1,p_{ji}}\}\\ \{(c_j+\gamma_j,\gamma_j)_{1,n}\}\;;\{(d_{ji}-\delta_{ji},\delta_{ji})_{1,q_{ji}}\}\end{matrix}\right.\right] \tag{5.186}$$

$$I_4''(ux) = I\left[ux\left|\begin{matrix}\{(k_j-\alpha_j,\alpha_j)_{1,m}\};\{(b_{ji}+\beta_{ji},\beta_{ji})_{1,p_{ji}}\}\\ \{(f_j+\gamma_j,\gamma_j)_{1,n}\}\;;\{(d_{ji}-\delta_{ji},\delta_{ji})_{1,q_{ji}}\}\end{matrix}\right.\right] \tag{5.187}$$

This gives final solution of the dual integral Equations (5.111) and (5.112).

SPECIAL CASES

i. Now we take $c_j = f_j$ $(j = 1,\ldots,n)$ and $g_j = k_j$ $(1,\ldots m)$ gives the integral Equations (5.184)–(5.187) and our solution reduces to,

$$\int_0^\infty I'(ux)a(u)\,du = \frac{1}{4}T(x) \tag{5.188}$$

ii. If $i = 1$ in Equations (5.187)–(5.190) we arrive at, in the form of H-function

$$\begin{aligned}
I_1''(ux) &= H\left[ux\left|\begin{matrix}\{(k_j-\alpha_j,\alpha_j)_{1,m}\};\{(b_j+\beta_j,\beta_j)_{1,p_j}\}\\ \{(f_j+\gamma_j,\gamma_j)_{1,n}\}\;;\{(d_j-\delta_j,\delta_j)_{1,q_j}\}\end{matrix}\right.\right]\\
I_2''(ux) &= H\left[ux\left|\begin{matrix}\{(k_j-\alpha_j,\alpha_j)_{1,m}\};\{(b_j+\beta_j,\beta_j)_{1,p_j}\}\\ \{(f_j+\gamma_j,\gamma_j)_{1,n}\}\;;\{(d_j-\delta_j,\delta_j)_{1,q_j}\}\end{matrix}\right.\right]\\
I_3''(ux) &= H\left[ux\left|\begin{matrix}\{(k_j-\alpha_j,\alpha_j)_{1,m}\};\{(b_j+\beta_j,\beta_j)_{1,p_j}\}\\ \{(f_j+\gamma_j,\gamma_j)_{1,n}\}\;;\{(d_j-\delta_j,\delta_j)_{1,q_j}\}\end{matrix}\right.\right]\\
I_4''(ux) &= H\left[ux\left|\begin{matrix}\{(k_j-\alpha_j,\alpha_j)_{1,m}\};\{(b_j+\beta_j,\beta_j)_{1,p_j}\}\\ \{(f_j+\gamma_j,\gamma_j)_{1,n}\}\;;\{(d_j-\delta_j,\delta_j)_{1,q_j}\}\end{matrix}\right.\right]
\end{aligned} \tag{5.189}$$

iii. Again we put in place of $\alpha j = \beta j = \delta j = \gamma j = 1$, then above set of equations are reduced to

$$I_1''(ux) = G\left[ux \middle| \begin{matrix} \{(k_j-1,\ 1)_{1,m}\};\{(b_j+1,\ 1)_{1,p}\} \\ \{(f_j+1,\ 1)_{1,n}\}\ \ ;\{(d_j-1,\ 1)_{1,q}\} \end{matrix}\right]$$

$$I_2''(ux) = G\left[ux \middle| \begin{matrix} \{(k_j-1,\ 1)_{1,m}\};\{(b_j+1,1)_{1,p}\} \\ \{(f_j+1,\ 1)_{1,m}\}\ \ ;\{(d_j-1,\ 1)_{1,q}\} \end{matrix}\right] \tag{5.190}$$

$$I_3''(ux) = G\left[ux \middle| \begin{matrix} \{(k_j-1,\ 1)_{1,m}\};\{(b_j+1,\ 1)_{1,p}\} \\ \{(f_j+1,\ 1)_{1,n}\}\ \ ;\{(d_j-1,\ 1)_{1,q}\} \end{matrix}\right]$$

$$I_4''(ux) = G\left[ux \middle| \begin{matrix} \{(k_j-1,\ 1)_{1,m}\};\{(b_j+1,\ 1)_{1,p}\} \\ \{(f_j+1,\ 1)_{1,n}\}\ \ ;\{(d_j-1,\ 1)_{1,q}\} \end{matrix}\right] \tag{5.191}$$

6 Integral Equations Involving Bessel–Maitland Functions

In this chapter, the authors have obtained a solution of an integral equation in which Bessel–Maitland function occurs as a part of kernel. Erdelyi–Kober fractional integral operators have been applied to transform the kernel in desired form so that the conditions of unsymmetrical Fourier kernels are satisfied. These results may be useful in many applications of integral equations [60]).

6.1 PRELIMINARIES

6.1.1 Fractional Integrals

The Weyl fractional integral is defined in the following form:

$$W^{\mu}\{f(t):p\}=\frac{1}{\Gamma(\mu)}\int_{p}^{\infty}(t-p)^{\mu-1}f(t)\,dt;\text{ where }\operatorname{Re}(\mu)>0 \tag{6.1}$$

The Laplace transform of $f(t)$ is denoted by $L[f(t)]$ is defined as

$$L[(f(t)):p]=\int_{0}^{\infty}e^{-pt}f(t)\,dt=F(p) \tag{6.2}$$

Here we stabilized a formula exhibiting a relationship between (6.1) and (6.2) which provides the more effective tools and allows the straightforward derivation of the Weyl fractional integral operators associated with Saxena's I-function, Fox's H-function and Meijer's G-function.

Fox has used the following generalized Erdelyi–Kober operators

$$[\gamma,\varepsilon:m]\,g(x)=\frac{m}{\Gamma\gamma}x^{-\varepsilon-\gamma m+m-1}\int_{0}^{x}(x^{m}-v^{m})^{\gamma-1}v^{\varepsilon}\,g(v)\,dv \tag{6.3}$$

and

$$[\gamma,\varepsilon:m]\,g(x)=\frac{m}{\Gamma\gamma}x^{\varepsilon}\int_{x}^{\infty}(v^{m}-x^{m})^{\gamma-1}v^{\varepsilon-\gamma m+m-1}\,g(v)\,dv \tag{6.4}$$

DOI: 10.1201/9781003517757-6

The operator $\Im$ exists if $g(x) \epsilon L_p(0,\infty)$, $p \geq 0$, $\gamma > 0$. $\varepsilon > \frac{1-p}{p}$. If in addition $g(x)$ can be differentiated sufficiently often, then the operator $\Im$ exists for both negative and positive value of γ. The operator $\Re$ exists if $g(x) \epsilon L_p(0,\infty)$, $p \geq 1$. If $g(x)$ can be differentiated sufficiently often, then $\Re$ exists if $m > 0$, $\varepsilon > -\frac{1}{p}$ while γ can take any negative or positive values.

The Beta function in the integral forms is most useful for the application of the above-mentioned $\Im$ and $\Re$ operators. These integrals are given as

$$\int_0^x \left(x^{\frac{1}{g_1}} - v^{\frac{1}{g_1}} \right)^{a_1-\tau_1-1} v^{\frac{\tau_1}{g_1}-1-\xi} dv = \frac{g_1\Gamma(a_1-\tau_1)\Gamma(\tau_1-g_1\xi)}{\Gamma(a_1-g_1\xi)} x^{\frac{a_1}{g_1}-\frac{1}{g_1}-\xi} \tag{6.5}$$

provided $a_1 > \tau_1$ and $\frac{\tau_1}{g_1} > c$, $(\xi = c + it)$.

$$\int_x^\infty \left(v^{\frac{1}{f_1}} - x^{\frac{1}{f_1}} \right)^{b_1-\mu_1-1} v^{\frac{1}{f_1}-\frac{b_1}{f_1}-\xi-1} dv = \frac{f_1\Gamma(b_1-\mu_1)\Gamma(\mu_1+f_1\xi)}{\Gamma(b_1+f_1\xi)} x^{-\frac{\mu_1}{f_1}-\xi} \tag{6.6}$$

provided $b_1 > \mu_1$ at the lower limit and $f_1\text{Re}(\xi) + \mu_1 > 0$ at the upper limit.

6.1.2 Mellin Transform

Let $f(t)$ be a function on the positive real axis $0 < t < \infty$. The Mellin transformation M is the operation mapping the function f into the function F defined on the complex plane by the relation

$$M\{f(t):s\} = F(s) = \int_0^\infty f(t)t^{s-1}dt \tag{6.7}$$

The function $F(s)$ is called the Mellin transform of $f(t)$.

6.1.3 Inversion Formula

The inversion formula of Mellin transform is given as

$$M^{-1}\{F(s):t\} = f(t) = \frac{1}{2\pi i}\int_{c-i\infty}^{c+i\infty} F(s)t^{-s}ds$$

where the integration is along a vertical line through $\text{Re}(s) = c$.

Elliptic integrals occur in a number of physical problems [61],[62] in the form of single and multiple integrals. Evans et al. [63] have treated the elliptic-type integrals

$$\Omega_j(k)=\int_0^{\pi}\left(1-k^2\cos\theta\right)^{-j-\frac{1}{2}}d\theta;\qquad j=0,1,2 \tag{6.8}$$

and $0\le k\le 1$.

Elliptic integrals (6.8) have been studied and generalized by many authors notably by Kalla et al. [64–66,68], Salman [67], Saxena and Bromberg [69] and others.

Some of the important generalizations of elliptic-type integrals (6.8) are as follows:

Kalla [64,65] introduced the generalization of the form:

$$R_\mu(k,\alpha,\gamma)=\int_0^{\pi}\frac{\cos^{2\alpha-1}(\theta/2)\sin^{2\gamma-2\alpha-1}(\theta/2)}{\left(1-k^2\cos\theta\right)^{\mu+\frac{1}{2}}}d\theta, \tag{6.9}$$

where $0\le k\langle 1,\ Re(\gamma)\rangle\mathrm{Re}(\alpha)>0,\mathrm{Re}(\mu)>\frac{-1}{2}$. Results for this generalization are also derived by Glasser and Kalla [70] and Al-Saqabi [71] defined and studied the generalization given by integral

$$B_\mu(k,m,v)=\int_0^{\pi}\frac{\cos^{2m}(\theta)\sin^{2v}(\theta)}{\left(1-k^2\cos\theta\right)^{\mu+\frac{1}{2}}}d\theta, \tag{6.10}$$

where $0\le k\langle 1; m\in\mathbb{N}_0,\ \mu\in\mathbb{C},\ \ \mathrm{Re}(\mu)\rangle\frac{-1}{2}$.

Asymptotic expansion of (6.10) has recently been discussed by Matera et al. [72]. The integral

$$\Lambda_v(\alpha,k)=\int_0^{\pi}\frac{\exp\left[\alpha\sin^2(\theta/2)\right]}{\left(1-k^2\cos\theta\right)^{v+\frac{1}{2}}}d\theta, \tag{6.11}$$

where $0\le k<1;\ \alpha,v\in\mathbb{R}$; presents another generalization of (6.8), given by Siddiqui [62].

Srivastava and Siddiqi [73] have given an interesting unification and extension of the families of elliptic-type integrals in the following form

$$\Lambda_{(\lambda,\mu)}^{(\alpha,\beta)}(\rho;k)=\int_0^{\pi}\frac{\cos^{2\alpha-1}(\theta/2)\sin^{2\beta-1}(\theta/2)}{\left(1-k^2\cos\theta\right)^{\mu+\frac{1}{2}}}\left[1-\rho\sin^2(\theta/2)\right]^{-\lambda}d\theta, \tag{6.12}$$

where $0\le k\langle 1,\ \mathrm{Re}(\alpha)\rangle 0,\mathrm{Re}(\beta)>0,\ \lambda,\mu\in\mathbb{C},\ |\rho|<1$.

The generalization of Equation (6.12) by means of the following integral and also obtained its asymptotic expansion [74]

$$\Lambda_{(\lambda,\gamma,\mu)}^{(\alpha,\beta)}(\rho,\delta;k)=\int_0^{\pi}\cos^{2\alpha-1}\left(\frac{\theta}{2}\right)\sin^{2\beta-1}\left(\frac{\theta}{2}\right)\left(1-k^2\cos\theta\right)^{-\mu-\frac{1}{2}}$$
$$\times\left(1-\rho\sin^2(\theta/2)\right)^{-\lambda}\left(1+\delta\cos^2(\theta/2)\right)^{-\gamma}d\theta \tag{6.13}$$

where $0\le k\langle 1,\ Re(\alpha)\rangle 0, Re(\beta)>0,\ \lambda,\mu,\gamma\in\mathbb{C}$ and either $|\rho|,|\delta|<1\ or\ \rho(or\ \delta)\in\mathbb{C}$ whenever $\lambda=m\ or\ \gamma=-m\ ,\ m\in\mathbb{N}_0$ respectively.

Kalla and Saxena [75] have studied families of elliptic-type integrals of the form

$$\Omega_{(\sigma_1,\dots\dots,\sigma_{n-2},:\delta,\mu)}^{(\alpha,\beta)}(\rho_1,\dots\dots\ \rho_{n-2},\delta;k)$$
$$=\int_0^{\pi}\cos^{2\alpha-1}\left(\frac{\theta}{2}\right)\sin^{2\beta-1}\left(\frac{\theta}{2}\right)\prod_{j=1}^{n-2}\left[1-\rho_j\sin^2\left(\frac{\theta}{2}\right)\right]^{-\sigma_j}$$
$$\times\left[1+\cos^2\left(\frac{\theta}{2}\right)\right]^{-\gamma}\left(1-k^2\cos\theta\right)^{-\mu-\frac{1}{2}}\ d\theta, \tag{6.14}$$

where $0\le k\langle 1,\ \mathrm{Re}(\alpha)\rangle 0,\mathrm{Re}(\beta)>0;\sigma_j\ (j=1,\dots\dots,n-2)\ ,\gamma,\mu\in\mathbb{C};$

$$\max\left\{|\rho_j|,\left|\frac{\delta}{1+\delta}\right|,\left|\frac{2k^2}{k^2-1}\right|\right\}<1.$$

In a recent paper an extension of Equation (6.14) has been done in the form [76]

$$\Omega_{(\sigma_1,\dots\dots,\sigma_m,\gamma;\tau_1,\dots\dots,\tau_n)}^{(\alpha,\beta)}(\rho_1,\dots\dots\dots,\rho_{n-2},\delta;\lambda_1,\dots\dots\dots,\lambda_n)$$
$$=\int_0^{\pi}\cos^{2\alpha-1}\left(\frac{\theta}{2}\right)\sin^{2\beta-1}\left(\frac{\theta}{2}\right)\prod_{i=1}^{m}\left[1-\rho_i\sin^2\left(\frac{\theta}{2}\right)\right]^{-\sigma_i}$$
$$\times\left[1+\delta\cos^2\left(\frac{\theta}{2}\right)\right]^{-\gamma}\prod_{j=1}^{n}\left(1-\lambda_j^2\cos\theta\right)^{-\tau_j}d\theta, \tag{6.15}$$

where $\min(\mathrm{Re}(\alpha),\mathrm{Re}(\beta))>0;|\lambda_j|<1;\ \sigma_i,\gamma,\tau_j\in\mathbb{C};$

$$\max\left\{|\rho_i|,\left|\frac{\delta}{1+\delta}\right|,\left|\frac{2\lambda_j^2}{\lambda_j^2-1}\right|\right\}<1\quad(i=1,\dots,m;j=1,\dots.,n).$$

Similarly, Chaurasia and Pandey [77] investigated a new family of unified and generalized elliptic-type integrals:

$$\bar{\Omega}^{(\alpha,\beta)}_{(\lambda_j,\tau_j)}\left((\rho_i),(\delta_i);k_j\right)=\bar{\Omega}^{(\alpha,\beta)}_{\lambda_1,\ldots,\lambda_N,\tau_1,\ldots,\tau_M}\left(\rho_1,\ldots\ldots,\rho_N;\delta_1,\ldots\ldots,\delta_N;k_1,\ldots\ldots,k_M\right)$$

$$=\int_0^{\pi}\cos^{2\alpha-1}\left(\frac{\theta}{2}\right)\sin^{2\beta-1}\left(\frac{\theta}{2}\right)\prod_{i=1}^{N}\left[1+\rho_i\sin^2\left(\frac{\theta}{2}\right)+\delta_i\cos^2\left(\frac{\theta}{2}\right)\right]^{-\lambda_i}$$

$$\times\prod_{j=1}^{M}\left(1-k_j^2\cos\theta\right)^{-\tau_j}d\theta, \tag{6.16}$$

where $\min\left(\text{Re}(\alpha),\text{Re}(\beta)\right)>0;\left|k_j\right|<1;\ \lambda_i,\tau_j\in\mathbb{C};$

$$\max\left\{\left|\rho_i\right|,\left|\delta_i\right|,\left|\frac{\delta_i-\rho_i}{1+\delta_i}\right|,\left|\frac{2k_j^2}{k_j^2-1}\right|\right\}<1\quad(i=1,\ldots,N;j=1,\ldots,M),$$

Which include most of the known generalized and unified families of elliptic-type integrals (including those discussed in (6.8) through (6.15). For more details see also earlier referred papers of [78], upon a closer examination of the above Equation (6.16). It can be seen that the family of elliptic-type integral $\bar{\Omega}^{(\alpha,\beta)}_{(\lambda_j,\tau_j)}\left((\rho_i),(\delta_i);k_j\right)$ can be put into the following form involving Euler-type integral

$$\bar{\Omega}^{(\alpha,\beta)}_{(\lambda_1,\ldots,\lambda_N;\tau_1,\ldots,\tau_M)}\left(\rho_1,\ldots\ldots,\rho_N,\delta_1,\ldots\ldots,\delta_N;k_1,\ldots\ldots,k_M\right)$$

$$=\prod_{j=1}^{M}\left(1-k_j^2\right)^{-\tau_j}\prod_{i=1}^{N}\left[1+\delta_i\right]^{-\lambda_i}\int_0^1 w^{\beta-1}(1-w)^{\alpha-1}$$

$$\times\prod_{j=1}^{M}\left(1-\frac{2wk_j^2}{k_j^2-1}\right)^{-\tau_j}\prod_{i=1}^{N}\left[1-\frac{(\delta_i-\rho_i)w}{1+\delta_i}\right]^{-\lambda_i}dw. \tag{6.17}$$

A two-variable generating function $F(z,t)$ possesses a formal power series representation in t and can be written in the following form

$$F(z,t)=\sum_{n=0}^{\infty}C_n f_n(z)t^n, \tag{6.18}$$

where each member of the set $\left\{f_n(z)\right\}_{n=0}^{\infty}$ is independent of t.

The I-function has relation with Fox's H-function and is also defined in the form

$$I(z)=I^{m,n}_{p_i,q_i;r}[z]=I^{m,n}_{p_i,q_i:r}\left[z\left|\begin{matrix}(a_j,\alpha_j)_{1,n}; & (a_{ji},\alpha_{ji})_{n+1,p_i}\\ (b_j,\beta_j)_{1,m}; & (b_{ji},\beta_{ji})_{m+1,q_i}\end{matrix}\right.\right]$$

$$=\sum_{k=0}^{\infty}\sum_{h=1}^{m}\frac{(-1)^k\prod_{j=1}^{m}\Gamma\left(b_j-\beta_j\left\{\frac{b_h+k}{\beta_h}\right\}\right)}{\beta_h k!\sum_{i=1}^{r}\left\{\prod_{j=m+1}^{q_i}\Gamma\left(1-b_{ji}+\beta_{ji}\left\{\frac{b_h+k}{\beta_h}\right\}\right)\right.} \tag{6.19}$$

$$\times\frac{\prod_{j=1}^{n}\Gamma\left(1-a_j+\alpha_{j\xi}\left\{\frac{b_h+k}{\beta_h}\right\}\right)z^{\frac{b_h+k}{\beta_h}}}{\left.\prod_{j=n+1}^{p_i}\Gamma\left(a_{ji}-\alpha_{ji}\left\{\frac{b_h+k}{\beta_h}\right\}\right)\right\}}$$

In this chapter, we have studied four new theorems associated with the two variables generating functions and I-function. Some new families of elliptic-type integrals can also be deduced with the help of the results obtained in this article. Such generalized and new families of elliptic-type integrals play an important role in the evaluation of a number of Euler-type integrals involving various generating functions. The basic idea of evolving the theorems discussed in the article is inspired by the previous research work of Mohammed [79], Saran [80] and Srivastava and Yeh [81].

6.1.4 Parseval's Theorem (Convolution Theorem)

Parseval's theorem for the Mellin transform is given as follows

$$\int_0^{\infty} g(tx)\,f(t)\,dt=\lim_{T\to\infty}\frac{1}{2\pi i}\int_{c-iT}^{c+iT}x^{-s}G(s)\,F(1-s)\,ds \tag{6.20}$$

Additional conditions, for the validity of (6.20), are that

$$F(s)\int L_p(c-i\infty,c+i\infty)\text{ and }x^{1-c}\,g(x)\int L_p(0,\infty)\,;\,p\geq 1,$$

where we denote by L_p the class of functions $g(x)$ such that $\int_0^{\infty}|g(x)|^p\,\frac{dx}{x}<\infty.$

6.1.5 Bessel–Maitland Function

The special function of the form defined by the series representation

$$J_{\mu}^{\lambda}(x)=\sum_{r=0}^{\infty}\frac{(-1)^r\,x^r}{\Gamma(1+\mu+\lambda r)} \tag{6.21}$$

is known as Wright generalized Bessel function or Bessel–Maitland function. Here $0 < \lambda < 1$ and $M \geq 0$.

6.1.6 The I-Function

As indicated earlier the I-function is the latest and most general form of hypergeometric functions and is defined in terms of following Mellin–Barnes type integral

$$\mathrm{I(z)} = I^{m,n}_{p_i,q_i;r}[\mathrm{z}] = I^{m,n}_{p_i,q_i;r}\left[z \left| \begin{matrix} (a_j,\alpha_j)_{1,n} \ ; & (a_{ji},\alpha_{ji})_{n+1,\ p_i} \\ (b_j,\beta_j)_{1,m} \ ; & (b_{ji},\beta_{ji})_{m+1,\ q_i} \end{matrix} \right. \right] \tag{6.22}$$

$$= \frac{1}{2\pi i}\int_L \phi(\xi) z^{\xi} d\xi$$

where

$$\phi(\xi) = \frac{\prod_{j=1}^{m}\Gamma(b_j - \beta_j\xi)\prod_{j=1}^{n}\Gamma(1-a_j+\alpha_j\xi)}{\sum_{i=1}^{r}\left[\prod_{j=m+1}^{q_i}\Gamma(1-b_{ji}+\beta_{ji}\xi)\prod_{j=n+1}^{p_i}\Gamma(a_{ji}-\alpha_{ji}\xi)\right]}$$

$p_i, q_i\ (i = 1,2,\ldots\ldots r)$, m and n are integers satisfying $0 \leq n \leq p_i$, $0 \leq \mathrm{m} \leq q_i\ (i = 1,2,\ldots\ldots r)$, r is finite, $\alpha_j, \beta_j, \alpha_{ji}, \beta_{ji}$ are real and positive and a_j, b_j, a_{ji}, b_{ji} are complex number such that

$$\alpha_k(b_h + v) \neq \beta_h(a_k - 1 - k) \text{ for } k, v = 0,1,2\ldots;\ h = 1,2,\ldots,m;\ i = 1,2,\ldots,r.$$

L is contour running form $\sigma - i\infty$ *to* $\sigma + i\infty$, where σ is real in the complex ξ-plane such that the points

$$\xi = \frac{(a_j - 1 - v)}{\alpha_j},\quad j = 1,2,\ldots n\ ;\ v = 0,1,2,\cdots$$

$$\xi = \frac{(b_j + v)}{\beta_j},\qquad j = 1,2\cdots,m\ ;\ v = 0,1,2,\cdots$$

lie to the left- and right-hand sides of L, respectively.

The I-function converges absolutely in ξ-plane if

i. $\lambda' > 0$, $|\arg z| < \frac{\pi}{2}\lambda'$

ii. $\lambda' \geq 0, |\arg z| \geq \frac{\pi}{2}\lambda', \operatorname{Re}(\mu'+1) < 0$

where

$$\lambda' = \sum_{j=1}^{n}\alpha_j + \sum_{j=1}^{m}\beta_j - \max_{1\leq i\leq r}\left[\sum_{j=n+1}^{p_i}\alpha_{ji} + \sum_{j=m+1}^{q_i}\beta_{ji}\right] \tag{6.23}$$

$$\mu' = \sum_{j=1}^{m}b_j + \sum_{j=1}^{n}a_j - \max_{1\leq i\leq r}\left[\sum_{j=n+1}^{p_i}a_j - \sum_{j=m+1}^{q_i}b_j + \frac{p_i}{2} - \frac{q_i}{2}\right] \tag{6.24}$$

If we put $r = 1$ in (6.22), it reduces Fox's H-function.

Following the results of Srivastava [83] (with z replaced by iz) are required in the present paper:

$$z^{\lambda}F\left[\begin{matrix}(a),(a');(c);(c');\\(b),(b');(d);(d');\end{matrix}\quad -x^2z^2,-y^2z^2\right] =$$

$$\sum_{n=0}^{\infty}\frac{(\lambda+2n)\Gamma(\lambda+n)}{n!}J_{\lambda+2n}(2z)F\left[\begin{matrix}-n,\lambda+n,(a),(a');(c);(c');\\(b),(b');(d);(d');\end{matrix}\quad x^2,y^2\right] \tag{6.25}$$

and

$$z^{\lambda}F\left[\begin{matrix}(a),(a');(c);(c');\\(b),(b');(d);(d');\end{matrix}\quad -x^2z^2,-y^2z^2\right]$$

$$= \Gamma(\lambda+1)\sum_{n=0}^{\infty}\frac{z^n}{2^n n!}J_{\lambda+n}(2z)F\left[\begin{matrix}-n,1+\lambda,(a),(a');(c);(c');\\(b),(b');(d);(d');\end{matrix}\quad x^2,y^2\right] \tag{6.26}$$

where $A+A'+C \leq B+B'+D$, $A+A'+C' \leq B+B'+D'$ and for all values of λ with possible exception of zero and negative integers. (a) represents the sequence of A parameters $a_1, a_2, \ldots, a_A$ and the convention will be retained throughout the chapter. The notation for double hypergeometric function and, for the sake of brevity, to an earlier one introduced by Appell and Kampe's Feriet [84].

6.2 THE INTEGRAL EQUATIONS

We consider the integral equation of the type

$$\int_0^\infty u^\alpha J_\mu^\lambda(ux)\, f(u)\,du = g(x),\ \ x>1 \tag{6.27}$$

Here α, λ *and* μ are arbitrary positive real numbers $(0<\lambda<1)$ and the kernel does not satisfy the condition for direct inversion. Hence we transform the same to Fox's H-function which makes the equation eligible to find the solution. Here *g*(*x*) is known and *f*(*x*) is to be determined. $J_\mu^\lambda(u)$ is Bessel–Maitland function defined in (6.21).

We have

$$M\left\{u^\alpha J_\mu^\lambda(u)\right\} = \frac{\Gamma(\alpha+\xi)}{\Gamma(1+\mu-\lambda\alpha-\lambda\xi)} \tag{6.28}$$

Applying Parseval's theorem of the Mellin transforms in (6.27) from (6.20), we obtain

$$\lim_{t\to\infty}\frac{1}{2\pi i}\int_{\sigma_0-it}^{\sigma_0+it}\frac{\Gamma(\alpha+\xi)}{\Gamma(1+\mu-\lambda\alpha-\lambda\xi)}x^{-\xi}F(1-\xi)\,d\xi = g(x),x>1 \tag{6.29}$$

where $\xi=\sigma_0+it$ and $F(\xi)$ is Mellin transform of *f*(*u*).

Now we shall use operator $\Re$ defined in (6.4) and the following definition of Beta function

$$\int_x^\infty\left(v^{\frac{1}{f_1}}-x^{\frac{1}{f_1}}\right)^{b_1-e_1-1} v^{\frac{1}{f_1}-\frac{b_1}{f_1}-1} = \frac{f_1\Gamma(b_1-e_1)\Gamma(e_1+f_1\xi)}{\Gamma(b_1+f_1\xi)}x^{-\frac{\mu_1}{f_1}-\xi} \tag{6.30}$$

For convergence of (6.30), we require $b_1>e_1$ at the lower limit and $f_1\mathrm{Re}(\ )+e_1>0$ at the upper limit. But when the fractional integral operator is introduced, some of these conditions may no longer be necessary.

Now we replace *x* by *v* in (6.29), multiply both the sides by

$$\int_x^\infty\left(v^{\frac{1}{f_1}}-x^{\frac{1}{f_1}}\right)^{b_1-e_1-1} v^{\frac{1}{f_1}-\frac{b_1}{f_1}-1}$$

and integrate with respect to *v* from *x* to ∞ and using (6.30). Consequently, we have

$$\begin{aligned}&\frac{1}{2\pi i}\int\frac{\Gamma(\alpha+\xi)\Gamma(e_1+f_1\xi)}{\Gamma(1+\mu-\lambda\alpha-\lambda\xi)\Gamma(b_1+f_1\xi)}x^{-\xi}F(1-\xi)\,d\xi\\&=\frac{1}{f_1\Gamma(b_1-e_1)}x^{\frac{\mu_1}{f_1}}\int_x^\infty\left(v^{\frac{1}{f_1}}-x^{\frac{1}{f_1}}\right)^{b_1-e_1-1}v^{\frac{1}{f_1}-\frac{b_1}{f_1}-1}g(v)\,dv\end{aligned} \tag{6.31}$$

Now, we introduce fractional integral operator $\Re$ defined in (6.4), we write

$$\left[b_1 - e_1, \frac{\mu_1}{f_1} : \frac{1}{f_1}\right] g(x) = {}_1[g(x)] \tag{6.32}$$

Again, if we apply the operator

$$\left[b_2 - e_2, \frac{e_2}{f_2} : \frac{1}{f_2}\right]$$

So that the Equation (6.31) becomes

$$\frac{1}{2\pi i}\int_L \frac{\Gamma(\alpha+\xi)\Gamma(e_1+f_1\xi)\Gamma(e_2+f_2\xi)}{\Gamma(1+\mu-\lambda\alpha-\lambda\xi)\Gamma(b_1+f_1\xi)\Gamma(b_2+f_2\xi)} x^{-\xi} F(1-\xi)\, d\xi =_{21} [g(x)], x > 1 \tag{6.33}$$

Continuing in the same manner up to m-times, we obtain

$$\frac{1}{2\pi i}\int_L \frac{\Gamma(\alpha+\xi)\prod_{k=1}^{m}\Gamma\{e_k+f_k\xi\}}{\Gamma(1+\mu-\lambda\alpha-\lambda\xi)\prod_{k=1}^{m}\Gamma\{b_k+f_k\xi\}} x^{-\xi} F(1-\xi)\, d\xi = \prod_{k=1}^{m}\{{}_k[g(x)]\}, x > 1 \tag{6.34}$$

where

$$g_1(x) = \prod_{k=1}^{m}\{{}_k[g(x)]\}, x > 1 \tag{6.35}$$

Now applying Parseval's theorem of Mellin transforms (6.20), we have obtained the transformation of the given equation (6.27) as

$$\int_0^\infty H^{m+1,0}_{m,m+2}\left[ux \,\middle|\, \begin{matrix} (\ldots\ldots), (b_k, f_k)_{1,m}, (\ldots\ldots) \\ (\alpha, 1), (e_k, f_k)_{1,m}, (1+\mu-\lambda\alpha, \lambda) \end{matrix}\right] f(u)\, du = g_1(x), x > 1 \tag{6.36}$$

In this way, our problem has been reduced to finding the solution of integral Equation (6.36). Now from (6.7) and (6.34), we get

$$G(\xi) = \frac{\Gamma(\alpha+\xi)\prod_{k=1}^{m}\Gamma\{e_k+f_k\xi\}}{\Gamma(1+\mu-\lambda\alpha-\lambda\xi)\prod_{k=1}^{m}\Gamma\{b_k+f_k\xi\}} F(1-\xi) \tag{6.37}$$

where $G(\xi)$ is the Mellin transform of $g_1(x)$ and the parameters e_k, b_k and f_k are chosen in such a way that the kernel fulfils the conditions for inversion.

From (6.37), we have

$$F(s)=\frac{\Gamma(1+\mu-\lambda\alpha-\lambda+\lambda s)\prod_{k=1}^{m}\Gamma\{b_k+f_k-f_k s\}}{\Gamma(\alpha+1-s)\prod_{k=1}^{m}\Gamma\{e_k+f_k-f_k s\}}G(1-s) \tag{6.38}$$

Taking Mellin inverse on both sides, we have

$$f(x)=\frac{1}{2\pi i}\int_L \frac{\Gamma(1+\mu-\lambda\alpha-\lambda+\lambda s)\prod_{k=1}^{m}\Gamma\{b_k+f_k-f_k s\}}{\Gamma(\alpha+1-s)\prod_{k=1}^{m}\Gamma\{e_k+f_k-f_k s\}}x^{-s}\, G(1-s)ds \tag{6.39}$$

Again, applying Parseval's theorem of defined in (6.20), we find that (6.39) takes the form

$$f(x)=\int_0^\infty H_{m,m+2}^{1,m}\left[ux\left|\begin{matrix}(\cdots\cdots),(b_k+f_k,f_k)_{1,m},(\cdots\cdots)\\(1+\mu-\lambda\alpha-\lambda,\lambda),(e_k+f_k,f_k)_{1,m},(\alpha+1,1)\end{matrix}\right.\right]g_1(u)du \tag{6.40}$$

Now we will extend the above result. If we take an integral equation involving summation of N-Bessel–Maitland functions instead of single function and apply the same process of the above result, then we will find the transformation of this equation in terms of summation of H-functions and solution in terms of I-function.

Theorem 6.1

If $f(x)$ is the solution of the integral equation

$$\int_0^\infty \sum_{j=1}^{N} u^{\alpha_j} J_{\mu_j}^{\lambda_j}(ux)f(u)du=g(x),\quad x>1 \tag{6.41}$$

where α,λ and μ are arbitrary real numbers. Then

$$f(x)=\int_0^\infty I_{m+N-1,m+N+1;r}^{N,m}\left[ux\left|\begin{matrix}(A_j,\gamma_j)_{1,m},(A_{ji},\gamma_{ji})_{1,N-1}\\(B_j,\delta_j)_{1,N},(B_{ji},\delta_{ji})_{1,m+1}\end{matrix}\right.\right]g_1(u)du \tag{6.42}$$

where

$$g_1(x)=\prod_{k=1}^{m}\left\{{}_k[g(x)]\right\},\ \left[b_k-e_k,\frac{\mu_k}{f_k}:\frac{1}{f_k}\right]g(x)={}_k[g(x)],\ x>1.$$

$A_j=1-b_k-f_k,\gamma_j=f_k,(j,k=1,2,...m);B_j=1-\lambda_j+\mu_j-\lambda_j\alpha_j$

$\delta_j=\lambda_j,(j=1,2,...,N);\ A_{ji}=1-\lambda_{ji}+\mu_{ji}-\lambda_{ji}\alpha_{ji},\gamma_{ji}=\lambda_{ji,}(j=1,2,...,N-1;i=1,2,...,r)B_{ji}$

$=1-e_{ki}-f_{ki},\delta_{ji}=f_{ki},(j,k=1,2,...m\ ;i=1,2,...,r)\ ;\ B_{m+1,i}=-\alpha_i,\delta_{m+1,i}=1,(i=1,2,...,r).$

provided

i. $\lambda'>0,\ |\arg x|<\frac{1}{2}\lambda'\pi$

ii. $\lambda'\geq 0,\ |\arg x|\leq\frac{1}{2}\lambda'\pi,\ \text{Re}(\mu'+1)<0$

where

$$\lambda'=\sum_{j=1}^{m}\gamma_j+\sum_{j=1}^{N}\delta_j-\max_{1\leq i\leq r}\left[\sum_{j=1}^{N-1}\gamma_{ji}+\sum_{j=1}^{m+1}\delta_{ji}\right]\tag{6.43}$$

and

$$\mu'=\sum_{j=1}^{N}B_j-\sum_{j=1}^{m}A_j-\min_{1\leq i\leq r}\left[\sum_{j=1}^{N-1}A_{ji}-\sum_{j=1}^{m+1}B_{ji}-1\right]\tag{6.44}$$

Proof. Applying Parseval's theorem of the Mellin transforms in (6.41) from (6.20), we obtain

$$\lim_{t\to\infty}\frac{1}{2\pi i}\int_{\sigma_0-it}^{\sigma_0+it}\sum_{j=1}^{N}\frac{\Gamma(\alpha_j+\xi)}{\Gamma(1+\mu_j-\lambda_j\alpha_j-\lambda_j\xi)}x^{-\xi}F(1-\xi)d\xi=\ g(x),x>1\tag{6.45}$$

where $\xi=\sigma_0+it$ and $F(\xi)$ is Mellin transform of $f(u)$.

Proceeding on the same lines of above and apply the second fractional integral operator $\Re$ defined in (6.4), we obtain the transformation of the Equation (6.41) as

$$\int_0^\infty\sum_{j=1}^{N}H_{jm+2,\,m}^{0,m+1}\left[ux\left|\begin{matrix}(\cdots\cdots),(b_k,f_k)_{1,m},(\cdots\cdots)\\(\alpha_j,1),(e_k,f_k)_{1,m},(1+\mu_j-\lambda_j\alpha,\lambda_j)\end{matrix}\right.\right]f(u)du=g_1(x)\tag{6.46}$$

where $g_1(x)$ is defined in (6.35). Proceeding on the similar lines as above. We obtain

$$f(x)=\frac{1}{2\pi i}\int_L \frac{\prod_{j=1}^{N}\Gamma(1+\mu_j-\lambda_j\alpha_j-\lambda_j+\lambda_j s)\prod_{k=1}^{m}\Gamma\{b_k+f_k-f_k s\}x^{-s}\;G(1-s)ds}{\sum_{i=1}^{r}\left[\Gamma(\alpha_i+1-s)\prod_{j=1}^{N-1}\Gamma(1+\mu_{ji}-\lambda_{ji}\alpha_{ji}-\lambda_{ji}+\lambda_{ji}s)\prod_{k=1}^{m}\Gamma\{e_{ki}+f_{ki}-f_{ki}s\}\right]} \tag{6.47}$$

Again, applying Parseval's theorem defined in (6.20), we finally obtain right-hand side of (6.42).

Lemma 6.1

If α,λ *and* μ are arbitrary positive real numbers $(0<\lambda_j<1)$ for $j=1\forall N$, and if

$$\int_0^\infty \sum_{j=1}^{N} u^{\alpha_j} J_{\mu_j}^{\lambda_j}(ux)f(u)du = g(x),\;\; 0<x<1 \tag{6.48}$$

Then

$$f(x)=\int_0^\infty I^{m+N,0}_{m+N-1,m+N+1;r}\left[ux\left|\begin{matrix}(a_k-g_k,g_k)_{1,m},(1-\lambda_{ji}+\mu_{ji}-\lambda_{ji}\alpha_{ji},\lambda_{ji})_{1,N-1}\\ (1-\lambda_j+\mu_j-\lambda_j\alpha_j,\lambda_j)_{1,N},(\tau_k-g_k,g_k)_{1,m},(-\alpha_j,1)\end{matrix}\right.\right]h_1(u)du \tag{6.49}$$

where

$$h_1(x)=\prod_{k=1}^{m}\{{}_k[g(x)]\},\;\; {}_k[g(x)]=\left[a_k-\tau_k,\,\tau_k g_k^{-1}-1:g_k^{-1}\right]g(x),\;\; 0<x<1.$$

provided $a_k>\tau_k$ and $\dfrac{\tau_k}{g_k}>c,\,(\xi=c+it),m>0,N>0,\,\mathrm{Re}(\tau_k-\min a_k)>0,$ $(k=1,2,\ldots,m)$ and other conditions of I-function are the same as given in (6.22).

Proof. Now, to establish the next inversion, we shall use the operator $\Im$, defined in (6.3) and the definition of Beta function (6.5). Further we replace x by v in (6.45) and multiply both sides by

$$\left(x^{\frac{1}{g_1}} - v^{\frac{1}{g_1}}\right)^{a_1-\tau_1-1} v^{\frac{\tau_1}{g_1}-1}$$

and integrate under the integral sign with respect to 0 to x and apply the same process as above we obtain the transformations of the integral Equation (6.48) given as

$$\int_0^\infty \sum_{j=1}^{N} H^{1,m}_{jm,m+2}\left[ux \left| \begin{matrix} (\ldots\ldots), (\tau_k, g_k)_{1,m}, (\ldots\ldots) \\ (\alpha_j, 1), (a_k, g_k)_{1,m}, (1+\mu_j - \lambda_j\alpha, \lambda_j) \end{matrix} \right.\right] f(u)\,du = h_1(x) \tag{6.50}$$

where

$$h_1(x) = \prod_{k=1}^{m} \{{}_k[g(x)]\}, \quad 0 < x < 1 \tag{6.51}$$

by proceeding on similar lines, we will obtain right-hand side of (6.49).

Particular case: If we put $N = 1$ in (6.48), then the solution (6.49) of the lemma reduces to the form

$$f(x) = \int_0^\infty H^{m+1,0}_{m,m+2}\left[ux \left| \begin{matrix} (\cdots\cdots), (a_k - g_k, g_k)_{1,m} \\ (1+\mu-\lambda\alpha-\lambda, \lambda), (\tau_k - g_k, g_k)_{1,m}, (-\alpha, 1) \end{matrix} \right.\right] h_1(u)\,du \tag{6.52}$$

where $h_1(x)$ is defined in (6.51). Provided $m > 0$, $\mathrm{Re}(\tau_k - \min a_k) > 0$, ($k$ = 1,2,..,m) and other conditions of I-function are same as given in (6.22) with $r = 1$.

Theorem 6.2

The following result holds true:

$$\int_0^\infty z^{\rho+\lambda-1}\sin 2z\ F\left[\begin{matrix}(a),(a');(c);(c');\\(b),(b');(d);(d');\end{matrix}\ -x^2z^2,-y^2z^2\right]I^{k,1}_{p_i,q_i;r}\left[\beta z^{2m}\left|\begin{matrix}\ldots\ldots,\ldots\ldots\\\ldots\ldots,\ldots\ldots\end{matrix}\right.\right]dz$$

$$=\sum_{n=0}^{\infty}\frac{(\lambda+2n)\Gamma(\lambda+n)}{2^{1+\rho-\lambda-2n}n!}F\left[\begin{matrix}-n,\lambda+n,(a),(a');(c);(c');\\(b),(b');(d);(d');\end{matrix}\ x^2,y^2\right]$$

$$\times I^{k+1,l+1}_{p_i+3,q_i+1;r}\left[\beta z^{-2m}\left|\begin{matrix}\left(\frac{1}{2}-n-\frac{\lambda}{2}-\frac{\rho}{2},m\right),\ldots\ldots\ldots,(1+2n+\lambda-\rho,2m),\left(1-n-\frac{\lambda}{2}-\frac{\rho}{2},m\right)\\\left(\frac{1}{2}-\rho,2m\right),\ldots\ldots\ldots\end{matrix}\right.\right]\tag{6.53}$$

where $A+A'+C\le B+B'+D,\ A+A'+C'\le B+B'+D'$,

$$R\left(\rho+\lambda+\frac{2mb_j}{\beta_j}\right)>-1\quad(j=1,\ldots,k)$$

$$R\left(\rho+\lambda+\frac{2m(a_j-1)}{\alpha_j}\right)<1\quad(j=1,\ldots,l)$$

$$M_2=\sum_{j=1}^{p_i}\alpha_{ji}-\sum_{j=1}^{q_i}\beta_{ji}\le 0$$

$$\sum_{j=1}^{l}\alpha_j-\sum_{j=l+1}^{p_i}\alpha_{ji}+\sum_{j=1}^{k}\beta_j-\sum_{j=k+1}^{q_i}\beta_{ji}=M_1>0$$

And $|\arg\beta|<1/2M_1\ \pi$.

$$\int_0^\infty z^{\rho+\lambda-1}\cos 2z\ F\left[\begin{matrix}(a),(a');(c);(c');\\(b),(b');(d);(d');\end{matrix}\ -x^2z^2,-y^2z^2\right]I^{k,1}_{p_i,q_i;r}\left[\beta z^{2m}\left|\begin{matrix}\ldots\ldots\ldots\\\ldots\ldots\ldots\end{matrix}\right.\right]dz$$

$$=\sum_{n=0}^{\infty}\frac{(\lambda+2n)\Gamma(\lambda+n)}{2^{1+\rho-\lambda-2n}n!}F\left[\begin{matrix}-n,\lambda+n,(a),(a');(c);(c');\\(b),(b');(d);(d');\end{matrix}\ x^2,y^2\right]$$

$$\times I^{k+1,l+1}_{p_i+3,q_i+1;r}\left[\beta z^{-2m}\left|\begin{matrix}\left(1-n-\frac{\lambda}{2}-\frac{\rho}{2},m\right),\ldots\ldots\ldots,(1+2n+\lambda-\rho,2m),\left(1-n-\frac{\lambda}{2}-\frac{\rho}{2},m\right)\\\left(\frac{1}{2}-\rho,2m\right),\ldots\ldots\ldots\end{matrix}\right.\right]\tag{6.54}$$

which is valid under the conditions

$$A+A'+C\leq B+B'+D,\ A+A'+C'\leq B+B'+D'$$

$$R\left(\rho+\lambda+\frac{2mb_j}{\beta_j}\right)>-1\quad (j=1,\ldots,k)$$

$$R\left(\rho+\lambda+\frac{2m(a_j-1)}{\alpha_j}\right)<1\quad (j=1,\ldots,l)$$

$M_1>0$, $M_2\leq 0$ and $|\arg\beta|<1/2M_1\ \pi$.

Proof: To prove (6.53) take the expression (6.25), multiply both sides by $f(z)$, integrate with respect to z between the limits 0 to ∞ and interchange the order of integration and summation of right-hand side, we get

$$\int_0^\infty z^\lambda F\left[\begin{matrix}(a),(a');(c);(c');\\(b),(b');(d);(d');\end{matrix}\ -x^2z^2,-y^2z^2\right]f(z)\,dz$$

$$=\sum_{n=0}^\infty\frac{(\lambda+2n)\Gamma(\lambda+n)}{n!}F\left[\begin{matrix}-n,\lambda+n,(a),(a');(c);(c');\\(b),(b');(d);(d');\end{matrix}\ x^2,y^2\right]\tag{6.55}$$

$$\times\int_0^\infty J_{\lambda+2n}(2z)f(z)\,dz,$$

For $A+A'+C\leq B+B'+D,\ A+A'+C'\leq B+B'+D'$

$$R(\lambda+\eta+1)>0$$

And

$$R(\lambda+\xi+1)<0$$

where $f(z)=O\left(|z|^{n}\right)$, for small z

And $f(z)=O\left(|z|^{\xi}\right)$, for large z

The change of integration and summation is justified [85], because

i. The series $\sum_{n=0}^{\infty} \frac{(\lambda+2n)\Gamma(\lambda+n)}{n!} F\left[\begin{matrix} -n,\lambda+n,(a),(a');(c);(c'); \\ (b),(b');(d);(d'); \end{matrix} \; x^2,y^2\right]$

Is uniformly convergent in $0 \le z \le N$, N being arbitrary.

ii. $f(z)$ is continuous function of z for all values of $z \ge z_0 > 0$;

iii. the integral on the left of (6.55) converges absolutely under the stated conditions.

Now on taking

$$f(z) = z^{\rho-1} \sin 2z I^{k,1}_{p_i,q_i;r}\left[\beta z^{2m} \left| \begin{matrix} \ldots\ldots,\ldots\ldots \\ \ldots\ldots,\ldots\ldots \end{matrix}\right.\right] dz$$

in (6.55), replacing I-function on the right by its equivalent contour integral, changing the order of integration which is justified due to the absolute convergence of the integrals, evaluating the inner integral with the help of Erdelyi [86] and interpreting it with definition of contour integral, we get (6.53).

If we take

$$f(z) = z^{\rho-1} \cos 2z I^{k,1}_{p_i,q_i;r}\left[\beta z^{2m} \left| \begin{matrix} \ldots\ldots,\ldots\ldots \\ \ldots\ldots,\ldots\ldots \end{matrix}\right.\right] dz$$

Proceed on the parallel lines as mentioned above and then in the light of the result of Truesdell [36], we obtain (6.54).

On considering the result (6.26), proceeding on the parallel lines as mentioned above and making use of the result of Erdelyi et al. [86], we get the following different forms of the integrals (6.53) and (6.54) as:

$$\int_0^\infty z^{\rho+\lambda-1} \sin 2z \; F\left[\begin{matrix} (a),(a');(c);(c'); \\ (b),(b');(d);(d'); \end{matrix} \; -x^2z^2,-y^2z^2\right] I^{k,1}_{p_i,q_i;r}\left[\beta z^{2m} \left| \begin{matrix} \ldots\ldots,\ldots\ldots \\ \ldots\ldots,\ldots\ldots \end{matrix}\right.\right] dz$$

$$= \frac{(\lambda+1)}{2^{1+\rho-\lambda}} \sum_{n=0}^{\infty} \frac{1}{2^n n!} F\left[\begin{matrix} -n,\lambda+n,(a),(a');(c);(c'); \\ (b),(b');(d);(d'); \end{matrix} \; x^2,y^2\right]$$

$$\times I^{k+1,l+1}_{p_i+3,q_i+1;r}\left[\beta z^{-2m} \left| \begin{matrix} \left(\frac{1}{2}-n-\frac{\lambda}{2}-\frac{\rho}{2},m\right),\ldots\ldots\ldots,(1+\lambda-\rho,2m),\left(1-n-\frac{\lambda}{2}-\frac{\rho}{2},m\right) \\ \left(\frac{1}{2}-\rho-n,2m\right),\ldots\ldots\ldots \end{matrix}\right.\right] \tag{6.56}$$

which is valid under the same conditions as (6.53) and

$$\int_0^\infty z^{\rho+\lambda-1}\cos 2z\, F\left[\begin{matrix}(a),(a');(c);(c');\\(b),(b');(d);(d');\end{matrix}\quad -x^2z^2,-y^2z^2\right] I^{k,1}_{p_i,q_i;r}\left[\beta z^{2m}\left|\begin{matrix}\ldots\ldots,\ldots\ldots\\\ldots\ldots,\ldots\ldots\end{matrix}\right.\right]dz$$

$$=\frac{(\lambda+1)}{2^{1+\rho-\lambda}}\sum_{n=0}^{\infty}\frac{1}{2^n n!}F\left[\begin{matrix}-n,\lambda+n,(a),(a');(c);(c');\\(b),(b');(d);(d');\end{matrix}\quad x^2,y^2\right]$$

$$\times I^{k+1,l+1}_{p_i+3,q_i+1;r}\left[\beta z^{-2m}\left|\begin{matrix}\left(1-n-\frac{\lambda}{2}-\frac{\rho}{2},m\right),\ldots\ldots\ldots,(1+\lambda-\rho,2m),\left(1-n-\frac{\lambda}{2}-\frac{\rho}{2},m\right)\\\left(\frac{1}{2}-\rho-n,2m\right),\ldots\ldots\ldots\end{matrix}\right.\right]$$

(6.57)

The conditions of validity for (6.57) are the same as for (6.54).

6.2.1 Particular Cases

For $a = b$ and $a' = b'$; the double hypergeometric function in the left breaks up into the product of two generalized hypergeometric functions and from (6.53) we thus get

$$\int_0^\infty z^{\rho+\lambda-1}\sin 2z\,{}_CF_D\left[\begin{matrix}(c);\\(d);\end{matrix}\quad -x^2z^2\right]{}_{C'}F_{D'}\left[\begin{matrix}(c');\\(d');\end{matrix}\quad -y^2z^2\right] I^{k,1}_{p_i,q_i;r}\left[\beta z^{2m}\left|\begin{matrix}\ldots\ldots,\ldots\ldots\\\ldots\ldots,\ldots\ldots\end{matrix}\right.\right]dz$$

$$=\sum_{n=0}^{\infty}\frac{(\lambda+2n)\Gamma(\lambda+n)}{2^{1+\rho-\lambda-2n}n!}F\left[\begin{matrix}-n,\lambda+n,(a),(a');(c);(c');\\(b),(b');(d);(d');\end{matrix}\quad x^2,y^2\right]$$

$$\times I^{k+1,l+1}_{p_i+3,q_i+1;r}\left[\beta z^{-2m}\left|\begin{matrix}\left(\frac{1}{2}-n-\frac{\lambda}{2}-\frac{\rho}{2},m\right),\ldots\ldots\ldots,(1+2n+\lambda-\rho,2m),\left(1-n-\frac{\lambda}{2}-\frac{\rho}{2},m\right)\\\left(\frac{1}{2}-\rho,2m\right),\ldots\ldots\ldots\end{matrix}\right.\right]$$

(6.58)

The conditions of validity for (6.58) are the same (with$A = B$, $A' = B'$) is given in (6.53).

On the other hand, since

$$F\left[\begin{matrix}(a),(a');(c);(c');\\(b),(b');(d);(d');\end{matrix}\quad x,y\right]={}_{A+A'+C}F_{B+B'+D}\left[\begin{matrix}(a),(a');(c);\\(b),(b');(d);\end{matrix}\quad x\right]$$

When $y = 0$

The special case $A = A' = B = B' = 0$ of (3) provides us

$$\int_0^\infty z^{\rho+\lambda-1} \sin 2z \, {}_C F_D \left[\begin{matrix} (c); \\ (d); \end{matrix} \; -x^2 z^2 \right] I^{k,1}_{p_i,q_i;r} \left[\beta z^{2m} \left| \begin{matrix} \ldots\ldots,\ldots\ldots \\ \ldots\ldots,\ldots\ldots \end{matrix} \right. \right] dz$$

$$= \sum_{n=0}^\infty \frac{(\lambda+2n)\Gamma(\lambda+n)}{2^{1+\rho-\lambda-2n} n!} \, {}_{C+2}F_D \left[\begin{matrix} -n, \lambda+n, (c); \\ (d); \end{matrix} \; x^2 \right]$$

$$\times I^{k+1,l+1}_{p_i+3,q_i+1;r} \left[\beta z^{-2m} \left| \begin{matrix} \left(\frac{1}{2}-n-\frac{\lambda}{2}-\frac{\rho}{2}, m\right), \ldots\ldots\ldots\ldots, (1+2n+\lambda-\rho, 2m), \left(1-n-\frac{\lambda}{2}-\frac{\rho}{2}, m\right) \\ \left(\frac{1}{2}-\rho, 2m\right), \ldots\ldots\ldots\ldots \end{matrix} \right. \right] \tag{6.59}$$

which is valid under the same conditions as for (3) with $A = A' = B = B' = C' = D' = 0$.

Further, with $C = 0$, $D = 1$, $d_1 = 1+\lambda$, $x = 1$, express ${}_0F_1$ as Bessel function, evaluate ${}_2F_1$ on the right with the help of Gauss's theorem [37, p. 49] and then on a closer examination we find

$$\int_0^\infty z^{\rho-1} J_\lambda(2z) \sin 2z \, I^{k,1}_{p_i,q_i;r} \left[\beta z^{2m} \left| \begin{matrix} \ldots\ldots,\ldots\ldots \\ \ldots\ldots,\ldots\ldots \end{matrix} \right. \right] dz$$

$$= \frac{1}{2^{1+\rho-\lambda}} I^{k+1,l+1}_{p_i+3,q_i+1;r} \left[\beta z^{-2m} \left| \begin{matrix} \left(\frac{1}{2}-n-\frac{\lambda}{2}-\frac{\rho}{2}, m\right), \ldots\ldots\ldots\ldots, (1+2n+\lambda-\rho, 2m), \left(1-n-\frac{\lambda}{2}-\frac{\rho}{2}, m\right) \\ \left(\frac{1}{2}-\rho, 2m\right), \ldots\ldots\ldots\ldots \end{matrix} \right. \right] \tag{6.60}$$

The conditions of validity for (6.60) are $M > 0$, $|\arg \beta| < 1/2M\ \pi$,

$$R\left(\rho+\lambda+\frac{2mb_j}{\beta_j}\right) > -1 \quad (j = 1, \ldots, k)$$

$$R\left(\rho+\lambda+\frac{2m(a_j-1)}{\alpha_j}\right) < \frac{1}{2} \quad (j = 1, \ldots, l)$$

Similar consequence of the integral (6.54) is the result

$$\int_0^\infty z^{\rho-1} J_\lambda(2z)\cos 2z I^{k,1}_{p_i,q_i;r}\left[\beta z^{2m}\left|\begin{matrix}\ldots\ldots,\ldots\ldots\\ \ldots\ldots,\ldots\ldots\end{matrix}\right.\right]dz$$

$$= \frac{1}{2^{1+\rho-\lambda}} I^{k+1,l+1}_{p_i+3,q_i+1;r}\left[\beta z^{-2m}\left|\begin{matrix}\left(1-\frac{\lambda}{2}-\frac{\rho}{2},m\right),\ldots\ldots\ldots,\left(\frac{1}{2}+\lambda-\rho,2m\right),\left(1-\frac{\lambda}{2}-\frac{\rho}{2},m\right)\\ \left(\frac{1}{2}-\rho,2m\right),\ldots\ldots\ldots\end{matrix}\right.\right] \tag{6.61}$$

where

$$M_1 > 0, \left|\arg\beta\right| < 1/2M_1\ \pi,$$

$$R\left(\rho+\lambda+\frac{2mb_j}{\beta_j}\right) > -1 \quad (j=1,\ldots,k)$$

$$R\left(\rho+\lambda+\frac{2m(a_j-1)}{\alpha_j}\right) < \frac{1}{2} \quad (j=1,\ldots,l)$$

It may be of interest to conclude with the remark that the integrals (6.58) and (6.59) provide some very interesting results on reducing some or all the functions occurring in the integrand on the left and it does not seem out of place to mention that especially in the light of the result of [10].

$${}_2F_3\left[\begin{matrix}\frac{a}{2}+\frac{b}{2},\frac{a}{2}+\frac{b}{2}-\frac{1}{2};\\ a,b,a+b-1;\end{matrix}\quad 4x\right] = {}_0F_1\left[\begin{matrix}-;\\ a;\end{matrix}\quad x\right]_0 F_1\left[\begin{matrix}-;\\ b;\end{matrix}\quad x\right]$$

And

$${}_2F_3\left[\begin{matrix}a,b-a;\\ b,\frac{b}{2};\frac{b}{2}+\frac{1}{2};\end{matrix}\quad x^2/4\right] = {}_1F_1\left[\begin{matrix}a;\\ b;\end{matrix}\quad -x\right]_1 F_1\left[\begin{matrix}a;\\ b;\end{matrix}\quad x\right]$$

Because by proper choice of parameters, ${}_0F_1$ can be reduced to a Bessel function and can also be transformed to ${}_1F_1$ by Kummer's second theorem [10, p. 126], the ${}_1F_1$ can also be reduced to a Whittaker function $M_{k,x}(x)$, Weber's parabolic cylinder function $D_n(x)$, Bessel function of the first kind $I_n(x)$, generalized Leguerre polynomial $L_n^\alpha(x)$ and Hermite polynomial $H_n(x)$.

Theorem 6.3

Consider the generating function $F(z,t)$ and I-function defined in (6.18) and (6.19) respectively, then

$$\int_0^1 w^{\alpha-1}(1-w)^{\gamma-\alpha-1}\prod_{i^*=1}^{R} I^{m^{(i^*)},n^{(i^*)}}_{p_i^{(i^*)},q_i^{(i^*)}:r^{(i^*)}}\left[(z_{i^*}w)^{\xi_{i^*}}\left|\begin{matrix}\left(a_j^{(i^*)},\alpha_j^{(i^*)}\right)_{1,n^{(i^*)}};\ \left(a_{ji}^{(i^*)},\alpha_{ji}^{(i^*)}\right)_{n^{(i^*)}+1,p_i^{(i^*)}}\\ \left(b_j^{(i^*)},\beta_j^{(i^*)}\right)_{1,m^{(i^*)}};\ \left(b_{ji}^{(i^*)},\beta_{ji}^{(i^*)}\right)_{m^{(i^*)}+1,q_i^{(i^*)}}\end{matrix}\right.\right]F\left[z,tw^{\eta}(1-w)^{\mu}\right]dw$$

$$=\Gamma(\gamma-\alpha)\sum_{n=0}^{\infty}C_n f_n(z)t^n(\gamma-\alpha)_{\mu n}\prod_{i^*=1}^{R} I^{m^{(i^*)},n^{(i^*)}+1}_{p_i^{(i^*)}+1,q_i^{(i^*)}+1:r^{(i^*)}}\left[z_{i^*}{}^{\xi_{i^*}}\left|\begin{matrix}(1-\alpha-\eta n,\xi_{i^*}),\left(a_j^{(i^*)},\alpha_j^{(i^*)}\right)_{1,n^{(i^*)}};\ \left(a_{ji}^{(i^*)},\alpha_{ji}^{(i^*)}\right)_{n^{(i^*)}+1,p_i^{(i^*)}}\\ \left(b_j^{(i^*)},\beta_j^{(i^*)}\right)_{1,m^{(i^*)}};\ \left(b_{ji}^{(i^*)},\beta_{ji}^{(i^*)}\right)_{m^{(i^*)}+1,q_i^{(i^*)}},(1-\gamma-\eta n-\mu n,\xi_{i^*})\end{matrix}\right.\right] \tag{6.62}$$

Provided that $\mathrm{Re}\left[\alpha+\xi_{i^*}\min\left(\dfrac{b_j^{(i^*)}}{\beta_j^{(i^*)}}\right)\right]>0$, $\mathrm{Re}\ [\gamma-\alpha]>0$, $\mathrm{Re}\ (\eta)>0$, $\mathrm{Re}\ (\mu)>0$ and $\left|\arg z_{i^*}{}^{\xi_{i^*}}\right|<\frac{1}{2}\pi B_i$, $j=1,2,\ldots,m$, $i=1,2,\ldots,r$, and $i^*=1,2,\ldots,R$,

where $B_i=\sum_{j=1}^{n}\alpha_j^{(i^*)}-\sum_{j=n+1}^{p_i}\alpha_j^{(i^*)}+\sum_{j=1}^{m}\beta_j^{(i^*)}-\sum_{j=m+1}^{q_i}\beta_j^{(i^*)}$.

Theorem 6.4

Consider the generating function $F(z,t)$ and I-function defined in (6.18) and (6.19) respectively, then

$$\int_0^1 w^{\alpha-1}(1-w)^{\gamma-\alpha-1}\prod_{i^*=1}^{R} I^{m^{(i^*)},n^{(i^*)}}_{p_i^{(i^*)},q_i^{(i^*)}:r^{(i^*)}}\left[(z_{i^*}(1-w))^{\xi_{i^*}}\left|\begin{matrix}\left(a_j^{(i^*)},\alpha_j^{(i^*)}\right)_{1,n^{(i^*)}};\ \left(a_{ji}^{(i^*)},\alpha_{ji}^{(i^*)}\right)_{n^{(i^*)}+1,p_i^{(i^*)}}\\ \left(b_j^{(i^*)},\beta_j^{(i^*)}\right)_{1,m^{(i^*)}};\ \left(b_{ji}^{(i^*)},\beta_{ji}^{(i^*)}\right)_{m^{(i^*)}+1,q_i^{(i^*)}}\end{matrix}\right.\right]F\left[z,tw^{\eta}(1-w)^{\mu}\right]dw$$

$$= \Gamma(\alpha)\sum_{n=0}^{\infty} C_n f_n(z) t^n (\alpha)_{\eta n} \prod_{i^*=1}^{R} I^{m^{(i^*)}, n^{(i^*)}+1}_{p_i^{(i^*)}+1, q_i^{(i^*)}+1: r^{(i^*)}}$$

$$\left[z_{i^*}^{\xi_{i^*}} \left| \begin{matrix} (1+\alpha-\mu n, \xi_{i^*}), \left(a_j^{(i^*)}, \alpha_j^{(i^*)}\right)_{1,n^{(i^*)}}; & \left(a_{ji}^{(i^*)}, \alpha_{ji}^{(i^*)}\right)_{n^{(i^*)}+1, p_i^{(i^*)}} \\ \left(b_j^{(i^*)}, \beta_j^{(i^*)}\right)_{1,m^{(i^*)}}; & \left(b_{ji}^{(i^*)}, \beta_{ji}^{(i^*)}\right)_{m^{(i^*)}+1, q_i^{(i^*)}}, (1-\gamma-\eta n-\mu n, \xi_{i^*}) \end{matrix} \right. \right] \tag{6.63}$$

Provided that $\mathrm{Re}\left[\gamma-\alpha+\xi_{i^*}\, min\left(\frac{b_j^{(i^*)}}{\beta_j^{(i^*)}}\right)\right] > 0$, $\mathrm{Re}[\alpha] > 0$, $\mathrm{Re}(\eta) > 0$, $\mathrm{Re}(\mu) > 0$ and $\left|\arg z_{i^*}^{\xi_{i^*}}\right| < \frac{1}{2}\pi B_i$, $j = 1, 2, \ldots, m$, $i = 1, 2, \ldots, r$, and $i^* = 1, 2, \ldots, R$,

where $B_i = \sum_{j=1}^{n} \alpha_j^{(i^*)} - \sum_{j=n+1}^{p_i} \alpha_j^{(i^*)} + \sum_{j=1}^{m} \beta_j^{(i^*)} - \sum_{j=m+1}^{q_i} \beta_j^{(i^*)}$.

Theorem 6.5

Consider the generating function $F(z,t)$ and I-function defined in (6.18) and (6.19) respectively, then

$$\int_0^1 w^{\alpha-1}(1-w)^{\gamma-\alpha-1} \prod_{i^*=1}^{R} I^{m^{(i^*)}, n^{(i^*)}}_{p_i^{(i^*)}, q_i^{(i^*)}: r^{(i^*)}}$$

$$\left[\left(z_{i^*}(1-w)w\right)^{\xi_{i^*}} \left| \begin{matrix} \left(a_j^{(i^*)}, \alpha_j^{(i^*)}\right)_{1,n^{(i^*)}}; & \left(a_{ji}^{(i^*)}, \alpha_{ji}^{(i^*)}\right)_{n^{(i^*)}+1, p_i^{(i^*)}} \\ \left(b_j^{(i^*)}, \beta_j^{(i^*)}\right)_{1,m^{(i^*)}}; & \left(b_{ji}^{(i^*)}, \beta_{ji}^{(i^*)}\right)_{m^{(i^*)}+1, q_i^{(i^*)}} \end{matrix} \right. \right] F\left[z, tw^{\eta}(1-w)^{\mu}\right] dw$$

$$= \sum_{n=0}^{\infty} C_n f_n(z) t^n \prod_{i^*=1}^{R} I^{m^{(i^*)}, n^{(i^*)}+2}_{p_i^{(i^*)}+2, q_i^{(i^*)}+1: r^{(i^*)}}$$

$$\left[z_{i^*}^{\xi_{i^*}} \left| \begin{matrix} (1-\alpha-\eta n, \xi_{i^*}), (1+\alpha-\gamma-\mu n, \xi_{i^*}), \left(a_j^{(i^*)}, \alpha_j^{(i^*)}\right)_{1,n^{(i^*)}}; & \left(a_{ji}^{(i^*)}, \alpha_{ji}^{(i^*)}\right)_{n^{(i^*)}+1, p_i^{(i^*)}} \\ \left(b_j^{(i^*)}, \beta_j^{(i^*)}\right)_{1,m^{(i^*)}}; & \left(b_{ji}^{(i^*)}, \beta_{ji}^{(i^*)}\right)_{m^{(i^*)}+1, q_i^{(i^*)}}, (1-\gamma-\eta n-\mu n, 2\xi_{i^*}) \end{matrix} \right. \right] \tag{6.64}$$

Provided that $\mathrm{Re}\left[\alpha+\xi_{i^*}\,\min\left(\frac{b_j^{(i^*)}}{\beta_j^{(i^*)}}\right)\right] > 0$, $\mathrm{Re}\left[\gamma-\alpha+\xi_{i^*}\,\min\left(\frac{b_j^{(i^*)}}{\beta_j^{(i^*)}}\right)\right] > 0$, $\mathrm{Re}(\eta) > 0$, $\mathrm{Re}(\mu) > 0$ and $\left|\arg z_{i^*}^{\xi_{i^*}}\right| < \frac{1}{2}\pi B_i$, $j = 1, 2, \ldots, m$, $i = 1, 2, \ldots, r$, and $i^* = 1, 2, \ldots, R$, where $B_i = \sum_{j=1}^{n} \alpha_j^{(i^*)} - \sum_{j=n+1}^{p_i} \alpha_j^{(i^*)} + \sum_{j=1}^{m} \beta_j^{(i^*)} - \sum_{j=m+1}^{q_i} \beta_j^{(i^*)}$.

Now we state another modification of the Theorem XVII, which can be used to obtain various new interesting generalizations of elliptic-type integrals.

Theorem 6.6

Consider the generating function $F(z,t)$ and I-function defined in (6.18) and (6.19) respectively, then

$$\int_0^1 w^{\beta-1}(1-w)^{\alpha-1} \prod_{i^*=1}^{R} I^{m^{(i^*)},n^{(i^*)}}_{p_i^{(i^*)},q_i^{(i^*)};r^{(i^*)}}$$

$$\left[(z_{i^*}w)^{\xi_{i^*}} \left| \begin{matrix} \left(a_j^{(i^*)},\alpha_j^{(i^*)}\right)_{1,n^{(i^*)}} ; & \left(a_{ji}^{(i^*)},\alpha_{ji}^{(i^*)}\right)_{n^{(i^*)}+1,p_i^{(i^*)}} \\ \left(b_j^{(i^*)},\beta_j^{(i^*)}\right)_{1,m^{(i^*)}} ; & \left(b_{ji}^{(i^*)},\beta_{ji}^{(i^*)}\right)_{m^{(i^*)}+1,q_i^{(i^*)}} \end{matrix} \right. \right]$$

$$\times \prod_{l=1}^{M}\left[1-\frac{2wk_l^2}{k_l^2-1}\right]^{-\tau_l} \prod_{l'=1}^{N}\left[1-\frac{(\delta_{l'}-\rho_{l'})w}{1+\delta_{l'}}\right]^{-\lambda_{l'}} F\left[z, tw^{\eta}(1-w)^{\mu}\right]dw$$

$$=\Gamma(\alpha)\sum_{n=0}^{\infty} C_n f_n(z) t^n (\alpha)_{\mu n} \prod_{l=1}^{M}\sum_{n_l=0}^{\infty} \frac{(\tau_l)_{n_l}}{n_l!}\left(\frac{2k_l^2}{k_l^2-1}\right)^{n_l}$$

$$\prod_{l'=1}^{N}\sum_{n_{l'}=0}^{\infty}\left(\frac{(\delta_{l'}-\rho_{l'})}{1+\delta_{l'}}\right)^{n_{l'}} \frac{(\lambda_{l'})_{n_{l'}}}{n_{l'}!} \prod_{i^*=1}^{R} I^{m^{(i^*)},n^{(i^*)}+1}_{p_i^{(i^*)}+1,q_i^{(i^*)}+1:r^{(i^*)}}$$

$$\left[z_{i^*}{}^{\xi_{i^*}} \left| \begin{matrix} (1-\beta-n_l-n_{l'}-\eta n,\xi_{i^*}),\left(a_j^{(i^*)},\alpha_j^{(i^*)}\right)_{1,n^{(i^*)}} ; & \left(a_{ji}^{(i^*)},\alpha_{ji}^{(i^*)}\right)_{n^{(i^*)}+1,p_i^{(i^*)}} \\ \left(b_j^{(i^*)},\beta_j^{(i^*)}\right)_{1,m^{(i^*)}} ; & \left(b_{ji}^{(i^*)},\beta_{ji}^{(i^*)}\right)_{m^{(i^*)}+1,q_i^{(i^*)}},(1-\alpha-\beta-n_l-n_{l'}-\eta n-\mu n,\xi_{i^*}) \end{matrix} \right. \right] \tag{6.65}$$

Provided that $\mathrm{Re}\left[\beta+\xi_{i^*}\min\left(\frac{b_j^{(i^*)}}{\beta_j^{(i^*)}}\right)\right]>0$, $\mathrm{Re}[\alpha]>0$, $\mathrm{Re}(\eta)>0$, $\mathrm{Re}(\mu)>0$; $\delta_{l'}, \rho_{l'}, \lambda_{l'}, \tau_l \in \mathbb{C}$; $|k_l|<1$ and max $\left\{|\rho_{l'}|,|\delta_{l'}|,\left|\frac{2k_l^2}{k_l^2-1}\right|,\left|\frac{(\delta_{l'}-\rho_{l'})}{1+\delta_{l'}}\right|<1\right\} l'=1,2\ldots.N,\ l=1,2,\ldots.,M$ and $\left|\arg z_{i^*}{}^{\xi_{i^*}}\right|<\frac{1}{2}\pi B_i$, $j=1,2,\ldots.,m$, $i=1,2,\ldots.,r$, and $i^*=1,2,\ldots.,R$,

where $B_i = \sum_{j=1}^{n} \alpha_j^{(i^*)} - \sum_{j=n+1}^{p_i} \alpha_j^{(i^*)} + \sum_{j=1}^{m} \beta_j^{(i^*)} - \sum_{j=m+1}^{q_i} \beta_j^{(i^*)}$.

Proof: Expressing $F[z,t]$ as a power series from (6.18) in the integral (6.62), changing the order of integration and summation, which is permissible due to uniform convergence of the series involved and using the definition (6.19) of the I-function in the evaluation of the resulting integral, we get the result (6.62), which proves Theorem 6.3. The proofs of Theorem 6.4–6.6 are similar to that of the Theorem 6.1.

6.2.2 Applications

Keeping in mind the usefulness of the theorems discussed in the last section, we are concerned here with some interesting applications, which demonstrate the generality of the results obtained in this article.

i. Consider the generating function [31]

$$F(z,t)=(1-zt)^{-\sigma}=\sum_{n=0}^{\infty}(\sigma)_n\frac{z^n t^n}{n!} \tag{6.66}$$

And by the use of the Theorem 6.3, under the stated condition, we get the following

$$\begin{aligned}&\int_0^1 w^{\alpha-1}(1-w)^{\gamma-\alpha-1}\prod_{i^*=1}^{R} I^{m^{(i^*)},n^{(i^*)}}_{p_i^{(i^*)},q_i^{(i^*)};r^{(i^*)}}\\&\left[(z_{i^*}w)^{\xi_{i^*}}\left|\begin{matrix}\left(a_j^{(i^*)},\alpha_j^{(i^*)}\right)_{1,n^{(i^*)}}; & \left(a_{ji}^{(i^*)},\alpha_{ji}^{(i^*)}\right)_{n^{(i^*)}+1,p_i^{(i^*)}}\\ \left(b_j^{(i^*)},\beta_j^{(i^*)}\right)_{1,m^{(i^*)}}; & \left(b_{ji}^{(i^*)},\beta_{ji}^{(i^*)}\right)_{m^{(i^*)}+1,q_i^{(i^*)}}\end{matrix}\right.\right]\left[1-ztw^{\eta}(1-w)^{\mu}\right]^{-\sigma}dw\\&-\Gamma(\gamma-\alpha)\sum_{n=0}^{\infty}\frac{(\sigma)_n z^n t^n(\gamma-\alpha)_{\mu n}}{n!}\prod_{i^*=1}^{R} I^{m^{(i^*)},n^{(i^*)}+1}_{p_i^{(i^*)}+1,q_i^{(i^*)}+1;r^{(i^*)}}\\&\left[z_{i^*}{}^{\xi_{i^*}}\left|\begin{matrix}(1-\alpha-\eta n,\xi_{i^*}),\left(a_j^{(i^*)},\alpha_j^{(i^*)}\right)_{1,n^{(i^*)}}; & \left(a_{ji}^{(i^*)},\alpha_{ji}^{(i^*)}\right)_{n^{(i^*)}+1,p_i^{(i^*)}}\\ \left(b_j^{(i^*)},\beta_j^{(i^*)}\right)_{1,m^{(i^*)}}; & \left(b_{ji}^{(i^*)},\beta_{ji}^{(i^*)}\right)_{m^{(i^*)}+1,q_i^{(i^*)}},(1-\gamma-\eta n-\mu n,\xi_{i^*})\end{matrix}\right.\right]\end{aligned} \tag{6.67}$$

When we put $w=\cos^2\left(\frac{\theta}{2}\right)$ and $\cos\theta=2\cos^2\left(\frac{\theta}{2}\right)-1$, then the Equation (6.67) gives the following generalization of the elliptic-type integral

$$\int_0^{\pi} \cos^{2\alpha-1}\left(\frac{\theta}{2}\right)\sin^{2\gamma-2\alpha-1}\left(\frac{\theta}{2}\right)\prod_{i^*=1}^{R} I^{m^{(i^*)},n^{(i^*)}}_{p_i^{(i^*)},q_i^{(i^*)}:r^{(i^*)}}\left[\left\{z_{i^*}\cos^2\left(\frac{\theta}{2}\right)\right\}^{\xi_{i^*}}\right.$$

$$\left.\begin{array}{l|l} & \left(a_j^{(i^*)},\alpha_j^{(i^*)}\right)_{1,n^{(i^*)}};\ \left(a_{ji}^{(i^*)},\alpha_{ji}^{(i^*)}\right)_{n^{(i^*)}+1,p_i^{(i^*)}} \\ & \left(b_j^{(i^*)},\beta_j^{(i^*)}\right)_{1,m^{(i^*)}};\ \left(b_{ji}^{(i^*)},\beta_{ji}^{(i^*)}\right)_{m^{(i^*)}+1,q_i^{(i^*)}}\end{array}\right]\left[1-zt\cos^{2\eta}\left(\frac{\theta}{2}\right)\sin^{2\mu}\left(\frac{\theta}{2}\right)\right]^{-\sigma} d\theta$$

$$= \Gamma(\gamma-\alpha)\sum_{n=0}^{\infty}\frac{(\sigma)_n z^n t^n (\gamma-\alpha)_{\mu n}}{n!}\prod_{i^*=1}^{R} I^{m^{(i^*)},n^{(i^*)}+1}_{p_i^{(i^*)}+1,q_i^{(i^*)}+1:r^{(i^*)}}$$

$$\left[z_{i^*}{}^{\xi_{i^*}}\left|\begin{array}{l}(1-\alpha-\eta n,\xi_{i^*}),\left(a_j^{(i^*)},\alpha_j^{(i^*)}\right)_{1,n^{(i^*)}};\ \left(a_{ji}^{(i^*)},\alpha_{ji}^{(i^*)}\right)_{n^{(i^*)}+1,p_i^{(i^*)}} \\ \left(b_j^{(i^*)},\beta_j^{(i^*)}\right)_{1,m^{(i^*)}};\ \left(b_{ji}^{(i^*)},\beta_{ji}^{(i^*)}\right)_{m^{(i^*)}+1,q_i^{(i^*)}},(1-\gamma-\eta n-\mu n,\xi_{i^*})\end{array}\right.\right]. \tag{6.68}$$

If we set $w=\sin^2\left(\frac{\theta}{2}\right)$ and using $\cos\theta = 1-2\sin^2\left(\frac{\theta}{2}\right)$ and $\sigma \to 0$ in (6.67), we get the following result:

$$\int_0^{\pi} \sin^{2\alpha-1}\left(\frac{\theta}{2}\right)\cos^{2\gamma-2\alpha-1}\left(\frac{\theta}{2}\right)\prod_{i^*=1}^{R} I^{m^{(i^*)},n^{(i^*)}}_{p_i^{(i^*)},q_i^{(i^*)}:r^{(i^*)}}$$

$$\left[\left\{z_{i^*}\sin^2\left(\frac{\theta}{2}\right)\right\}^{\xi_{i^*}}\left|\begin{array}{l}\left(a_j^{(i^*)},\alpha_j^{(i^*)}\right)_{1,n^{(i^*)}};\ \left(a_{ji}^{(i^*)},\alpha_{ji}^{(i^*)}\right)_{n^{(i^*)}+1,p_i^{(i^*)}} \\ \left(b_j^{(i^*)},\beta_j^{(i^*)}\right)_{1,m^{(i^*)}};\ \left(b_{ji}^{(i^*)},\beta_{ji}^{(i^*)}\right)_{m^{(i^*)}+1,q_i^{(i^*)}}\end{array}\right.\right] d\theta \tag{6.69}$$

$$= \Gamma(\gamma-\alpha)\prod_{i^*=1}^{R} I^{m^{(i^*)},n^{(i^*)}+1}_{p_i^{(i^*)}+1,q_i^{(i^*)}+1:r^{(i^*)}}$$

$$\left[z_{i^*}{}^{\xi_{i^*}}\left|\begin{array}{l}(1-\alpha,\xi_{i^*}),\left(a_j^{(i^*)},\alpha_j^{(i^*)}\right)_{1,n^{(i^*)}};\ \left(a_{ji}^{(i^*)},\alpha_{ji}^{(i^*)}\right)_{n^{(i^*)}+1,p_i^{(i^*)}} \\ \left(b_j^{(i^*)},\beta_j^{(i^*)}\right)_{1,m^{(i^*)}};\ \left(b_{ji}^{(i^*)},\beta_{ji}^{(i^*)}\right)_{m^{(i^*)}+1,q_i^{(i^*)}},(1-\gamma,\xi_{i^*})\end{array}\right.\right].$$

It can be seen that the above elliptic-type integral (6.67) also provides generalization to a number of new families of elliptic-type integrals, which also generalizes known families of elliptic integrals. Also using the generating function (6.66) and applying the Theorem 6.4, under the stated conditions, we have obtained the following new family of elliptic-type integrals, which also generalizes known families of elliptic-type integrals.

$$\int_0^1 w^{\beta-1}(1-w)^{\alpha-1}\prod_{i^*=1}^{R} I^{m^{(i^*)},n^{(i^*)}}_{p_i^{(i^*)},q_i^{(i^*)}:r^{(i^*)}}\left[(z_{i^*}w)^{\xi_{i^*}}\left|\begin{matrix}\left(a_j^{(i^*)},\alpha_j^{(i^*)}\right)_{1,n^{(i^*)}};\ \left(a_{ji}^{(i^*)},\alpha_{ji}^{(i^*)}\right)_{n^{(i^*)}+1,p_i^{(i^*)}}\\ \left(b_j^{(i^*)},\beta_j^{(i^*)}\right)_{1,m^{(i^*)}};\ \left(b_{ji}^{(i^*)},\beta_{ji}^{(i^*)}\right)_{m^{(i^*)}+1,q_i^{(i^*)}}\end{matrix}\right.\right]$$

$$\times\prod_{l=1}^{M}\left[1-\frac{2wk_l^2}{k_l^2-1}\right]^{-\tau_l}\prod_{l'=1}^{N}\left[1-\frac{(\delta_{l'}-\rho_{l'})w}{1+\delta_{l'}}\right]^{-\lambda_{l'}}\left[1-ztw^{\eta}(1-w)^{\mu}\right]^{-\sigma}dw$$

$$=\Gamma(\alpha)\sum_{n=0}^{\infty}(\sigma)_n z^n t^n(\alpha)_{\mu n}\prod_{l=1}^{M}\sum_{n_l}^{\infty}\frac{(\tau_l)_{n_l}}{n_l!}\left(\frac{2k_l^2}{k_l^2-1}\right)^{n_l}$$

$$\prod_{l'=1}^{N}\sum_{n_{l'}=0}^{\infty}\left(\frac{(\delta_{l'}-\rho_{l'})}{1+\delta_{l'}}\right)^{n_{l'}}\frac{(\lambda_{l'})_{n_{l'}}}{n_{l'}!}\prod_{i^*=1}^{R} I^{m^{(i^*)},n^{(i^*)}+1}_{p_i^{(i^*)}+1,q_i^{(i^*)}+1:r^{(i^*)}}$$

$$\left[z_{i^*}^{\xi_{i^*}}\left|\begin{matrix}(1-\beta-n_l-n_{l'}-\eta n,\xi_{i^*}),\left(a_j^{(i^*)},\alpha_j^{(i^*)}\right)_{1,n^{(i^*)}};\ \left(a_{ji}^{(i^*)},\alpha_{ji}^{(i^*)}\right)_{n^{(i^*)}+1,p_i^{(i^*)}}\\ \left(b_j^{(i^*)},\beta_j^{(i^*)}\right)_{1,m^{(i^*)}};\ \left(b_{ji}^{(i^*)},\beta_{ji}^{(i^*)}\right)_{m^{(i^*)}+1,q_i^{(i^*)}},(1-\alpha-\beta-n_l-n_{l'}-\eta n-\mu n,\xi_{i^*})\end{matrix}\right.\right] \tag{6.70}$$

For $\sigma \to 0$ and $w=\sin^2\left(\frac{\theta}{2}\right)$, the Equality (6.70) gives the following explicit representation of generalized family of the elliptic-type integral.

$$\int_0^1 \cos^{2\alpha-1}\left(\frac{\theta}{2}\right)\sin^{2\beta-1}\left(\frac{\theta}{2}\right)\prod_{i^*=1}^{R} I^{m^{(i^*)},n^{(i^*)}}_{p_i^{(i^*)},q_i^{(i^*)}:r^{(i^*)}}\left[\left(z_{i^*}\sin^2\left(\frac{\theta}{2}\right)\right)^{\xi_{i^*}}\left|\begin{matrix}\left(a_j^{(i^*)},\alpha_j^{(i^*)}\right)_{1,n^{(i^*)}};\ \left(a_{ji}^{(i^*)},\alpha_{ji}^{(i^*)}\right)_{n^{(i^*)}+1,p_i^{(i^*)}}\\ \left(b_j^{(i^*)},\beta_j^{(i^*)}\right)_{1,m^{(i^*)}};\ \left(b_{ji}^{(i^*)},\beta_{ji}^{(i^*)}\right)_{m^{(i^*)}+1,q_i^{(i^*)}}\end{matrix}\right.\right]$$

$$\times\prod_{l=1}^{M}\left[1-k_l^2\cos\theta\right]^{-\tau_l}\prod_{l'=1}^{N}\left[1+\rho_{l'}\sin^2\left(\frac{\theta}{2}\right)+\delta_{l'}\cos^2\left(\frac{\theta}{2}\right)\right]^{-\lambda_{l'}}d\theta$$

$$=\Gamma(\alpha)\prod_{l=1}^{M}\left[1-k_l^2\right]^{-\tau_l}\prod_{l'=1}^{N}\left[1+\delta_{l'}\right]^{-\lambda_{l'}}\prod_{l=1}^{M}\sum_{n_l=0}^{\infty}\frac{(\tau_l)_{n_l}}{n_l!}\left(\frac{2k_l}{k_l^2-1}\right)^{n_l}$$

$$\times\prod_{l'=1}^{N}\sum_{n_{l'}=0}^{\infty}\left(\frac{(\delta_{l'}-\rho_{l'})}{1+\delta_{l'}}\right)^{n_{l'}}\frac{(\lambda_{l'})_{n_{l'}}}{n_{l'}!}\prod_{i^*=1}^{R} I^{m^{(i^*)},n^{(i^*)}+1}_{p_i^{(i^*)}+1,q_i^{(i^*)}+1:r^{(i^*)}}$$

$$\left[z_{i^*}^{\xi_{i^*}}\left|\begin{matrix}(1-\beta-n_l-n_{l'},\xi_{i^*}),\left(a_j^{(i^*)},\alpha_j^{(i^*)}\right)_{1,n^{(i^*)}};\ \left(a_{ji}^{(i^*)},\alpha_{ji}^{(i^*)}\right)_{n^{(i^*)}+1,p_i^{(i^*)}}\\ \left(b_j^{(i^*)},\beta_j^{(i^*)}\right)_{1,m^{(i^*)}};\ \left(b_{ji}^{(i^*)},\beta_{ji}^{(i^*)}\right)_{m^{(i^*)}+1,q_i^{(i^*)}},(1-\alpha-\beta-n_l-n_{l'},\xi_{i^*})\end{matrix}\right.\right] \tag{6.71}$$

ii. Consider the generating function [90]

$$F(z,t)=(1-z_1t)^{-\alpha_1}(1-z_2t)^{-\alpha_2}=\sum_{n=0}^{\infty}g_n^{\alpha_1,\alpha_2}(z_1,z_2)t^n, \tag{6.72}$$

where $g_n^{\alpha_1,\alpha_2}$ is the Lagrange polynomial defined by

$$g_n^{\alpha_1,\alpha_2}(x,y)=\sum_{n=0}^{\infty}\frac{(\alpha_1)_r(\alpha_2)_{n-r}}{r!(n-r)!}x^r y^{n-r}. \tag{6.73}$$

By the application of the Theorem 6.3, under the stated condition, we get

$$\int_0^1 w^{\alpha-1}(1-w)^{\gamma-\alpha-1}\prod_{i^*=1}^{R} I^{m^{(i^*)},n^{(i^*)}}_{p_i^{(i^*)},q_i^{(i^*)};r^{(i^*)}}\left[(z_{i^*}w)^{\xi_{i^*}}\left|\begin{matrix}\left(a_j^{(i^*)},\alpha_j^{(i^*)}\right)_{1,n^{(i^*)}}; & \left(a_{ji}^{(i^*)},\alpha_{ji}^{(i^*)}\right)_{n^{(i^*)}+1,p_i^{(i^*)}}\\ \left(b_j^{(i^*)},\beta_j^{(i^*)}\right)_{1,m^{(i^*)}}; & \left(b_{ji}^{(i^*)},\beta_{ji}^{(i^*)}\right)_{m^{(i^*)}+1,q_i^{(i^*)}}\end{matrix}\right.\right]\sum_{j^*=1}^{2}\left[1-z_{j^*}tw^{\eta}(1-w)^{\mu}\right]^{-\sigma_{j^*}}dw$$

$$=\Gamma(\gamma-\alpha)\sum_{n=0}^{\infty}g_n^{\sigma_1,\sigma_2}(z_1,z_2)t^n(\gamma-\alpha)_{\mu n}\prod_{i^*=1}^{R} I^{m^{(i^*)},n^{(i^*)}+1}_{p_i^{(i^*)}+1,q_i^{(i^*)}+1:r^{(i^*)}}$$

$$\left[z_{i^*}{}^{\xi_{i^*}}\left|\begin{matrix}(1-\alpha-\eta n,\xi_{i^*}),\left(a_j^{(i^*)},\alpha_j^{(i^*)}\right)_{1,n^{(i^*)}}; & \left(a_{ji}^{(i^*)},\alpha_{ji}^{(i^*)}\right)_{n^{(i^*)}+1,p_i^{(i^*)}}\\ \left(b_j^{(i^*)},\beta_j^{(i^*)}\right)_{1,m^{(i^*)}}; & \left(b_{ji}^{(i^*)},\beta_{ji}^{(i^*)}\right)_{m^{(i^*)}+1,q_i^{(i^*)}},(1-\gamma-\eta n-\mu n,\xi_{i^*})\end{matrix}\right.\right]. \tag{6.74}$$

Also, by the application of Theorem 6.6 under the stated conditions and by the use of the generating relation (6.72), we get

$$\int_0^1 w^{\beta-1}(1-w)^{\alpha-1}\prod_{i^*=1}^{R} I^{m^{(i^*)},n^{(i^*)}}_{p_i^{(i^*)},q_i^{(i^*)};r^{(i^*)}}\left[(z_{i^*}w)^{\xi_{i^*}}\left|\begin{matrix}\left(a_j^{(i^*)},\alpha_j^{(i^*)}\right)_{1,n^{(i^*)}}; & \left(a_{ji}^{(i^*)},\alpha_{ji}^{(i^*)}\right)_{n^{(i^*)}+1,p_i^{(i^*)}}\\ \left(b_j^{(i^*)},\beta_j^{(i^*)}\right)_{1,m^{(i^*)}}; & \left(b_{ji}^{(i^*)},\beta_{ji}^{(i^*)}\right)_{m^{(i^*)}+1,q_i^{(i^*)}}\end{matrix}\right.\right]$$

$$\times\prod_{l=1}^{M}\left[1-\frac{2wk_l^2}{k_l^2-1}\right]^{-\tau_l}\prod_{l'=1}^{N}\left[1-\frac{(\delta_{l'}-\rho_{l'})w}{1+\delta_{l'}}\right]^{-\lambda_{l'}}\sum_{j^*=1}^{2}\left[1-z_{j^*}tw^{\eta}(1-w)^{\mu}\right]^{-\sigma_{j^*}}dw$$

$$= \Gamma(\alpha)\sum_{n=0}^{\infty} g_n^{\sigma_1,\sigma_2}(z_1,z_2)\,t^n(\alpha)_{\mu n}\prod_{l=1}^{M}\sum_{n_l=0}^{\infty}\frac{(\tau_l)_{n_l}}{n_l\,!}\left(\frac{2k_l^2}{k_l^2-1}\right)^{n_l}$$

$$\prod_{l'=1}^{N}\sum_{n_{l'}=0}^{\infty}\left(\frac{(\delta_{l'}-\rho_{l'})}{1+\delta_{l'}}\right)^{n_{l'}}\frac{(\lambda_{l'})_{n_{l'}}}{n_{l'}\,!}\prod_{i^*=1}^{R} I_{p_i^{(i^*)}+1,q_i^{(i^*)}+1:r^{(i^*)}}^{m^{(i^*)},n^{(i^*)}+1}\Bigg[$$

$$z_{i^*}^{\xi_{i^*}}\left|\begin{matrix}(1-\beta-n_l-n_{l'}-\eta n,\xi_{i^*}),\left(a_j^{(i^*)},\alpha_j^{(i^*)}\right)_{1,n^{(i^*)}}; & \left(a_{ji}^{(i^*)},\alpha_{ji}^{(i^*)}\right)_{n^{(i^*)}+1,p_i^{(i^*)}} \\ \left(b_j^{(i^*)},\beta_j^{(i^*)}\right)_{1,m^{(i^*)}}; & \left(b_{ji}^{(i^*)},\beta_{ji}^{(i^*)}\right)_{m^{(i^*)}+1,q_i^{(i^*)}},(1-\alpha-\beta-n_l-n_{l'}-\eta n-\mu n,\xi_{i^*})\end{matrix}\right.\Bigg] \tag{6.75}$$

iii. Consider the well-known generating function

$$F(z,t)=e^{-zt}=\sum_{n=0}^{\infty}\frac{(-1)^n z^n t^n}{n!} \tag{6.76}$$

By the use of Theorem 6.3, under the stated conditions, we get

$$\int_0^1 w^{\alpha-1}(1-w)^{\gamma-\alpha-1}\prod_{i^*=1}^{R} I_{p_i^{(i^*)},q_i^{(i^*)}:r^{(i^*)}}^{m^{(i^*)},n^{(i^*)}}$$

$$\left[(z_{i^*}w)^{\xi_{i^*}}\left|\begin{matrix}\left(a_j^{(i^*)},\alpha_j^{(i^*)}\right)_{1,n^{(i^*)}}; & \left(a_{ji}^{(i^*)},\alpha_{ji}^{(i^*)}\right)_{n^{(i^*)}+1,p_i^{(i^*)}} \\ \left(b_j^{(i^*)},\beta_j^{(i^*)}\right)_{1,m^{(i^*)}}; & \left(b_{ji}^{(i^*)},\beta_{ji}^{(i^*)}\right)_{m^{(i^*)}+1,q_i^{(i^*)}}\end{matrix}\right.\right]e^{-zt}\left[w^{\eta}(1-w)^{\mu}\right]dw$$

$$=\Gamma(\gamma-\alpha)\sum_{n=0}^{\infty}\frac{t^n(\gamma-\alpha)_{\mu n}(-z)^n}{n!}\prod_{i^*=1}^{R} I_{p_i^{(i^*)}+1,q_i^{(i^*)}+1:r^{(i^*)}}^{m^{(i^*)},n^{(i^*)}+1}$$

$$\left[z_{i^*}^{\xi_{i^*}}\left|\begin{matrix}(1-\alpha-\eta n,\xi_{i^*}),\left(a_j^{(i^*)},\alpha_j^{(i^*)}\right)_{1,n^{(i^*)}}; & \left(a_{ji}^{(i^*)},\alpha_{ji}^{(i^*)}\right)_{n^{(i^*)}+1,p_i^{(i^*)}} \\ \left(b_j^{(i^*)},\beta_j^{(i^*)}\right)_{1,m^{(i^*)}}; & \left(b_{ji}^{(i^*)},\beta_{ji}^{(i^*)}\right)_{m^{(i^*)}+1,q_i^{(i^*)}},(1-\gamma-\eta n-\mu n,\xi_{i^*})\end{matrix}\right.\right] \tag{6.77}$$

Also by application of Theorem 6.6, under the stated conditions, and by the use of the generating function (iii), we get the following useful integral

$$\int_0^1 w^{\beta-1}(1-w)^{\alpha-1}\prod_{i^*=1}^{R} I^{m^{(i^*)},n^{(i^*)}}_{p_i^{(i^*)},q_i^{(i^*)};r^{(i^*)}}\left[(z_{i^*}w)^{\xi_{i^*}}\left|\begin{matrix}\left(a_j^{(i^*)},\alpha_j^{(i^*)}\right)_{1,n^{(i^*)}};\ \left(a_{ji}^{(i^*)},\alpha_{ji}^{(i^*)}\right)_{n^{(i^*)}+1,p_i^{(i^*)}}\\ \left(b_j^{(i^*)},\beta_j^{(i^*)}\right)_{1,m^{(i^*)}};\ \left(b_{ji}^{(i^*)},\beta_{ji}^{(i^*)}\right)_{m^{(i^*)}+1,q_i^{(i^*)}}\end{matrix}\right.\right]$$

$$\times\prod_{l=1}^{M}\left[1-\frac{2wk_l^2}{k_l^2-1}\right]^{-\tau_l}\prod_{l'=1}^{N}\left[1-\frac{(\delta_{l'}-\rho_{l'})w}{1+\delta_{l'}}\right]^{-\lambda_{l'}} e^{-zt}\left[w^{\eta}(1-w)^{\mu}\right]dw$$

$$=\Gamma(\alpha)\sum_{n=0}^{\infty}\frac{(-z)^n t^n(\alpha)_{\mu n}}{n!}\prod_{l=1}^{M}\sum_{n_l=0}^{\infty}\frac{(\tau_l)_{n_l}}{n_l!}\left(\frac{2k_l^2}{k_l^2-1}\right)^{n_l}$$

$$\prod_{l'=1}^{N}\sum_{n_{l'}=0}^{\infty}\left(\frac{(\delta_{l'}-\rho_{l'})}{1+\delta_{l'}}\right)^{n_{l'}}\frac{(\lambda_{l'})_{n_{l'}}}{n_{l'}!}\prod_{i^*=1}^{R} I^{m^{(i^*)},n^{(i^*)}+1}_{p_i^{(i^*)}+1,q_i^{(i^*)}+1:r^{(i^*)}}$$

$$\left[z_{i^*}^{\xi_{i^*}}\left|\begin{matrix}(1-\beta-n_l-n_{l'}-\eta n,\xi_{i^*}),\left(a_j^{(i^*)},\alpha_j^{(i^*)}\right)_{1,n^{(i^*)}};\ \left(a_{ji}^{(i^*)},\alpha_{ji}^{(i^*)}\right)_{n^{(i^*)}+1,p_i^{(i^*)}}\\ \left(b_j^{(i^*)},\beta_j^{(i^*)}\right)_{1,m^{(i^*)}};\ \left(b_{ji}^{(i^*)},\beta_{ji}^{(i^*)}\right)_{m^{(i^*)}+1,q_i^{(i^*)}},(1-\alpha-\beta-n_l-n_{l'}-\eta n-\mu n,\xi_{i^*})\end{matrix}\right.\right] \tag{6.78}$$

It can be seen that the theorems discussed in this article provide generalizations to a number of known elliptic-type integrals. Furthermore, these theorems have also their applications in formulation of various new elliptic-type integrals, and some use in representing various forms of Euler-type integrals in terms of different generalized functions.

6.2.3 Special Case

i. If we take $r = 1$ in result (6.62) through (6.65), results (6.68) through (6.71), (6.74), (6.75), (6.76), (6.77) and (6.78) are reduced to the well-known Fox H-function recently obtained by Chaurasia and Meghwal [82].
ii. If we put $r = 1, \alpha_j^{(i^*)} = \alpha_j^{(i^*)} = \alpha_j^{(i^*)} = \alpha_j^{(i^*)} = \alpha$ in results (6.62) through (6.65), results (6.68) through (6.71), (6.74), (6.75), (6.76), (6.77) and (6.78) reduces to a Meijer's G-function recently obtained by Chaurasia and Meghwal [82].

Laplace Transform of I-Function: By the definition of Laplace transform [87,88] we get

$$L[(f(t)):p]=\int_0^\infty e^{-pt}I(t)\,dt$$

$$=\int_0^\infty e^{-pt}\frac{1}{2\pi w}\int_L\left\{\frac{\prod_{j=1}^{m}\Gamma(b_j-\beta_j s)\prod_{j=1}^{n}\Gamma(1-a_j+\alpha_j s)}{\sum_{i=1}^{r}\left\{\prod_{j=m+1}^{q_i}\Gamma(1-b_{ji}+\beta_j s)\prod_{j=n+1}^{p_i}\Gamma(a_{ji}-\alpha_{ji}s)\right\}}\right\}t^s\,ds\,dt$$

By changing the order of integration

$$= \frac{1}{2\pi w}\left[\int_L \left\{ \frac{\prod_{j=1}^{m}\Gamma(b_j-\beta_j s)\prod_{j=1}^{n}\Gamma(1-a_j+\alpha_j s)}{\sum_{i=1}^{r}\left\{\prod_{j=m+1}^{q_i}\Gamma(1-b_{ji}+\beta_j s)\prod_{j=n+1}^{p_i}\Gamma(a_{ji}-\alpha_{ji}s)\right\}} \right\} \int_0^\infty e^{-pt} t^s \, dt \, ds\right]$$

$$= \frac{1}{2\pi w}\int_L \left\{ \frac{\prod_{j=1}^{m}\Gamma(b_j-\beta_j s)\prod_{j=1}^{n}\Gamma(1-a_j+\alpha_j s)\Gamma(s+1)}{\sum_{i=1}^{r}\left\{\prod_{j=m+1}^{q_i}\Gamma(1-b_{ji}+\beta_j s)\prod_{j=n+1}^{p_i}\Gamma(a_{ji}-\alpha_{ji}s)\right\} p^{s+1}} \right\} ds$$

$$O(p) = \frac{1}{2\pi w}\int_L \frac{\theta(s)\Gamma(s+1)}{p^{s+1}} ds = \hat{I}(p) \tag{6.79}$$

where

$$\theta(s) = \frac{\prod_{j=1}^{m}\Gamma(b_j-\beta_j s)\prod_{j=1}^{n}\Gamma(1-a_j+\alpha_j s)}{\sum_{i=1}^{r}\left\{\prod_{j=m+1}^{q_i}\Gamma(1-b_{ji}+\beta_j s)\prod_{j=n+1}^{p_i}\Gamma(a_{ji}-\alpha_{ji}s)\right\}} \tag{6.80}$$

Now we shall establish a theorem involving Laplace transform of Saxena's I-function

Theorem 6.7

If the integral

$$O(p) = \frac{1}{2\pi w}\int_L \frac{\theta(s)\Gamma(s+1)t^s}{p^{s+1}} ds \tag{6.81}$$

where $\theta(s)$ is given by (6.80) Then under the assumption of absolute convergence

$$W^{\mu}\left\{t^{-\lambda}I(t): p\right\} = p^{\mu-\lambda-1}\frac{1}{2\pi w}\int_L \frac{\theta(s)\Gamma(1-\mu+\lambda+s)}{p^s\Gamma(1+\lambda+s)} ds$$

$$= p^{\mu-\lambda-1} I^{m,n+1}_{p_i+1,q_i+1:r}\left[\frac{1}{p}\left| \begin{matrix} (\mu-\lambda,1),(a_j,\alpha_j)_{1,n}; & (a_{ji},\alpha_{ji})_{n+1,p_i} \\ (b_j,\beta_j)_{1,m}; & (b_{ji},\beta_{ji})_{m+1,q_i},(-\lambda,1) \end{matrix}\right.\right] \tag{6.82}$$

Proof: We have [89,90].

$$L\left[t^{\mu-1}(t+a)^{-\lambda}:p\right]=\Gamma(\mu)\,p^{(\lambda-\mu-1)/2}p^{(\lambda-\mu-1)/2}e^{ap/2}W_{k,m}(ap)$$
$$=\phi(ap) \tag{6.83}$$

where$k=\dfrac{(1-\mu-\lambda)}{2}$, $m=\dfrac{(\mu-\lambda)}{2}$

and $W_{k,m}$ is usual Whittaker function

$$L\left[t^{-\lambda}(t-a)^{\mu-1}\;H(t-a):p\right]=e^{-ap}\phi(ap) \tag{6.84}$$

where $H(t)$ is Heaviside's function

Now applying the operational pair (6.2) and (6.84) in the Parseval-Goldstein Theorem, for Laplace transform and changing the order of integration we get

$$\int_a^{\infty}t^{-\lambda}(t-a)^{\mu-1}\;I(t)\,dt=\int_0^{\infty}t^{-\lambda}(t-a)^{\mu-1}\;g(t)H(t-a)\,dt$$
$$=\int_0^{\infty}e^{-ap}\phi(at)I(t)\,dt \tag{6.85}$$

Now, substituting $I(t)$ and $\phi(at)$ from definition of I-function and (6.83) respectively in (6.85) and upon performing the indicated integration with the help of Erdelyi et al. [89] we get our desired results.

6.2.4 Special Cases

i. By setting $r=1$ then I-function reduces to Fox's H-function [24] and in this case, the results reduce

$$W^{\mu}\left\{t^{-\lambda}H(t):p\right\}$$
$$=p^{\mu-\lambda-1}H_{p+1,q+1}^{m,n+1}\left[\frac{1}{p}\left|\begin{matrix}(\mu-\lambda,1), & (a_j,\alpha_j)_{1,p}\\ (b_j,\beta_j)_{1,m}, & (-\lambda,1)\end{matrix}\right.\right] \tag{6.86}$$

ii. By setting $r = 1$, $\alpha_j = 1, \beta_j = 1$ then I-function reduces to Meijer's G-function [11] and in this case, results reduce

$$W^{\mu}\left\{t^{-\lambda}G(t):p\right\}$$

$$= p^{\mu-\lambda-1}G_{p+1,q+1}^{m,n+1}\left[\frac{1}{p}\left|\begin{matrix}\mu-\lambda, & a_1,\ldots,a_p \\ b_1,\ldots,b_q, & -\lambda\end{matrix}\right.\right] \tag{6.87}$$

Also considering particular values of parameters the results can be converted into corresponding results in the form of series.

7 Associated Generating Function q and (p,q)-Analogue of I-Function

Recent developments in the theory of basic hypergeometric functions have gained interest due to the introduction of new generalizations like Saxena's I-function. This chapter gives certain results associated with the associated generating function.

7.1 (P,Q)-ANALOGUE OF I-FUNCTION WITH (P,Q)-DERIVATIVE

The basic number occurs in the theory of two parameter quantum algebras and has also been introduced in combinatorics by Chakrabarti and Jagannathan [47]. Around the same time, Brodimas et al. [48] and Arik et al. [49] also independently introduced the $(p; q)$-number in the physics literature, but in a very much less detailed manner. The (p, q)-identities thus derived, with doubling of the number of parameters, offer more choices for applications. It has been observed that many of the q-results can be generalized directly to $(p; q)$-results. If we have the $(p; q)$-results, the q-results can be obtained more easily by mere substitutions for the parameters instead of any limiting process as required in the usual q-theory. This also provides a new look for the q-identities.

The q-deformed algebra and their generalization to (p, q)-analogue have attracted much attention of the researchers to increase the accessibility of different dimensions of (p, q)-analogue algebra. The main reason is that these topics stand for real life problems, in mathematics and physics, later to the theory of quantum calculus.

Here, the attention is towards defining the (p, q)-analogue of I-function with $(p; q)$-derivative by using the $(p; q)$ generalization of the Gamma and Beta functions [46].

From the theory of basic hypergeometric series, some basic definitions are given below:

The q-shifted factorial is given by

$$(a;q)_n = \begin{cases} 1, n=0 \\ (1-a)(1-aq)\ldots\left(1-aq^{n-1}\right), n=1,2,3,\ldots \end{cases} \tag{7.1}$$

with $(a_1,a_2,\ldots,a_k;q)_n = (a_1;q)_n (a_2;q)_n, \ldots, (a_k;q)_n$

DOI: 10.1201/9781003517757-7

The *q*-gamma function was first introduced by Thomae and later by Jackson.

The q-analogue of Gamma function which is defined by Jackson [50] is given by

$$\Gamma_q(x)=\frac{(q;q)_\infty(1-q)^{1-x}}{\left(q^x;q\right)_\infty},\ 0<q<1$$

Jackson gave the general definition which is given below

$$\int_a^b f(t)d_q(t)=\int_0^b f(t)d_q(t)-\int_0^a f(t)d_q(t)$$

where

$$\int_0^a f(t)d_q(t)=a(1-q)\sum_{n=0}^{\infty} f\left(aq^n\right)q^n$$

Jackson also defined an integral i.e. $\int_0^\infty f(t)d_q(t)=(1-q)\sum_{n=-\infty}^{\infty} f\left(q^n\right)q^n$

Sadjang [51] introduced the so-called the (p, q)-powers as follows:

$$(x\Theta a)_{p,q}^n=(x-a)(px-aq)\left(p^2x-aq^2\right)\ldots\left(xp^{n-1}-aq^{n-1}\right). \tag{7.2}$$

$$(x\oplus a)_{p,q}^n=(x+a)(px+aq)\left(p^2x+aq^2\right)\ldots\left(xp^{n-1}+aq^{n-1}\right). \tag{7.3}$$

These definitions are extended to

$$(a\Theta b)_{p,q}^\infty=\prod_{k=0}^{\infty}\left(ap^k-q^kb\right) \tag{7.4}$$

$$(a\oplus b)_{p,q}^\infty=\prod_{k=0}^{\infty}\left(ap^k+q^kb\right) \tag{7.5}$$

Let x be a complex number, the (p, q)-Gamma function is defined by [52]

$$\Gamma_{p,q}(x)=\frac{(p\Theta q)_{p,q}^\infty}{(p^x\Theta q^x)_{p,q}^\infty}(p-q)^{1-x},0<q<p \tag{7.6}$$

And if we put $p=1$, then $\Gamma_{p,q}$ reduces to Γ_q.

The (p, q)-Gamma function fulfils the following fundamental relation

$$\Gamma_{p,q}(x+1)=[x]_{p,q}\Gamma_{p,q}(x).$$

If n is a non-negative integer, it follows from above that

$$\Gamma_{p,q}(x+1)=[x]_{p,q}!.$$

It can be also easily seen from the definition that

$$\Gamma_{p,q}(n+1)=\frac{(p\Theta q)_{p,q}^{n}}{(p-q)_{p,q}^{n}}$$

Sadjang [52] also defined the (p, q)-Beta function as

$$B_{p,q}(x,y)=\frac{\Gamma_{p,q}(x)\Gamma_{p,q}(y)}{\Gamma_{p,q}(x+y)}. \tag{7.7}$$

The (p, q)-derivative of the function $f(x)$ is defined as follows:

$$D_{p,q}f(x)=\frac{f(px)-f(qx)}{(p-q)x},\ x\neq 0$$

and $D_{p,q}f(0)=f^{'}(0)$, provided that $f(x)$ is differentiable at $x=0$.

The (p, q)-numbers $[n]_{p,q}$ and (p,q) factorials $[n]_{p,q}!$ are defined as:

$$[n]_{p,q}=\frac{p^{n}-q^{n}}{p-q}$$

and $[n]_{p,q}!=[1]_{p,q}.[2]_{p,q}.\ \dots\ [n]_{p,q}$ respectively.
Also it happens that $D_{p,q}\left(x^{n}\right)=[n]_{p,q}\,x^{n-1}$.

Remark 1: $D_{p,q}f(x)$ reduces to Hahn Derivative $d_{q}f(x)$ *iff* $p\to 1$.

Remark 2: $[n]_{p,q}=[n]_{q}$ (Hahn Basic Number) iff $p\to 1$.

where $[n]_{q}=\dfrac{1-q^{n}}{1-q},\ q\neq 1.$

NOTE

$$D_{p\cdot q}^{n}\left(x^{\mu}\right)=\frac{\Gamma_{p,q}(\mu+1)}{\Gamma_{p,q}(\mu-n+1)}X^{\mu-n},\mathrm{Re}(\mu)+1>0 \tag{7.8}$$

$$I_{p\cdot q}^{n}\left(X^{\mu}\right)=\frac{\Gamma_{p,q}(\mu+1)}{\Gamma_{p,q}(\mu+n+1)}X^{\mu+n},\mathrm{Re}(\mu)+1>0 \tag{7.9}$$

Saxena [53] introduced the following basic analogue of I-function in terms of the Mellin–Barnes type basic contour integral as:

$$I(z) = I_{A_i, B_i; r}^{m,n}\left[z; q \left| \begin{matrix} (a_j, \alpha_j)_{1,n}; (a_{ji}, \alpha_{ji})_{n+1, A_i} \\ (b_j, \beta_j)_{1,m}; (b_{ji}, \beta_{ji})_{m+1, B_i} \end{matrix} \right. \right]$$

$$= \frac{1}{2\pi i} \int_L \frac{\prod_{j=1}^{m} G\left(q^{b_j - \beta_j s}\right) \prod_{j=1}^{n} G\left(q^{1 - a_j + \alpha_j s}\right)}{\sum_{i=1}^{r} \left[\prod_{j=m+1}^{B_i} G\left(q^{1 - b_{ji} + \beta_{ji} s}\right) \prod_{j=n+1}^{A_i} G\left(q^{a_{ji} - \alpha_{ji} s}\right) G\left(q^{s}\right) G\left(q^{1-s}\right) \sin \pi s \right]} \pi z^s ds \tag{7.10}$$

where α_j, β_j, α_{ji}, β_{ji} are real and positive, a_j, b_j, a, b_{ji} are complex numbers and

$$G\left(q^{\alpha}\right) = \prod_{n=0}^{\infty} \left(1 - q^{\alpha + n}\right)^{-1} = \frac{1}{\left(q^{\alpha}; q\right)_{\infty}} \tag{7.11}$$

where L is contour of integration running from $-i\infty$ to $i\infty$ in such a manner so that all poles of $G\left(q^{b_j - \beta_j s}\right)$; $1 \le j \le m$ *are* to right of the path and those of $G\left(q^{1 - a_j + \alpha_j s}\right)$; $1 \le j \le n$, *are to left*. The integral converges if Re [s log (x) – log sin $\pi s\} < 0$, for large values of |s| on the contour L.

Setting $r = 1$, $A_i = A$, $B_i = B$, in Equation (7.10) we get q-analogue of H-function defined by Saxena et al. [53] as follows:

$$H_{A,B}^{m,n}\left[z; q \left| \begin{matrix} (a_j, \alpha_j)_{1,A} \\ (b_j, \beta_j)_{1,B} \end{matrix} \right. \right]$$

$$= \frac{1}{2\pi i} \int_L \frac{\prod_{j=1}^{m} G\left(q^{b_j - \beta_j s}\right) \prod_{j=1}^{n} G\left(q^{1 - a_j + \alpha_j s}\right)}{\prod_{j=m+1}^{B} G\left(q^{1 - b_j + \beta_j s}\right) \prod_{j=n+1}^{A} G\left(q^{a_j - \alpha_j s}\right) G\left(q^{s}\right) G\left(q^{1-s}\right) \sin \pi s} \pi z^s ds \tag{7.12}$$

Further if we put $\alpha_j = \beta_j = 1$, Equation (7.12) reduces to the basic analogue of Meijer's G-function given by Saxena et al. [53]

$$G_{A,B}^{m,n}\left[z; q \left| \begin{matrix} a_1, a_2, \ldots, a_A \\ b_1, b_2, \ldots, b_B \end{matrix} \right. \right]$$

$$= \frac{1}{2\pi i} \int_L \frac{\prod_{j=1}^{m} G\left(q^{b_j - s}\right) \prod_{j=1}^{n} G\left(q^{1 - a_j + s}\right)}{\prod_{j=m+1}^{B} G\left(q^{1 - b_j + s}\right) \prod_{j=n+1}^{A} G\left(q^{a_j - s}\right) G\left(q^{s}\right) G\left(q^{1-s}\right) \sin \pi s} \pi z^s ds. \tag{7.13}$$

(A) Pathak and Jain [54] defined the (p, q)-Gamma function as follows

$$\Gamma_{p,q}(x) = \frac{((p,q);(p,q))_\infty (p-q)^{1-x}}{\left(\left(p^x,q^x\right);(p,q)\right)_\infty}, \quad 0<q<p \tag{7.14}$$

When $x=n+1$ with a non-negative integer this definition reduces to

$$\Gamma_{p,q}(n+1) = [n]_{p,q}! \tag{7.15}$$

And

$$\Gamma_{p,q}(x+1) = [x]_{p,q}\Gamma_{p,q}(x)$$

We can deduce that we shall make use of (p,q)-Gamma function for (p,q)-analogue of I-function.

$$I^{m,n}_{A_i,B_i+1:r}\left[z(p-q)^{\sum_{t=1}^{m}\beta_t-\sum_{t=1}^{n}\alpha_t+\sum_{i=1}^{r}\left\{\sum_{t=m+1}^{qi}\beta_{ti}-\sum_{t=n+1}^{pi}\alpha_{ti}\right\}};(p,q) \left| \begin{matrix} (a_j,\alpha_j)_{1,n}; & (a_{ji},\alpha_{ji})_{n+1,A_i} \\ (b_j,\beta_j)_{1,m}; & (b_{ji},\beta_{ji})_{n+1,B_i},(1,1) \end{matrix} \right. \right]$$

$$= \frac{1}{2\pi i}\int_L \frac{\prod_{j=1}^{m}G\left(p^{b_j-\beta_j s},\ q^{b_j-\beta_j s}\right)\prod_{j=1}^{n}G\left(p^{1-a_j+\alpha_j s},\ q^{1-a_j+\alpha_j s}\right)(p-q)^{s\left\{\sum_{t=1}^{m}\beta_t-\sum_{t=1}^{n}\alpha_t+\sum_{i=1}^{r}\left\{\sum_{t=m+1}^{qi}\beta_{ti}-\sum_{t=n+1}^{pi}\alpha_{ti}\right\}\right.}}{\sum_{i=1}^{r}\left[\prod_{j=m+1}^{B_i}G\left(p^{1-b_{ji}+\beta_{ji}s},\ q^{1-b_{ji}+\beta_{ji}s}\right)\prod_{j=n+1}^{A_i}G\left(p^{a_{ji}-\alpha_{ji}s},\ q^{a_{ji}-\alpha_{ji}s}\right)G\left(p^s,\ q^{\ s}\right)G\left(p^{1-s},\ q^{1-s}\right)\sin\pi s\right]}\pi z^s\,ds \tag{7.16}$$

On multiplying Equation (7.9) by:

$$(p-q)^{\sum_{t=1}^{n}a_t-\sum_{t=1}^{m}b_t+m+n-1+\sum_{i=1}^{r}\left\{\sum_{t=n+1}^{Ai}\alpha_{ti}-\sum_{t=m+1}^{Bi}\beta_{ti}-A_i\right\}} G(p,q)^{\sum_{t=1}^{n}A_t+B_t-(m+n-1)}$$

and then making use of identity

$$\Gamma_{p,q}(x) = \frac{G\left(p^x,q^x\right)}{(p-q)^{x-1}G(p,q)}, \left|\frac{q}{p}\right|<1. \tag{7.17}$$

where

$$G\left(p^x,q^x\right) = \prod_{n=0}^{\infty}\left(p^{\alpha+n}-q^{\alpha+n}\right)^{-1}$$

$$= \frac{1}{\left((p^\alpha,q^\alpha);(p,q)\right)_\infty}$$

The right-hand side of Equation (7.16) becomes

$$= \frac{1}{2\pi i}\int_L \frac{\prod_{j=1}^{m}\Gamma_{p,q}(b_j-\beta_j s)\prod_{j=1}^{n}\Gamma_{p,q}(1-a_j+\alpha_j s)\pi z^s ds}{\sum_{i=1}^{r}\left[\prod_{j=m+1}^{B_i}\Gamma_{p,q}(1-b_{ji}+\beta_{ji}s)\prod_{j=n+1}^{A_i}\Gamma_{p,q}(a_{ji}-\alpha_{ji}s)\Gamma_{p,q}(s)\Gamma_{p,q}(1-s)\sin\pi s\right]}$$

7.1.1 Special Cases

Taking $r=1$, in the above equation we will get the (*p*,*q*)-analogue of Fox's H-function as follows:

$$H_{A,B}^{m,n}\left[z;(p,q)\left|\begin{matrix}(a_j,\alpha_j)_{1,A}\\(b_j,\beta_j)_{1,B}\end{matrix}\right.\right]$$

$$= \frac{1}{2\pi i}\int_L \frac{\prod_{j=1}^{m}\Gamma_{p,q}(b_j-\beta_j s)\prod_{j=1}^{n}\Gamma_{p,q}(1-a_j+\alpha_j s)\pi z^s ds}{\prod_{j=m+1}^{B}\Gamma_{p,q}(1-b_j+\beta_j s)\prod_{j=n+1}^{A}\Gamma_{p,q}(a_j-\alpha_j s)\Gamma_{p,q}(s)\Gamma_{p,q}(1-s)\sin\pi s} \tag{7.18}$$

If we take $\alpha_j=\beta_j=1$, in Equation (7.18) we will get the (*p*,*q*)-analogue of Meijer's G-function defined by Pathak et al. [54] as follows:

$$G_{p,q}{}_{A,B}^{m,n}\left[z;(p,q)\left|\begin{matrix}a_1,a_2,\ldots,a_A\\b_1,b_2,\ldots,b_B\end{matrix}\right.\right]$$

$$= \frac{1}{2\pi i}\int_L \frac{\prod_{j=1}^{m}\Gamma_{p,q}(b_j-s)\prod_{j=1}^{m}\Gamma_{p,q}(1-a_j+s)}{\prod_{j=m+1}^{B}\Gamma_{p,q}(1-b_j+s)\prod_{j=n+1}^{A}\Gamma_{p,q}(a_j-s)\Gamma_{p,q}(s)\Gamma_{p,q}(1-s)\sin\pi s} \tag{7.19}$$

(B) In this section, we will evaluate the (*p*,*q*)-derivative operator involving (*p*,*q*)-analogue of I-function.

Theorem 7.1

Let the parameters A_i, B_i are non-negative integers satisfying the inequality $0\le n\le A$, $0\le m\le B_i$ and a_j, b_j, a_{ji}, b_{ji} are complex numbers, then

$$zD_{(p,q)}\left[z^{1-a_1}I^{m,n}_{A_i,B_i\,;r}\left[z;(p,q)\left|\begin{matrix}(a_j,1)_{1,n}, & (a_{ji},1)_{n+1,\,A_i}\\(b_j,1)_{1,m}, & (b_{ji},\,1)_{m+1,\,B_i}\end{matrix}\right.\right]\right.$$

$$=z^{1-a_1}I^{m,n}_{A_i,B_i\,;r}\left[z;(p,q)\left|\begin{matrix}(a_1-1,1)(aj,1)_{2,\,n},\ldots. & (a_{ji},1)_{n+1,\,A_i}\\(b_j,1)_{1,m}, & (b_{ji},\,1)_{m+1,\,B_i}\end{matrix}\right.\right] \tag{7.20}$$

where $z\neq 0$, $0<q<p$ and $\omega=\sqrt{-1}$.

Proof: To prove theorem (7.20) when $a_1\geq 0$, we apply Equation (7.8) so that

Left-hand side

$$=zD_{(p,q)}\left[z^{1-a_1}\frac{1}{2\pi\omega}\int_L\frac{\prod_{j=1}^{m}\Gamma_{p,q}(b_j-s)\prod_{j=1}^{n}\Gamma_{p,q}(1-a_j+s)\pi z^s ds}{\sum_{i=1}^{r}\left\{\prod_{j=m+1}^{B_i}\Gamma_{p,q}(1-b_{ji}+S)\prod_{j=n+1}^{A_i}\Gamma_{p,q}(a_{ji}-s)\Gamma_{p,q}(s)\Gamma_{p,q}(1-s)\sin\pi s\right.}\right.$$

$$=z\frac{1}{2\pi\omega}\int_L\frac{\prod_{j=1}^{m}\Gamma_{p,q}(b_j-s)\prod_{j=1}^{n}\Gamma_{p,q}(1-a_j+s)\pi D_{p,q}\left[z^{1-a_1+s}\right]ds}{\sum_{i=1}^{r}\left\{\prod_{j=m+1}^{B_i}\Gamma_{p,q}(1-b_{ji}+S)\prod_{j=n+1}^{A_i}\Gamma_{p,q}(a_{ji}-s)\Gamma_{p,q}(s)\Gamma_{p,q}(1-s)\sin\pi s\right.}$$

$$=\frac{1}{2\pi\omega}\int_L\frac{\prod_{j=1}^{m}\Gamma_{p,q}(b_j-s)\prod_{j=1}^{n}\Gamma_{p,q}(1-a_j+s)\pi[1-a_1+s]_{p,q}z^{1-a_1+s}ds}{\sum_{i=1}^{r}\left\{\prod_{j=m+1}^{B_i}\Gamma_{q}(1-b_{ji}+S)\prod_{j=n+1}^{A_i}\Gamma_{q}(a_{ji}-s)\Gamma_{q}(s)\Gamma_{q}(1-s)\sin\pi s\right.}$$

Since,

$$\Gamma_{p,q}(1+a)=\frac{p^a-q^a}{p-q}\Gamma_{p,q}(a)=[a]_{p,q}\Gamma_{p,q}(a)$$

$$[a]_{p,q}\Gamma_{p,q}(a)=\Gamma_{p,q}(a+1)$$

Therefore $[1-a_1+s]_{p,q}\Gamma_{p,q}(1-a_1+s)=\Gamma_{p,q}(1-(a_1-1)+s)$

Thus

$$\text{L.H.S.}=\frac{1}{2\pi\omega}\int_L\frac{\prod_{j=1}^{m}\Gamma_{p,q}(b_j-s)\,\Gamma_{p,q}(1-(a_1-1)+s)\prod_{j=2}^{n}\Gamma_{p,q}(1-a_j+s)\pi Z^s Z^{1-a_1}ds}{\sum_{i=1}^{r}\left\{\prod_{j=m+1}^{B_i}\Gamma_{p,q}(1-b_{ji}+S)\prod_{j=n+1}^{A_i}\Gamma_{p,q}(a_{ji}-s)\Gamma_{p,q}(s)\Gamma_{p,q}(1-s)\sin\pi s\right.}$$

which implies,

$$zD_{p,q}\left[z^{1-a_1}I^{m,n}_{A_i,B_i;r}\left[z;(p,q)\middle|\begin{matrix}(a_j,1)_{1,n}, & (a_{ji},1)_{n+1,A_i}\\(b_j,1)_{1,m}, & (b_{ji},1)_{m+1,B_i}\end{matrix}\right]\right]$$

$$=z^{1-a_1}I^{m,n}_{A_i,B_i;r}\left[z;(p,q)\middle|\begin{matrix}(a_1-1,1)(aj,1)_{2,n}, & (a_{ji},1)_{n+1,A_i}\\(b_j,1)_{1,m}, & (b_{ji},1)_{m+1,B_i}\end{matrix}\right].$$

Hence the result.

Theorem 7.2

Let the parameters A_i, B_i are non-negative integers satisfying the inequality $0\le n\le A_i$, $0\le m\le B_i$ and $\alpha_j,\beta_j,\alpha_{ji},\beta_{ji}$ are positive real numbers and a_j,b_j,a_{ji},b_{ji} are complex numbers, then

$$D^{\mu}_{p,q}\left[I^{m,n}_{A_i,B_i;r}\left(px^{\lambda};(p,q)\middle|\begin{matrix}(a_j,\alpha_j)_{1,n};(a_{ji},\alpha_{ji})_{n+1,A_i}\\(b_j,\beta_j)_{1,m};(b_{ji},\beta_{ji})_{m+1,B_i}\end{matrix}\right)\right]$$
$$=x^{-\mu}I^{m,n+1}_{A_i,B_i+1;r}\left(px^{\lambda};(p,q)\middle|\begin{matrix}(0,\lambda)(a_j,\alpha_j)_{1,n};(a_{ji},\alpha_{ji})_{n+1,A_i}\\(b_j,\beta_j)_{1,m};(b_{ji},\beta_{ji})_{m+1,B_i}(\mu,\lambda)\end{matrix}\right) \tag{7.21}$$

where $z\ne 0$, $0<q<p$ and $\omega=\sqrt{-1}$.

Proof: To prove theorem (7.21) when $\lambda\ge 0$, we apply Equation (7.8)

$$D^{\mu}_{p,q}\left[I^{m,n}_{A_i,B_i;r}\left(px^{\lambda};(p,q)\middle|\begin{matrix}(a_j,\alpha_j)_{1,n};(a_{ji},\alpha_{ji})_{n+1,A_i}\\(b_j,\beta_j)_{1,m};(b_{ji},\beta_{ji})_{m+1,B_i}\end{matrix}\right)\right]$$

$$=\frac{1}{2\pi i}\int_L\frac{\prod_{j=1}^{m}\Gamma_{p,q}(b_j-\beta_js)\prod_{j=1}^{n}\Gamma_{p,q}(1-a_j+\alpha_js)\pi D^{\mu}_{p,q}\left[(px^{\lambda})^s\right]ds.}{\sum_{i=1}^{r}\left\{\prod_{j=m+1}^{B_i}\Gamma_{p,q}(1-b_{ji}+\beta_{ji}s)\prod_{j=n+1}^{A_i}\Gamma_{p,q}(a_{ji}-\alpha_{ji}s)\Gamma_{p,q}(s)\Gamma_{p,q}(1-s)\sin\pi s\right\}}$$

That is,

$$=\frac{1}{2\pi i}\int_L \frac{\prod_{j=1}^{m}\Gamma_{p,q}(b_j-\beta_j s)\prod_{j=1}^{n}\Gamma_{p,q}(1-a_j+\alpha_j s)\pi\Gamma_{p,q}(1-0+\lambda s)p^s x^{\lambda s-\mu}ds.}{\sum_{i=1}^{r}\left\{\prod_{j=m+1}^{B_i}\Gamma_{p,q}(1-b_{ji}+\beta_{ji}s)\prod_{j=n+1}^{A_i}\Gamma_{p,q}(a_{ji}-\alpha_{ji}s)\Gamma_{p,q}(1-\mu+\lambda s))\Gamma_{p,q}(s)\Gamma_{p,q}(1-s)\sin\pi s\right\}}$$

$$=x^{-\mu}I^{m,n+1}_{A_i,B_i+1;r}\left(px^{\lambda};(p,q)\left|\begin{matrix}(0,\lambda)(a_j,\alpha_j)_{1,n};(a_{ji},\alpha_{ji})_{n+1,A_i}\\(b_j,\beta_j)_{1,m};(b_{ji},\beta_{ji})_{m+1,B_i}(\mu,\lambda)\end{matrix}\right.\right).$$

Hence the result.

7.1.2 Special Cases

Put $r=1$ in (7.20) and (7.21) we get the (p,q)-analogue of H-function as mentioned below:

$$zD_{p,q}\left[z^{1-a_1}H^{m,n}_{P,Q}\left[z;(p,q)\left|\begin{matrix}(a_j,1)_{1,P}\\(b_j,1)_{1,Q}\end{matrix}\right.\right]\right]$$

$$=z^{1-a_1}H^{m,n}_{P,Q}\left[z;(p,q)\left|\begin{matrix}(a_1-1,1)(aj,1)_{2,P}\\(b_j,1)_{1,Q}\end{matrix}\right.\right] \tag{7.22}$$

$$D^{\mu}_{p,q}\left[H^{m,n}_{P,Q}\left(px^{\lambda};(p,q)\left|\begin{matrix}(a_j,A_j)_{1,P}\\(b_j,B_j)_{1,Q}\end{matrix}\right.\right)\right]$$

$$=x^{-\mu}H^{m,n+1}_{P,Q+1}\left(px^{\lambda};(p,q)\left|\begin{matrix}(0,\lambda)(a_j,A_j)_{1,P}\\(b_j,B_j)_{1,Q}(\mu,\lambda)\end{matrix}\right.\right) \tag{7.23}$$

Put $p=1$, the above results of (p,q)-analogue change into well-known results of the basic analogue of H-function as mentioned below:

$$zD_q\left[z^{1-a_1}H^{m,n}_{P,Q}\left[z;q\left|\begin{matrix}(a_j,1)_{1,P}\\(b_j,1)_{1,Q}\end{matrix}\right.\right]\right]$$

$$=z^{1-a_1}H^{m,n}_{P,Q}\left[z;q\left|\begin{matrix}(a_1-1,1)(aj,1)_{2,P}\\(b_j,1)_{1,Q}\end{matrix}\right.\right] \tag{7.24}$$

$$D_q^{\mu}\left[H_{P,Q}^{m,n}\left(x^{\lambda};q\left|\begin{matrix}(a_j,A_j)_{1,P}\\(b_j,B_j)_{1,Q}\end{matrix}\right.\right)\right]$$

$$=x^{-\mu}H_{P,Q+1}^{m,n+1}\left(x^{\lambda};q\left|\begin{matrix}(0,\lambda)(a_j,A_j)_{1,P}\\(b_j,B_j)_{1,Q}(\mu,\lambda)\end{matrix}\right.\right) \tag{7.25}$$

The results proved here give some contribution to the (p,q)-algebra and are believed to be new and fruitful for (p,q) analogue and are likely to find certain applications to the solution of the (p,q)-integral equations involving various (p,q)-hypergeometric functions.

7.2 GENERATING FUNCTIONS FOR I-FUNCTION AND THEIR Q-ANALOGUE SATISFYING TRUESDELL'S DESCENDING F_Q-EQUATION

The q-analogue of I-function has been introduced by Saxena et al. [53] in terms of Mellin–Barnes type basic contour integral as:

$$I\ (z)=I_{A_i,\ B_i;r}^{m,n}\left[z;q\left|\begin{matrix}(a_j,\ \alpha_j)_{1,n};(a_{ji},\ \alpha_{ji})_{n+1,A_i}\\(b_j,\ \beta_j)_{1,m};(b_{ji},\ \beta_{ji})_{m+1,B_i}\end{matrix}\right.\right]=$$

$$\frac{1}{2\pi i}\int_L\frac{\prod_{j=1}^{m}G\left(q^{b_j-\beta_j s}\right)\prod_{j=1}^{n}G\left(q^{1-a_j+\alpha_j s}\right)}{\sum_{i=1}^{r}\left[\prod_{j=m+1}^{B_i}G\left(q^{1-b_{ji}+\beta_{ji}s}\right)\prod_{j=n+1}^{A_i}G\left(q^{a_{ji}-\alpha_{ji}s}\right)G\left(q^{s}\right)G\left(q^{1-s}\right)\sin\pi s\right]}\pi z^{s}ds \tag{7.26}$$

where $\alpha_j,\ \beta_j,\alpha_{ji},\ \beta_{ji}$ are real and positive, $a_j,\ b_j,a,\ b_{ji}$ are complex numbers and

$$G\ (q^{\alpha})=\prod_{n=0}^{\infty}\left(1-q^{\alpha+n}\right)^{-1}=\frac{1}{\left(q^{\alpha};q\right)_{\infty}}$$

where L is contour of integration running from $-i\infty$ to $i\infty$ in such a manner so that all poles of $G\left(q^{b_j-\beta_j s}\right)$; $1\le j\le m$ are to right of the path and those of $G\left(q^{1-a_j+\alpha_j s}\right)$; $1\le j\le n$, are to left. The integral converges if Re [s log (x) – log sin πs} < 0, for large values of |s| on the contour L.

Setting $r=1,\ A_i=A,\ B_i=B,$ in Equation (7.26) we get q-analogue of H-function defined by Saxena et al. [53] as follows:

$$H_{A,B}^{m,n}\left[z;q \,\middle|\, \begin{matrix}(a_j,\alpha_j)_{1,A}\\(b_j,\beta_j)_{1,B}\end{matrix}\right]$$

$$=\frac{1}{2\pi i}\int_L \frac{\prod_{j=1}^{m} G\left(q^{b_j-\beta_j s}\right)\prod_{j=1}^{n} G\left(q^{1-a_j+\alpha_j s}\right)}{\prod_{j=m+1}^{B} G\left(q^{1-b_j+\beta_j s}\right)\prod_{j=n+1}^{A} G\left(q^{a_j-\alpha_j s}\right)G\left(q^{s}\right)G\left(q^{1-s}\right)\sin\pi s}\pi z^s ds \tag{7.27}$$

Further if we put $\alpha_j=\beta_j=1$, Equation (7.27) reduces to the basic analogue of Meijer's G-function given by Saxena et al. [53]

$$G_{A,B}^{m,n}\left[z;q \,\middle|\, \begin{matrix}a_1,a_2,\ldots,a_A\\b_1,b_2,\ldots,b_B\end{matrix}\right]$$

$$=\frac{1}{2\pi i}\int_L \frac{\prod_{j=1}^{m} G\left(q^{b_j-s}\right)\prod_{j=1}^{n} G\left(q^{1-a_j+s}\right)}{\prod_{j=m+1}^{B} G\left(q^{1-b_j+s}\right)\prod_{j=n+1}^{A} G\left(q^{a_j-s}\right)G\left(q^{s}\right)G\left(q^{1-s}\right)\sin\pi s}\pi z^s ds \tag{7.28}$$

Farooq et al. [55] defined the basic analogue of I-function in terms of Gamma function as follows

$$I_{q,A_i,B_i;R}^{m,n}\left[z;q \,\middle|\, \begin{matrix}(a_j,\alpha_j)_{1,n}\,(a_{ji},\alpha_{ji})_{n+1,A_i}\\(b_j,\beta_j)_{1,m}\,(b_{ji},\beta_{ji})_{m+1,B_i}\end{matrix}\right]$$

$$=\frac{1}{2\pi i}\int_L \frac{\prod_{j=1}^{m}\Gamma_q(b_j-\beta_j s)\prod_{j=1}^{n}\Gamma_q(1-a_j+\alpha_j s)}{\sum_{i=1}^{R}\left\{\prod_{j=m+1}^{B_i}\Gamma_q(1-b_{ji}+\beta_{ji}s)\prod_{j=n+1}^{A_i}\Gamma_q(a_{ji}-\alpha_{ji}s)\right\}\Gamma_q(s)\Gamma_q(1-s)\sin\pi s}\pi z^s\, ds \tag{7.29}$$

In his effort towards achieving the unification of special functions, Truesdell [36] has put forward a theory that yielded a number of results for special functions satisfying the so-called Truesdell's F_q-equation. Agrawal [56] extended this theory further and derived results for descending F_q-equation. He obtained various properties like orthogonality, Rodrigue's and Schalafli's formulae for F_q-equation, which turn out to be special functions.

The function $F(z,\alpha)$ is said to satisfy the descending F-equation if

$$D_z^r F(z,\alpha)=F(z,\alpha-r) \tag{7.30}$$

For $F(z,\alpha)$ satisfying descending F-equation, Agrawal [57] has obtained the following generating functions

$$F(z+y,\alpha)=\sum_{n=0}^{\infty} y^n \frac{F(z,\alpha-n)}{n!} \tag{7.31}$$

The q-derivative of Equation (7.30) can be written in the following manner:

$$D_{q,z}^{r} F(z,\alpha)=F(z,\alpha+r) \tag{7.32}$$

To obtain the main result of this chapter, we will make use of the following results which we have obtained on multiplication formulae for q-analogue of Gamma functions:

$$\prod_{k=0}^{m-1}\Gamma_q\left(\frac{\alpha-r+k}{m}\right)=\frac{(1-q)^r q^{\frac{r}{m}(r-2\alpha+1)}}{(-1)^r (q^{1-\alpha};q)_r}\prod_{k=0}^{m-1}\Gamma_q\left(\frac{\alpha+k}{m}\right) \tag{7.33}$$

$$\prod_{k=0}^{m-1}\Gamma_q\left(1-\frac{\alpha-r+k}{m}\right)=\frac{(q^{1-\alpha};q)_r}{(1-q)^r}\prod_{k=0}^{m-1}\Gamma_q\left(1-\frac{\alpha+k}{m}\right) \tag{7.34}$$

In the results that follow, by $\Delta(\mu,\alpha)$ we shall mean the array of μ parameters

$$\frac{\alpha}{\mu},\frac{\alpha+1}{\mu},\ldots,\frac{\alpha+\mu-1}{\mu};\quad (\mu=1,2,3,\ldots) \tag{7.35}$$

And $(\Delta(\mu,\alpha),\beta)s$ tan ds for

$$\left(\frac{\alpha}{\mu},\beta\right),\left(\frac{\alpha+1}{\mu},\beta\right),\ldots,\left(\frac{\alpha+\mu-1}{\mu},\beta\right) \tag{7.36}$$

7.2.1 Generating Functions for I-Function

Theorem 7.3

Consider $p,q,t_i,s_i \in \mathbb{N}_0$ such that $0\le q\le t_i, 0\le p\le s_i\,(i=1,\ldots.r)$ r is finite, $e_j, f_j, e_{ji}, f_{ji} \in \mathbb{N}$ and $a_j, b_j, a_{ji}, b_{ji}, \lambda \in \mathbb{C}$, then the following result holds true:

$$\sum_{n=0}^{\infty}\binom{\lambda+n-1}{n} I_{t_i+1,s_i+1:r}^{p+1,q}\left[x \left| \begin{matrix} (a_j,e_j)_{1,q}; & (a_{ji},e_{ji})_{q+1,t_i},(\lambda+n,0) \\ (\lambda+n,1)(b_j,f_j)_{1,p}; & (b_{ji},f_{ji})_{p+1,s_i} \end{matrix}\right.\right] y^n$$

$$= (1-y)^{-n} I_{t_i+1,s_i+1:r}^{p+1,q}\left[x(1-y) \left| \begin{matrix} (a_j,e_j)_{1,q}; & (a_{ji},e_{ji})_{q+1,t_i},(\lambda,0) \\ (\lambda,1)(b_j,f_j)_{1,p}; & (b_{ji},f_{ji})_{p+1,s_i} \end{matrix}\right.\right] \tag{7.37}$$

Proof: Consider L.H.S of the (7.12), and denote it by K, then we have

$$K = \sum_{n=0}^{\infty}\binom{\lambda+n-1}{n} I_{t_i+1,s_i+1:r}^{p+1,q}\left[x \left| \begin{matrix} (a_j,e_j)_{1,q}; & (a_{ji},e_{ji})_{q+1,t_i},(\lambda+n,0) \\ (\lambda+n,1)(b_j,f_j)_{1,p}; & (b_{ji},f_{ji})_{p+1,s_i} \end{matrix}\right.\right] y^n \tag{7.38}$$

By using the Mellin–Barnes type contour integral of I-function, we get

$$K = \sum_{n=0}^{\infty}\binom{\lambda+n-1}{n}\left\{\frac{1}{2\pi i}\int_L x^{\xi}\phi_{t_i+1,s_i+1:r}^{p+1,q}(\xi)\,d\xi\right\} y^n \tag{7.39}$$

Then using value of $\phi_{t_{i+1},s_{i+1}:r}^{p+1,q}(\xi)$, we have

$$K = \sum_{n=0}^{\infty}\binom{\lambda+n-1}{n}\left\{\frac{1}{2\pi i}\int_L x^{\xi}\frac{\prod_{j=1}^{p}\Gamma(b_j-f_j\xi)\prod_{j=1}^{p}\Gamma(1-a_j-e_j\xi)\Gamma(\lambda+n-\xi)}{\sum_{i=1}^{r}\left[\prod_{j=p+1}^{s_i}\Gamma(1-b_{ji}+f_{ji}\xi)\prod_{j=q+1}^{\eta}\Gamma(a_{ji}-e_{ji}\xi)\right]\Gamma(\lambda+n)}d\xi\right\} y^n \tag{7.40}$$

On interchanging order of summation and integration, we get

$$K = \frac{1}{2\pi i}\int_L \frac{\prod_{j=1}^{p}\Gamma(b_j-f_j\xi)\prod_{j=1}^{p}\Gamma(1-a_j-e_j\xi)x^{\xi}}{\sum_{i=1}^{r}\left[\prod_{j=p+1}^{s_i}\Gamma(1-b_{ji}+f_{ji}\xi)\prod_{j=q+1}^{\eta}\Gamma(a_{ji}-e_{ji}\xi)\right]}d\xi \tag{7.41}$$

$$\times\left\{\sum_{n=0}^{\infty}\binom{\lambda+n-1}{n}\frac{\Gamma(\lambda+n-\xi)y^n}{\Gamma(\lambda+n)}\right\}$$

After using some calculations, we have

$$K = \frac{1}{2\pi i}\int_L \frac{\prod_{j=1}^{p}\Gamma(b_j - f_j\xi)\prod_{j=1}^{p}\Gamma(1-a_j-e_j\xi)x^{\xi}}{\sum_{i=1}^{r}\left[\prod_{j=p+1}^{s_i}\Gamma(1-b_{ji}+f_{ji}\xi)\prod_{j=q+1}^{r_i}\Gamma(a_{ji}-e_{ji}\xi)\right]}d\xi \tag{7.42}$$

$$\times\left\{\sum_{n=0}^{\infty}\binom{\lambda+n-\xi-1}{n}\frac{\Gamma(\lambda-\xi)y^n}{\Gamma(\lambda)}\right\}$$

By using the following binomial identity

$$\sum_{n=0}^{\infty}\binom{\xi+n-1}{n}y^n = (1-y)^{-\xi} \tag{7.43}$$

we get,

$$K = \frac{1}{2\pi i}\int_L \frac{\prod_{j=1}^{p}\Gamma(b_j - f_j\xi)\prod_{j=1}^{p}\Gamma(1-a_j-e_j\xi)x^{\xi}}{\sum_{i=1}^{r}\left[\prod_{j=p+1}^{s_i}\Gamma(1-b_{ji}+f_{ji}\xi)\prod_{j=q+1}^{r_i}\Gamma(a_{ji}-e_{ji}\xi)\right]}d\xi \tag{7.44}$$

$$\times(1-y)^{\xi-\lambda}\frac{\Gamma(\lambda-\xi)}{\Gamma(\lambda)}$$

On re-arranging the terms, we get

$$K = (1-y)^{-\lambda}\frac{1}{2\pi i}\int_L \frac{\prod_{j=1}^{p}\Gamma(b_j - f_j\xi)\prod_{j=1}^{p}\Gamma(1-a_j-e_j\xi)\Gamma(\lambda-\xi)\left(x(1-y)\right)^{\xi}}{\sum_{i=1}^{r}\left[\prod_{j=p+1}^{s_i}\Gamma(1-b_{ji}+f_{ji}\xi)\prod_{j=q+1}^{r_i}\Gamma(a_{ji}-e_{ji}\xi)\right]\Gamma(\lambda)}d\xi \tag{7.45}$$

Then, using the definition of I-function, we get our desired result.

$$\sum_{n=0}^{\infty}\binom{\lambda+n-1}{n}I_{t_i+1,s_i+1:r}^{p+1,q}\left[x\left|\begin{matrix}(a_j,e_j)_{1,q}; & (a_{ji},e_{ji})_{q+1,t_i},(\lambda+n,0)\\ (\lambda+n,1)(b_j,f_j)_{1,p}; & (b_{ji},f_{ji})_{p+1,s_i}\end{matrix}\right.\right]y^n$$

$$=(1-y)^{-n}I_{t_i+1,s_i+1:r}^{p+1,q}\left[x(1-y)\left|\begin{matrix}(a_j,e_j)_{1,q}; & (a_{ji},e_{ji})_{q+1,t_i},(\lambda,0)\\ (\lambda,1)(b_j,f_j)_{1,p}; & (b_{ji},f_{ji})_{p+1,s_i}\end{matrix}\right.\right] \tag{7.46}$$

Corollary 7.1

The following result for H-function holds true:

$$\sum_{n=0}^{\infty}\binom{\lambda+n-1}{n} H_{t+1,s+1}^{p+1,q}\left[x\left|\begin{matrix}(a_j,e_j)_{1,q}; & (a_j,e_j)_{q+1,t},(\lambda+n,0)\\ (\lambda+n,1)(b_j,f_j)_{1,p}; & (b_j,f_j)_{p+1,s}\end{matrix}\right.\right]y^n$$

$$=(1-y)^{-n} H_{t+1,s+1}^{p+1,q}\left[x(1-y)\left|\begin{matrix}(a_j,e_j)_{1,q}; & (a_j,e_j)_{q+1,t},(\lambda,0)\\ (\lambda,1)(b_j,f_j)_{1,p}; & (b_j,f_j)_{p+1,s}\end{matrix}\right.\right] \tag{7.47}$$

Proof: If we set $r=1$, where $t_i=t,\ s_i=s,$ and $e_{ji}=e_j,\ f_{ji}=f_j$, $a_{ji}=a_j\ b_{ji}=b_j$ in the proof of above Theorem 7.3, then I-function reduces to H-function and we get our desired result.

Corollary 7.2

The following result for G-function holds true:

$$\sum_{n=0}^{\infty}\binom{\lambda+n-1}{n} G_{t+1,s+1}^{p+1,q}\left[x\left|\begin{matrix}a_1,a_2\ldots\ldots\ldots & .a_t,\lambda+n\\ \lambda+n,\ b_1,b_2\ldots\ldots\ldots & b_s\end{matrix}\right.\right]$$

$$y^n=(1-y)^{-n} G_{t+1,s+1}^{p+1,q}\left[x(1-y)\left|\begin{matrix}a_1,a_2\ldots\ldots\ldots & .a_t,\lambda\\ \lambda+n,\ b_1,b_2\ldots\ldots\ldots & b_s\end{matrix}\right.\right] \tag{7.48}$$

Proof: If we substitute$r=1,e_j=1,\ j=1,2,\ldots t$ and $e_j=0,\ j=t+1$

$$f_j=1,\quad j=1,2,\ldots,s+1,$$

where $t_i=t,\ s_i=s,$ and $e_{ji}=e_j,\ f_{ji}=f_j$, $a_{ji}=a_j\ b_{ji}=b_j$ in the proof of above Theorem 7.3, then I-function reduces to G-function and we get our desired result.

Theorem 7.4

Assume $p,q,t_i,s_i\in\mathbb{N}_0$ such that $0\le q\le t_i, 0\le p\le s_i(i=1,\ldots.r)$ r is finite, $e_j,f_j,e_{ji},f_{ji}\in\mathbb{N}$ and $a_j,b_j,a_{ji},b_{ji},\lambda\in\mathbb{C}$, then the following result holds true:

$$\sum_{q=0}^{\infty} Q_{q,p}^{u,v}(w_1,w_2)\frac{z^q}{q!}=\frac{(1+\xi)^{u+1}}{(1-v\xi)}\psi\left[w_1(-\xi)^p\right]I_{t_i,s_i:r}^{p,q}\left[w_2(1+\xi)\left|\begin{matrix}(a_j,e_j)_{1,q}; & (a_{ji},e_{ji})_{q+1,t_i}\\ (b_j,f_j)_{1,p}; & (b_{ji},f_{ji})_{p+1,s_i}\end{matrix}\right.\right] \tag{7.49}$$

where $= z(1+\xi)^{v+1}, \psi[z]=\sum_{q=0}^{\infty}\Omega_q z^q$, z^q is an arbitrary complex sequence and

$$Q_{q,p}^{u,v}(w_1,w_2)$$

$$=\sum_{k=0}^{[q/p]}(-q)_{pk}\,\Omega_k w_1^k I_{t_i+1,s_i+1:r}^{p,q+1}\left[w_2\left|\begin{matrix}(-u-(v+1)q,1)(a_j,e_j)_{1,q}; & (a_{ji},e_{ji})_{q+1,t_i}\\ (b_j,f_j)_{1,p}; & (b_{ji},f_{ji})_{p+1,s_i},(-u-vq-pk,1)\end{matrix}\right.\right]$$

Proof: Consider the definition of $Q_{q,p}^{u,v}(w_1,w_2)$, we have

$$Q_{q,p}^{u,v}(w_1,w_2)$$

$$=\sum_{k=0}^{[q/p]}(-q)_{pk}\,\Omega_k w_1^k I_{t_i+1,s_i+1:r}^{p,q+1}\left[w_2\left|\begin{matrix}(-u-(v+1)q,1)(a_j,e_j)_{1,q}; & (a_{ji},e_{ji})_{q+1,t_i}\\ (b_j,f_j)_{1,p}; & (b_{ji},f_{ji})_{p+1,s_i},(-u-vq-pk,1)\end{matrix}\right.\right] \tag{7.50}$$

By using Mellin–Barnes type contour integral of I-function, we get

$$Q_{q,p}^{u,v}(w_1,w_2)$$

$$=\sum_{k=0}^{[q/p]}(-q)_{pk}\,\Omega_k w_1^k\left\{\frac{1}{2\pi i}\int_L\frac{\Gamma(1+u+(v+1)q+\xi)}{\Gamma(1+u+vq+pk+\xi)}w_2^{\xi}\phi_{t_i,s_i:r}^{p,q}(\xi)\,d\xi\right. \tag{7.51}$$

where $\phi_{t_i,s_i:r}^{p,q}(\xi)$ is defined as follows

$$\phi_{t_i,s_i:r}^{p,q}(\xi)=\frac{\prod_{j=1}^{p}\Gamma(b_j-f_j\xi)\prod_{j=1}^{p}\Gamma(1-a_j-e_j\xi)}{\sum_{i=1}^{r}\left[\prod_{j=p+1}^{s_i}\Gamma(1-b_{ji}+f_{ji}\xi)\prod_{j=q+1}^{t_i}\Gamma(a_{ji}-e_{ji}\xi)\right]}$$

On interchanging order of summation and integration, we get

$$Q_{q,p}^{u,v}(w_1,w_2)$$

$$= \frac{1}{2\pi i}\int_L w_2^{\xi}\phi_{t_i,s_i:r}^{p,q}(\xi)\left\{\sum_{k=0}^{[q/p]}\frac{\Gamma(1+u+(v+1)q+\xi)}{\Gamma(1+u+vq+pk+\xi)}(-q)_{pk}\,\Omega_k w_1^k\right\}d\xi \tag{7.52}$$

Then consider the left-hand side of the (7.24), and denote it by M, we have

$$M=\sum_{q=0}^{\infty}Q_{q,p}^{u,v}(w_1,w_2)\frac{z^q}{q!} \tag{7.53}$$

By using the Equations (7.27) and (7.28), we get

$$M=\sum_{q=0}^{\infty}\left\{\frac{1}{2\pi i}\int_L w_2^{\xi}\phi_{t_i,s_i:r}^{p,q}(\xi)\left\{\sum_{k=0}^{[q/p]}\frac{\Gamma(1+u+(v+1)q+\xi)}{\Gamma(1+u+vq+pk+\xi)}(-q)_{pk}\,\Omega_k w_1^k\right\}d\xi\right\}\frac{z^n}{n!} \tag{7.54}$$

After changing the order of summation and integration with some calculations, we have

$$M=\frac{1}{2\pi i}\int_L w_2^{\xi}\phi_{t_i,s_i:r}^{p,q}(\xi)\left\{\sum_{q=0}^{\infty}\binom{u+\xi+(v+1)q}{q}z^q\sum_{k=0}^{[q/p]}\frac{(-q)_{pk}\,\Omega_k w_1^k}{(1+u+vq+\xi)_{pk}}\right\} \tag{7.55}$$

Then using the result discussed in Srivastava [59], we have

$$\sum_{q=0}^{\infty}\binom{u+(v+1)q}{q}z^q\sum_{k=0}^{[q/p]}\frac{(-q)_{pk}\,\mathrm{C}_k w_1^k}{(1+u+vq)_{pk}\,k!}=\frac{(1+\xi)^{u+1}}{(1-v\xi)}\sum_{k=0}^{\infty}C_k\frac{\left[w_1(-\xi)^p\right]^k}{k!} \tag{7.56}$$

With $u=u+\xi$ and $\mathrm{C}_k=k!\Omega_k,\ k\ge o$, using (7.30), we get

$$M=\frac{(1+\xi)^{u+1}}{(1-v\xi)}\sum_{k=0}^{\infty}\Omega_k w_1^k(-\xi)^{pk}\times\frac{1}{2\pi i}\int_l \phi_{t_i,s_i:r}^{p,q}(\xi)w_2^{\xi}(1+\xi)^{\xi}\,d\xi \tag{7.57}$$

Then using the definition of Saxena I-function, we get our desired result.

$$\sum_{q=0}^{\infty} Q_{q,p}^{u,v}(w_1,w_2)\frac{z^q}{q!}=\frac{(1+\xi)^{u+1}}{(1-v\xi)}\psi\left[w_1(-\xi)^p\right]I_{t_i,s_i:r}^{p,q}\left[w_2(1+\xi)\left|\begin{matrix}(a_j,e_j)_{1,q};&(a_{ji},e_{ji})_{q+1,t_i}\\(b_j,f_j)_{1,p};&(b_{ji},f_{ji})_{p+1,s_i}\end{matrix}\right.\right] \tag{7.58}$$

Corollary 7.3

The following bilateral generating relation for H-function holds true:

$$\sum_{q=0}^{\infty} Q_{q,p}^{u,v}(w_1,w_2)\frac{z^q}{q!}=\frac{(1+\xi)^{u+1}}{(1-v\xi)}\psi\left[w_1(-\xi)^p\right]H_{t,s}^{p,q}\left[w_2(1+\xi)\left|\begin{matrix}(a_j,e_j)_{1,q};&(a_j,e_j)_{q+1,t}\\(b_j,f_j)_{1,p};&(b_j,f_j)_{p+1,s}\end{matrix}\right.\right] \tag{7.59}$$

where $= z(1+\xi)^{v+1}, \psi[z]=\sum_{q=0}^{\infty}\Omega_q z^q$, z^q is an arbitrary complex sequence and

$$Q_{q,p}^{u,v}(w_1,w_2)$$

$$=\sum_{k=0}^{[q/p]}(-q)_{pk}\,\Omega_k w_1^k H_{t+1s+1}^{p,q+1}\left[w_2\left|\begin{matrix}(-u-(v+1)q,1)(a_j,e_j)_{1,q};&(a_j,e_j)_{q+1,t}\\(b_j,f_j)_{1,p};&(b_j,f_j)_{p+1,s},(-u-vq-pk,1)\end{matrix}\right.\right]$$

Corollary 7.4

The following bilateral generating relation for G-function holds true:

$$\sum_{q=0}^{\infty} Q_{q,p}^{u,v}(w_1,w_2)\frac{z^q}{q!}=\frac{(1+\xi)^{u+1}}{(1-v\xi)}\psi\left[w_1(-\xi)^p\right]G_{t,s}^{p,q}\left[w_2(1+\xi)\left|\begin{matrix}a_1,a_2,\ldots a_t\\b_1,b_2,\ldots .b_s\end{matrix}\right.\right] \tag{7.60}$$

where $= z(1+\xi)^{v+1}, \psi[z]=\sum_{q=0}^{\infty}\Omega_q z^q$, z^q is an arbitrary complex sequence and

$$Q_{q,p}^{u,v}(w_1,w_2)$$

$$=\sum_{k=0}^{[q/p]}(-q)_{pk}\,\Omega_k w_1^k G_{t+1s+1}^{p,q+1}\left[w_2\left|\begin{matrix}-u-(v+1)q,a_1,a_2,\ldots a_t\\b_1,b_2,\ldots .b_s,-u-vq-pk\end{matrix}\right.\right]$$

7.3 GENERATING FUNCTIONS FOR Q-ANALOGUE OF I-FUNCTION

(A) Jain et al. [58] obtained various forms of I-function which satisfy Truesdell's descending F-equation and hence in this connection, we established in this section the different forms of q-analogue of I-function which satisfy Truesdell's descending F_q-equation:

$$\text{I.}\quad \left(q^{\frac{\alpha}{2}\left[\frac{1-\rho}{\rho}\right]}z\right)^{\alpha-1} I^{m,n}_{p_i,q_i,l}\left[\frac{q^{\alpha h(\lambda+1)}}{z^{h\lambda}};q\,\middle|\,\begin{matrix}\{\Delta(\lambda,\alpha),h\},\{(a_j,\alpha_j)_{\lambda+1,n}\},\{(a_{ji},\alpha_{ji})_{n+1,p_i-\rho}\},\{\Delta(\rho,\alpha),h\}\\ \{(b_j,\beta_j)_{1,m}\},\{(b_{ji},\beta_{ji})_{m+1,q_i-\rho}\},\{\Delta(\rho,\alpha),h\}\end{matrix}\right] \tag{7.61}$$

$$\text{II.}\quad \frac{\left(q^{\frac{-\alpha}{2}\left[\frac{1+\lambda}{\lambda}\right]}z\right)^{\alpha-1}}{(1-q)^{\alpha}} I^{m,n}_{p_i,q_i,l}\left[\frac{q^{\alpha h(\lambda-1)}}{z^{h\lambda}};q\,\middle|\,\begin{matrix}\{\Delta(2\lambda,2\alpha),h\},\{(a_j,\alpha_j)_{2\lambda+1,n}\},\{(a_{ji},\alpha_{ji})_{n+1,p_i}\}\\ \left\{\Delta(\lambda,\alpha+\frac{1}{2}),h\right\},\{(b_j,\beta_j)_{\lambda+1,m}\},\{(b_{ji},\beta_{ji})_{m+1,q_i}\}\end{matrix}\right] \tag{7.62}$$

$$\text{III.}\quad \frac{\left(q^{\frac{-\alpha}{2(\alpha-1)}\left[\frac{3\alpha+1}{3\lambda}+\alpha-1\right]}z\right)^{\alpha-1}}{(1-q)^{\alpha}}$$

$$I^{m,n}_{p_i,q_i,l}\left[\frac{q^{\alpha h(\lambda+1)}}{z^{h\lambda}};q\,\middle|\,\begin{matrix}\{\Delta(3\lambda,3\alpha),h\},\{(a_j,\alpha_j)_{3\lambda+1,n}\},\{(a_{ji},\alpha_{ji})_{n+1,p_i}\}\\ \left\{\Delta(\lambda,\alpha+\frac{2}{3}),h\right\},\{(b_j,\beta_j)_{\lambda+1,m}\},\{(b_{ji},\beta_{ji})_{m+1,q_i-\lambda}\},\left\{\Delta(\lambda,\alpha+\frac{1}{3}),h\right\}\end{matrix}\right] \tag{7.63}$$

$$\text{IV.}\quad \left(q^{\frac{-\alpha}{2}}z\right)^{\alpha-1} e^{\pi i\alpha} I^{m,n}_{p_i,q_i,l}\left[\frac{q^{\alpha h\lambda}}{z^{h\lambda}};q\,\middle|\,\begin{matrix}\{\Delta(\lambda,\alpha),h\},\{(a_j,\alpha_j)_{\lambda+1,n}\},\{(a_{ji},\alpha_{ji})_{n+1,p_i}\}\\ \{(b_j,\beta_j)_{1,m}\},\{(b_{ji},\beta_{ji})_{m+1,q_i}\}\end{matrix}\right] \tag{7.64}$$

$$\text{V.}\quad \left(q^{\frac{\alpha}{2}\left(\frac{1-\lambda}{\lambda}\right)}z\right)^{\alpha-1} I^{m,n}_{p_i,q_i,l}\left[\frac{q^{h\alpha(\lambda-1)}}{z^{h\lambda}};q\,\middle|\,\begin{matrix}\{(a_j,\alpha_j)_{1,n}\},\{(a_{ji},\alpha_{ji})_{n+1,p_i-\lambda}\},\{\Delta(\lambda,\alpha),h\}\\ \{(b_j,\beta_j)_{1,m}\},\{(b_{ji},\beta_{ji})_{m+1,q_i}\}\end{matrix}\right] \tag{7.65}$$

VI. $$\left(q^{\frac{\alpha}{2}\left(\frac{1-\lambda}{\lambda}-\frac{1}{\rho}\right)}z\right)^{\alpha-1}$$

$$e^{\pi i\alpha} I^{m,n}_{p_i,q_i,l}\left[\frac{q^{\alpha h\lambda}}{z^{h\lambda}};q\left|\begin{matrix}\{\Delta(\rho,\alpha),h\}\left\{(a_j,\alpha_j)_{\rho+1,n}\right\},\left\{(a_{ji},\alpha_{ji})_{n+1,p_i-\lambda}\right\},\{\Delta(\lambda,\alpha),h\}\\ \{\Delta(\rho,\alpha),h\}\left\{(b_j,\beta_j)_{\rho+1,m}\right\},\left\{(b_{ji},\beta_{ji})_{m+1,q_i}\right\}\end{matrix}\right.\right] \tag{7.66}$$

Assuming that form (7.61) is $A(z,\alpha)$, replacing q-analogue of I-function by its definition (7.29) and then interchanging order of integration and differentiation, which is justified under the conditions of convergence, we observe

$$D^r_{q,z}A(z,\alpha)$$

$$=\frac{1}{2\pi i}\frac{\prod_{j=1}^{m}\Gamma_q(b_j-\beta_j s)\prod_{k=0}^{\lambda-1}\Gamma_q\left(1-\frac{\alpha+k}{\lambda}-hs\right)\prod_{j=\lambda+1}^{n}\Gamma_q(1-a_j+\alpha_j s)\,q^{\frac{\alpha(\alpha-1)(1-\rho)}{2\rho}}q^{\alpha h(\lambda+1)s}D^r_{q,z}\left\{z^{\alpha-1-h\lambda s}\right\}\pi ds}{\sum_{i=1}^{l}\left\{\prod_{j=m+1}^{q_i-\rho}\Gamma_q(1-b_{ji}+\beta_{ji}s)\prod_{k=0}^{\rho-1}\Gamma_q\left(1-\frac{\alpha+k}{\rho}+hs\right)\prod_{j=n+1}^{p_i-\rho}\Gamma_q(a_{ji}-\alpha_{ji}s)\prod_{k=0}^{\rho-1}\Gamma_q\left(\frac{\alpha+k}{\rho}-hs\right)\right\}\Gamma_q(s)\Gamma_q(1-s)\sin\pi s}$$

Now results (7.33) and (7.34) lead to two very important identities

$$\prod_{k=0}^{\lambda-1}\Gamma_q\left(\frac{\alpha+k}{\lambda}-hs\right)=\frac{(-1)^r\left(q^{1-(\alpha-h\lambda s)};q\right)_r}{(1-q)^r\,q^{\frac{r}{2m}(r-2(\alpha-h\lambda s)+1)}}\prod_{k=0}^{\lambda-1}\Gamma_q\left(\frac{\alpha-r+k}{\lambda}-hs\right). \tag{7.68}$$

$$\prod_{k=0}^{\lambda-1}\Gamma_q\left(1-\frac{\alpha+k}{\lambda}+hs\right)=\frac{(1-q)^r}{(q^{1-(\alpha-h\lambda s)};q)_r}\prod_{k=0}^{\lambda-1}\Gamma_q\left(1-\frac{\alpha-r+k}{\lambda}+hs\right) \tag{7.69}$$

Using these identities (7.68) and (7.69) we see that (7.67) takes the form

$$D^r_{q,z}A(z,\alpha)=[A(z,\alpha-r)] \tag{7.70}$$

This is Truesdell's form of descending F-equation.

Similarly, forms (7.62) to (7.66) can be shown to satisfy Truesdell's descending F-equation.

(B) In this section we employ forms (7.61) to (7.66), to establish the following generating functions for q-analogue of I-function using Truesdell's descending F-equation technique:

I. $$(1+q^{\alpha})^{\alpha-1} I_{p_i,q_i,l}^{m,n}\left[(1+q^{\alpha})^{-h\lambda}x;q\left|\begin{matrix}\{\Delta(\lambda,\alpha),h\},\{(a_j,\alpha_j)_{\lambda+1,n}\},\{(a_{ji},\alpha_{ji})_{n+1,p_i-\rho}\},\{\Delta(\rho,\alpha),h\}\\ \{(b_j,\beta_j)_{1,m}\},\{(b_{ji},\beta_{ji})_{m+1,q_i-\rho}\},\{\Delta(\rho,\alpha),h\}\end{matrix}\right.\right]$$

$$=\sum_{r=0}^{\infty}\frac{\left(q^{(-\alpha+r/2+1/2)(\frac{1-\rho}{\rho})+\alpha}\right)^r}{r!} I_{p_i,q_i,l}^{m,n}$$

$$\left[q^{-rh(\lambda+1)}x;q\left|\begin{matrix}\{\Delta(\lambda,\alpha-r),h\},\{(a_j,\alpha_j)_{\lambda+1,n}\},\{(a_{ji},\alpha_{ji})_{n+1,p_i-\rho}\},\{\Delta(\rho,\alpha-r),h\}\\ \{(b_j,\beta_j)_{1,m}\},\{(b_{ji},\beta_{ji})_{m+1,q_i-\rho}\},\{\Delta(\rho,\alpha-r),h\}\end{matrix}\right.\right] \tag{7.71}$$

II. $$(1+q^{\alpha})^{\alpha-1} I_{p_i,q_i,l}^{m,n}\left[(1+q^{\alpha})^{-h\lambda}x;q\left|\begin{matrix}\{\Delta(2\lambda,2\alpha),h\},\{(a_j,\alpha_j)_{2\lambda+1,n}\},\{(a_{ji},\alpha_{ji})_{n+1,p_i}\}\\ \left\{\Delta(\lambda,\alpha+\frac{1}{2}),h\right\},\{(b_j,\beta_j)_{\lambda+1,m}\},\{(b_{ji},\beta_{ji})_{m+1,q_i}\}\end{matrix}\right.\right]$$

$$=\sum_{r=0}^{\infty}\frac{\left(q^{\left(\alpha-\frac{1}{2}-\frac{r}{2}\right)\left(\frac{1+\lambda}{\lambda}\right)+\alpha}(1-q)\right)^r}{r!} I_{p_i,q_i,l}^{m,n}$$

$$\left[q^{rh(1-\lambda)}x;q\left|\begin{matrix}\{\Delta(2\lambda,2\alpha-2r),h\},\{(a_j,\alpha_j)_{2\lambda+1,n}\},\{(a_{ji},\alpha_{ji})_{n+1,p_i}\}\\ \left\{\Delta(\lambda,\alpha-r+\frac{1}{2}),h\right\},\{(b_j,\beta_j)_{\lambda+1,m}\},\{(b_{ji},\beta_{ji})_{m+1,q_i}\}\end{matrix}\right.\right] \tag{7.72}$$

III. $$(1+q^{\alpha})^{\alpha} I_{p_i,q_i,l}^{m,n}$$

$$\left[(1+q^{\alpha})^{-h\lambda}x;q\left|\begin{matrix}\{\Delta(3\lambda,3\alpha),h\},\{(a_j,\alpha_j)_{3\lambda+1,n}\},\{(a_{ji},\alpha_{ji})_{n+1,p_i}\}\\ \left\{\Delta(\lambda,\alpha+\frac{2}{3}),h\right\},\{(b_j,\beta_j)_{\lambda+1,m}\},\{(b_{ji},\beta_{ji})_{m+1,q_i-\lambda}\},\left\{\Delta(\lambda,\alpha+\frac{1}{3}),h\right\}\end{matrix}\right.\right]$$

$$= \sum_{r=0}^{\infty} \frac{\left(q^{\left(2\alpha+\frac{\alpha}{\lambda}-\frac{r}{2\lambda}-\frac{r}{2}+\frac{1}{6\lambda}+\frac{1}{2}\right)}(1-q) \right)^{r}}{r!} I_{p_i,q_i,l}^{m,n}$$

$$\left[q^{-rh(1+\lambda)}x;q \left| \begin{matrix} \{\Delta(3\lambda,3\alpha-3r),h\},\{(a_j,\alpha_j)_{3\lambda+1,n}\},\{(a_{ji},\alpha_{ji})_{n+1,p_i}\} \\ \{\Delta(\lambda,\alpha-r+\frac{2}{3}),h\},\{(b_j,\beta_j)_{\lambda+1,m}\},\{(b_{ji},\beta_{ji})_{m+1,q_i-\lambda}\},\{\Delta(\lambda,\alpha-r+\frac{1}{3}),h\} \end{matrix} \right. \right] \tag{7.73}$$

IV. $$(1+q^{\alpha})^{\alpha-1} I_{p_i,q_i,l}^{m,n} \left[(1+q^{\alpha})^{-h\lambda}x;q \left| \begin{matrix} \{\Delta(\lambda,\alpha),h\},\{(a_j,\alpha_j)_{\lambda+1,n}\},\{(a_{ji},\alpha_{ji})_{n+1,p_i}\} \\ \{(b_j,\beta_j)_{1,m}\},\{(b_{ji},\beta_{ji})_{m+1,q_i}\} \end{matrix} \right. \right]$$

$$= \sum_{r=0}^{\infty} \frac{(-1)^r \left(q^{2\alpha+\frac{r}{2}-\frac{1}{2}} \right)^r}{r!} I_{p_i,q_i,l}^{m,n} \left[q^{-rh\lambda}x;q \left| \begin{matrix} \{\Delta(\lambda,\alpha-r),h\},\{(a_j,\alpha_j)_{\lambda+1,n}\},\{(a_{ji},\alpha_{ji})_{n+1,p_i}\} \\ \{(b_j,\beta_j)_{1,m}\},\{(b_{ji},\beta_{ji})_{m+1,q_i}\} \end{matrix} \right. \right] \tag{7.74}$$

V. $$(1+q^{\alpha})^{\alpha-1} I_{p_i,q_i,l}^{m,n} \left[(1+q^{\alpha})^{-h\lambda}x;q \left| \begin{matrix} \{(a_j,\alpha_j)_{1,n}\},\{(a_{ji},\alpha_{ji})_{n+1,p_i-\lambda}\},\{\Delta(\lambda,\alpha),h\} \\ \{(b_j,\beta_j)_{1,m}\},\{(b_{ji},\beta_{ji})_{m+1,q_i}\} \end{matrix} \right. \right]$$

$$= \sum_{r=0}^{\infty} \frac{\left(q^{\left(\frac{1-\lambda}{\lambda}\right)\left[-\alpha+\frac{r}{2}+\frac{1}{2}\right]+\alpha} \right)^r}{r!}$$

$$I_{p_i,q_i,l}^{m,n} \left[q^{rh(1-\lambda)}x;q \left| \begin{matrix} \{(a_j,\alpha_j)_{1,n}\},\{(a_{ji},\alpha_{ji})_{n+1,p_i-\lambda}\},\{\Delta(\lambda,\alpha-r),h\} \\ \{(b_j,\beta_j)_{1,m}\},\{(b_{ji},\beta_{ji})_{m+1,q_i}\} \end{matrix} \right. \right] \tag{7.75}$$

VI. $$(1+q^{\alpha})^{\alpha-1} I_{p_i,q_i,l}^{m,n}$$

$$\left[(1+q^{\alpha})^{-h\lambda}x;q \left| \begin{matrix} \{\Delta(\rho,\alpha),h\}\{(a_j,\alpha_j)_{\rho+1,n}\},\{(a_{ji},\alpha_{ji})_{n+1,p_i-\lambda}\},\{\Delta(\lambda,\alpha),h\} \\ \{\Delta(\rho,\alpha),h\}\{(b_j,\beta_j)_{\rho+1,m}\},\{(b_{ji},\beta_{ji})_{m+1,q_i}\} \end{matrix} \right. \right]$$

$$= \sum_{r=0}^{\infty} \frac{(-1)^r \left(q^{\left(\frac{1}{\lambda}-\frac{1}{\rho}-1\right)\left[-\alpha+\frac{r}{2}+\frac{1}{2}\right]+\alpha} \right)^r}{r!}$$

$$I_{p_i,q_i,l}^{m,n}\left[q^{-rh\lambda}x;q \left| \begin{matrix} \{\Delta(\rho,\alpha-r),h\}\{(a_j,\alpha_j)_{\rho+1,n}\},\{(a_{ji},\alpha_{ji})_{n+1,p_i-\lambda}\},\{\Delta(\lambda,\alpha-r),h\} \\ \{\Delta(\rho,\alpha-r),h\}\{(b_j,\beta_j)_{\rho+1,m}\},\{(b_{ji},\beta_{ji})_{m+1,q_i}\} \end{matrix} \right. \right] \tag{7.76}$$

Proof: To establish (7.71) we substitute the form (7.61) in Truesdell's descending F-Equation (7.31) and replace z by $\frac{y}{q^{\alpha}}$ and $\frac{q^{\alpha h+2\alpha h\lambda}}{y^{h\lambda}}$ by x in succession to get the required result. Similarly, result (7.73) can be proved by substituting the form (7.63) in Truesdell's descending F-equation (7.6) and using the same replacement.

To establish (7.72) we substitute the form (7.62) in Truesdell's descending F-Equation (7.31) and replace z by $\frac{y}{q^{\alpha}}$ and $\frac{q^{2\alpha h\lambda-\alpha h}}{y^{h\lambda}}$ by x in succession to get the required result. Similarly, result (7.75) can be proved by substituting the form (7.65) in Truesdell's descending F-Equation (7.31) and using the same replacement.

To establish (7.74) we substitute the form (7.76) in Truesdell's descending F-Equation (7.31) and replace z by $\frac{y}{q^{\alpha}}$ and $\frac{q^{2\alpha h\lambda}}{y^{h\lambda}}$ by x in succession to get the required result. Similarly, result (7.76) can be proved by substituting the form (7.66) in Truesdell's descending F-Equation (7.31) and using the same replacement.

7.3.1 Special Cases

These results yield as special cases of certain generating function for q-analogue of Fox's H-function and q-analogue of Meijer's G-function.

(i): If we take $l=1$ then the series (7.71) reduces to generating function of q-analogue of Fox's H-function.

$$({}^1 H_{P,Q}^{m,n}\left[(1+q^{\alpha})^{-h\lambda} x;q \left| \begin{matrix} \{\Delta(\lambda,\alpha),h\},(a_j,\alpha_j),\{\Delta(\rho,\alpha),h\} \\ (b_j,\beta_j),\{\Delta(\rho,\alpha),h\} \end{matrix} \right. \right]$$

$$= \sum_{r=0}^{\infty} \frac{\left(q^{(-\alpha+r/2+1/2)\left(\frac{1-\rho}{\rho}\right)+\alpha} \right)^r}{r!} H_{P,Q}^{m,n}\left[q^{-rh(\lambda+1)}x;q \left| \begin{matrix} \{\Delta(\lambda,\alpha-r),h\},(a_j,\alpha_j),\{\Delta(\rho,\alpha-r),h\} \\ (b_j,\beta_j),\{\Delta(\rho,\alpha-r),h\} \end{matrix} \right. \right] \tag{7.77}$$

Again taking $\alpha_j = \beta_j = 1$ *and* $h = 1$ in (7.77), it gives Meijer's G-function as:

$$({}^{1}H_{P,Q}^{m,n}\left[\left(1+q^{\alpha}\right)^{-\lambda} x;q \,\middle|\, \begin{matrix} \Delta(\lambda,\alpha), a_j, \Delta(\rho,\alpha) \\ b_j, \Delta(\rho,\alpha) \end{matrix}\right]$$

$$= \sum_{r=0}^{\infty} \frac{\left(q^{(-\alpha+r/2+1/2)\left(\frac{1-\rho}{\rho}\right)+\alpha}\right)^{r}}{r!} H_{P,Q}^{m,n}\left[q^{-r(\lambda+1)}x;q \,\middle|\, \begin{matrix} \Delta(\lambda,\alpha-r), a_j, \Delta(\rho,\alpha-r) \\ b_j, \Delta(\rho,\alpha-r) \end{matrix}\right] \tag{7.78}$$

Similarly, (7.66) to (7.76) can be employed to yield apparently new and interesting results for q-analogue of Fox's H-function and Meijer's G-function.

8 Function Applications

In this chapter, we have given some applications and results of I-function in related areas. This includes finite integral transform of n-variables having product of Jacobi polynomials with their weight functions as a kernel. This transform is known as Jacobi transform of n-variables.

8.1 JACOBI TRANSFORM OF I-FUNCTION

In this section, we have given some properties and established the inversion formula and recurrence relations of the transform.

We define Jacobi Transform of n-variables in a function $h\,(x_1, x_2, x_3, \ldots\ldots x_n)$ according to the definition of Jacobi transform of two variables [90]

$$J\left[h\left(x_1,x_2,x_3,\ldots x_n\right)\right]=\int_{-1}^{1}\int_{-1}^{1}\ldots n\ times\int_{-1}^{1}\left(1-x_1\right)^{\alpha_{11}}\left(1+x_1\right)^{\beta_{11}}\left(1-x_2\right)^{\alpha_{12}}$$

$$\left(1+x_2\right)^{\beta_{12}}\ldots\ldots\left(1-x_n\right)^{\alpha_{1n}}\left(1+x_n\right)^{\beta_{1n}}P_{m1}^{\alpha_{11},\beta_{11}}\left(x_1\right) \tag{8.1}$$

$$h\left(x_1,x_2,x_3,\ldots\ldots x_n\right)\ P_{m_2}^{\alpha_{12},\beta_{12}}\left(x_2\right)\ldots\ldots P_{m_n}^{\alpha_{1n},\beta_{1n}}\left(\hat{x}_n\right)dx_1dx_2\ldots\ldots dx_n$$

This can be represented as

$$J\left\{h\left(x_1,x_2,\ldots x_n\right)\right\}=f\left(\begin{matrix}\alpha_{11},\beta_{11},\alpha_{13},\beta_{13}\\ ;m_1,m_2\ldots m_n\\ \alpha_{12},\beta_{12},\ldots\alpha_{1n},\beta_{1n}\end{matrix}\right) \tag{8.2}$$

where the function h $(x_1, x_2, \ldots\ldots x_n)$ is defined for values, $-1\le x_1, x_2, \ldots\ldots x_n \le 1$; $\alpha_{11}, \alpha_{12}, \ldots\ldots \alpha_{1n} > -1, \beta_{11}, \beta_{12}, \ldots\ldots \beta_{1n} > -1$.

where $P_l^{\mu,\nu}(x)$ is Jacobi polynomial of nth order, which is represented as

$$P_l^{\mu,\nu}\left(x\right)=\left(\begin{matrix}l+\mu\\ l\end{matrix}\right){}_2F_1\left[\begin{matrix}-l,l+\mu+\nu+1; & \\ & \frac{1}{2}-\frac{1}{2}x\\ \mu+1 \qquad ; & \end{matrix}\right] \tag{8.3}$$

$$=\left(-1\right)^n\left(\begin{matrix}l+\nu\\ l\end{matrix}\right){}_2F_1\left[\begin{matrix}-l,l+\mu+\nu+1; & \\ & \frac{1}{2}-\frac{1}{2}x\\ \nu+1 \qquad ; & \end{matrix}\right] \tag{8.4}$$

DOI: 10.1201/9781003517757-8

Re $(\mu) > -1$, Re $(\nu) > -1$.

The Jacobi polynomial $P_l^{\mu,\nu}(x)$defined in Equations (8.3) and (8.4) is a solution of differential equation.

$$\left(1-x^2\right)\frac{\partial^2 w}{\partial x^2}+\left[v-\mu-(\mu+v+2)z\right]\frac{\partial w}{\partial x}+l(l+\mu+v+1)w=0 \tag{8.5}$$

Orthogonal relation for, $P_l^{\mu,\nu}(x)$ is [94]

$$\int_{-1}^{1}(1-x)^{\mu}(1+x)^{\nu}P_l^{\mu,\nu}(x)P_k^{\mu,\nu}(x)dx=0,1\neq \mathrm{k}$$
$$=\delta_l, l=k, \tag{8.6}$$

$$\delta_l=\frac{2^{\mu+\nu+1}\Gamma(l+\mu+1)\Gamma(l+\nu+1)}{l(\mu+\nu+2l+1)\Gamma(l+\mu+\nu+1)} \tag{8.7}$$

The self-adjoint differential form for $P_l^{\mu,\nu}(x)$

$$\frac{d}{dx}\left[(1-x)^{\mu+1}(1+x)^{\nu+1}\frac{d}{dx}P_l^{\mu,\nu}(x)\right]$$
$$+l(l+\mu+\nu+1)(1-x)^{\mu}(1+x)^{\nu}P_l^{\mu,\nu}(x)=0 \tag{8.8}$$

8.1.1 The Inversion Formula for n –Variables

Now we establish the inversion formula of the nth Jacobi transform.

Let

$$h(x_1,x_2,x_2,\ldots,x_n)=\sum_{m=0}^{\infty}\sum_{n=0}^{\infty}A_{m_1,m_2,m_3,\ldots,m_n}P_{m_1}^{(\alpha_{11},\beta_{11})}(x_1)P_{m_2}^{(\alpha_{12},\beta_{12})}(x_2)$$
$$\times P_{m_3}^{(\alpha_{13},\beta_{13})}(x_3),\ldots,P_{m_n}^{(\alpha_{1n},\beta_{1n})}(x_n) \tag{8.9}$$

where $A_{m1,m2,m3\ldots\ldots mn}$ is constant independent of $x_1, x_2, \ldots\ldots x_n$ then multiplying Equation (8.9) by

$$(1-x_1)^{\alpha_{11}}(1+x_1)^{\beta_{11}}(1-x_2)^{\alpha_{12}}(1+x_2)^{\beta_{12}}\ldots\ldots(1-x_n)^{\alpha_{1n}}(1+x_n)^{\beta_{1n}}$$
$$P_{m_1}^{(\alpha_{11},\beta_{11})}(x_1)P_{m_2}^{(\alpha_{12},\beta_{12})}(x_2)\ldots\ldots P_{m_n}^{(\alpha_{1n},\beta_{1n})}(x_n)$$

and integrating both the sides with respect to $x_1, x_2, \ldots\ldots x_n$ from −1 to 1 under the sign of summation, we find

$$A_{m_1,m_2,\ldots\ldots mn}=\frac{1}{\delta_{m_1},\delta_{m_2},\ldots\ldots\delta_{m_n}}f\begin{pmatrix}\alpha_{11},\beta_{11}\\ m_1,m_2,\ldots.m_n\\ \alpha_{12},\beta_{12}\end{pmatrix} \tag{8.10}$$

δ_l is defined in Equation (8.7).

Substituting the value of $A_{m1,\,m2,\,.....mn}$ in Equation (8.9) we get,

$$h(x_1,x_2,....x_n)=\sum_{m_1=0}^{\infty}\sum_{m_2=0}^{\infty}..n-\text{times}\sum_{m_n=0}\frac{1}{\delta_{m_1},\delta_{m_2},.....\delta_{m_n}}P_{m_1}^{\alpha_{11},\beta_{11}}(x_1)$$
$$\times P_{m_2}^{\alpha_{12},\beta_{12}}(x_2)....P_{m_n}^{\alpha_{1n},\beta_{1n}}(x_n)f\begin{pmatrix}\alpha_{11},\beta_{11}\\ m_1,m_2,....m_n\\ \alpha_{12},\beta_{12}\end{pmatrix} \tag{8.11}$$

$$-1\le x_1,x_2,....x_n\le 1,\quad \alpha_{11},\alpha_{12},....\alpha_{1n}>-1,\quad \beta_{11},\beta_{12},.....,\beta_{1n}>-1$$

8.1.2 n-Variables Jacobi Transform of Differential Form

We establish a theorem which gives n-Jacobi transform of differential form of $h(x_1, x_2, x_3,......,x_n)$.

Theorem 8.1

Now we assume that

i. The function $h(x_1,x_2,x_3,...,x_n)$ and its partial derivatives
$\frac{\partial}{\partial x_1}h(x_1,x_2,x_2,....,x_n),\frac{\partial}{\partial x_2}h(x_1,x_2,x_2,....,x_n)$ and
$\frac{\partial}{\partial x_n}h(x_1,x_2,x_2,....,x_n)$ are bounded and almost continuous
in $-1\le x_1,x_2,x_3,....,x_n\le 1$

ii. The second-order partial derivatives

$$\frac{\partial^2}{\partial x_1^{\,2}}h(x_1,x_2,x_2,....,x_n),\frac{\partial^2}{\partial x_2^{\,2}}h(x_1,x_2,x_2,....,x_n)\text{ and}$$

$$\frac{\partial^2}{\partial x_n^{\,2}}h(x_1,x_2,x_2,....,x_n)\text{ are bounded and integrable}$$

therefore $C\left[h(x_1,x_2,x_3,....,x_n)\right]$ denotes the differential form

iii. $$\lim_{x_1\to\pm 1}(1-x_1)^{\alpha_{11}+1}(1+x_1)^{\beta_{11}+1}h(x_1x_2....x_n)=\lim_{x_1\to\pm 1}(1-x_1)^{\alpha_{11}+1}$$

$$(1+x_1)^{\beta_{11}+1}\frac{\partial}{\partial x}h(x_1x_2....x_n)=0$$

$$\lim_{x_2 \to \pm 1} (1-x_2)^{\alpha_{12}+1}(1+x_2)^{\beta_{12}+1} h(x_1x_2.....x_n) = \lim_{x_1 \to \pm 1} (1-x_2)^{\alpha_{12}+1}$$

$$(1+x_2)^{\beta_{12}+1} \frac{\partial}{\partial x_2} h(x_1x_2.....x_n) = 0$$

n times, ….

$$\lim_{x_n \to \pm 1} (1-x_n)^{\alpha_{1n}+1}(1+x_n)^{\beta_{1n}+1} h(x_1x_2.....x_n) = \lim_{x_n \to \pm 1} (1-x_n)^{\alpha_{1n}+1}$$

$$(1+x_n)^{\beta_{1n}+1} \frac{\partial}{\partial x_n} h(x_1x_2.....x_n) = 0$$

and

$$\begin{aligned} C\left[h(x_1,x_2,x_3,\ldots \mathrm{x}_n)\right] = {} & (1-x_1)^{-\alpha_{11}}(1+x_1)^{-\beta_{11}} \\ & \frac{\partial}{\partial x_1}\left[(1-x_1)^{\alpha_{11}+1}(1+x_1)^{\beta_{11}+1}\frac{\partial}{\partial x_1} h(x_1,x_2,x_3,\ldots,x_n)\right] \\ & +(1-x_2)^{-\alpha_{12}}(1+x_2)^{-\beta_{12}} \\ & \frac{\partial}{\partial x_2}(1-x_2)^{\alpha_{12}+1}\left((1+x_2)^{\beta_{12}+1}\right)\frac{\partial}{\partial x_2} h(x_1,x_2,x_3,\ldots,x_n) \\ & +\ldots n-\text{times} \\ & +(1-x_2)^{-\alpha_{1n}}(1+x_2)^{-\beta_{1n}} \\ & \frac{\partial}{\partial x_n}\left[(1-x_n)^{\alpha_{1n}+1}\left((1+x_n)^{\beta_{n}+1}\right)\frac{\partial}{\partial x_n} h(x_1,x_2,x_3,\ldots,x_n)\right] \end{aligned} \tag{8.12}$$

Then

$$\begin{aligned} & J\left\{c\left[h(x_1,x_2,\ldots..,x_n)\right]\right\} \\ & = -\left[m_1(m_1+\alpha_{11}+\beta_{11}+1)+m_2(m_2+\alpha_{12}+\beta_{12}+1)\ldots+m_n(m_n+\alpha_{1n}+\beta_{1n}+1)\right] \\ & \times f\begin{pmatrix} \alpha_{11},\beta_{11}\ldots. \\ m_1,m_2,\ldots.m_n \\ \alpha_{12},\beta_{12}\ldots. \end{pmatrix} \end{aligned} \tag{8.13}$$

$$-1 \le x_1,x_2,\ldots\ldots.x_n \le 1, \quad \alpha_{11},\alpha_{12},\ldots.\alpha_{1n} > -1, \quad \beta_{11},\beta_{12},\ldots..,\beta_{1n} > -1$$

Proof: We have to find the solution of n-variables Jacobi transform we assume that the function $h(x_1, x_2, \ldots, x_n)$ and its first and second-order partial derivatives are bounded and almost continuous.

$$J\left[C\left\{h(x_{11},x_{12},x_{13},\ldots x_{1n})\right\}\right]=\int_{-1}^{1}\int_{-1}^{1}\ldots n \text{ times}\int_{-1}^{1}(1-x_2)^{\alpha_{12}}(1+x_2)^{\beta_{12}}\ldots(1-x_n)^{\alpha_{1n}}(1+x_n)^{\beta_{1n}}$$

$$P_{m_z}^{(\alpha_{11},\beta_{11})}(x_1)P_{m_2}^{(\alpha_{12},\beta_{12})}(x_2),\ldots P_{m_n}^{(\alpha_{1n},\beta_{1n})}(x_n)$$

$$\times\frac{\partial}{\partial x_1}[(1-x_1)^{\alpha_{11}+1}(1+x_1)^{\beta_{11}+1}\frac{\partial}{\partial x_1}h(x_1,x_2,x_3,\ldots,x_n)]$$

$$dx_1dx_2,\ldots,dx_n$$

$$+\int_{-1}^{1}\int_{-1}^{1}\ldots n \text{ times}\int_{-1}^{1}(1-x_1)^{\alpha_{11}}(1+x_1)^{\beta_{11}}\ldots(1-x_n)^{\alpha_{1n}}$$

$$(1+x_n)^{\beta_{1n}}P_{m_1}^{(\alpha_{11},\beta_{11})}(x_1)$$

$$P_{m_2}^{(\alpha_{12},\beta_{12})}(x_2),\ldots P_{m_n}^{(\alpha_{1n},\beta_{1n})}(x_n)$$

$$\times\frac{\partial}{\partial x_2}[(1-x_2)^{\alpha_{12}+1}(1+x_2)^{\beta_{12}+1}\frac{\partial}{\partial x_2}h(x_1,x_2,x_3,\ldots,x_n)]$$

$$dx_1dx_2,\ldots,d_{xn}$$

$$+\int_{-1}^{1}\int_{-1}^{1}\ldots n \text{ times}\int_{-1}^{1}(1-x_1)^{\alpha_{11}}(1+x_1)^{\beta_{11}}(1-x_2)^{\alpha_{12}}(1+x_2)\ldots..$$

$$(1-x_{n-1})^{\alpha_{12}}(1+x_{n-1})^{\beta_{12}}$$

$$P_{m_1}^{(\alpha_{11},\beta_{11})}(x_1)P_{m_2}^{(\alpha_{12},\beta_{12})}(x_2),\ldots P_{m_n}^{(\alpha_{1n},\beta_{1n})}(x_n)$$

$$\frac{\partial}{\partial x_n}\left[(1-x_n)^{\alpha_{1(n)}+1}(1+x_n)^{\beta_{1(n)}+1}\times\frac{\partial}{\partial x_n}h(x_1,x_2,x_3,\ldots,x_n)\right]$$

$$dx_ndx_1,\ldots,dx_n \tag{8.14}$$

$$\lim_{x_1\to\pm1}(1-x_1)^{\alpha_{11}+1}(1+x_1)^{\beta_{11}+1}h(x_1,x_2\ldots x_n)=\lim_{x_1\to\pm1}(1-x_1)^{\alpha_{11}+1}$$

$$(1+x_1)^{\beta_{11}+1}\frac{\partial}{\partial x}h(x_1,x_2\ldots x_n)=0$$

$$\lim_{x_2 \to \pm 1} (1-x_2)^{\alpha_{12}+1}(1+x_2)^{\beta_{12}+1} h(x_1x_2\ldots x_n) = \lim_{x_1 \to \pm 1} (1-x_2)^{\alpha_{12}+1}$$

$$(1+x_2)^{\beta_{12}+1} \frac{\partial}{\partial x_2} h(x_1x_2\ldots x_n) = 0$$

n times

$$\lim_{x_n \to \pm 1} (1-x_n)^{\alpha_{1n}+1}(1+x_n)^{\beta_{1n}+1} h(x_1x_2\ldots x_n) = \lim_{x_n \to \pm 1} (1-x_n)^{\alpha_{1n}+1}$$

$$(1+x_n)^{\beta_{1n}+1} \frac{\partial}{\partial x_n} h(x_1x_2\ldots x_n) = 0$$

Making use of (8.8) we integrate right-hand side of (8.13) by parts separately with respect to x_1, $x_{2,}$ x_3.....x_n, hence to get,

$$J\left[c\left\{h(x_1,x_2,\ldots x_n)\right\}\right] = -\left[m_1(m_1+\alpha_{11}+\beta_{11}+1)+m_2(m_2+\alpha_{12}+\beta_{12}+1)\right.$$

$$\left.+m_3(m_3+\alpha_{13}+\beta_{13}+1)+\ldots+m_n(m_n+\alpha_{1n}+\beta_{1n}+1)\right]$$

$$\times f\begin{bmatrix} \alpha_{11},\beta_{11},\alpha_{13},\beta_{13} \\ ;m_1,m_2\ldots m_n \\ \alpha_{12},\beta_{12},\ldots\alpha_{1n},\beta_{1n} \end{bmatrix} \tag{8.15}$$

This theorem will be useful in solving many partial differential equations in n-variables.

8.2 *N*-VARIABLES JACOBI TRANSFORM OF I-FUNCTION AND ITS SPECIAL CASES

The n-variables Jacobi transform of I- function can be defined as

$$J\left\{ I\frac{(1-x_1)(1-x_2)(1-x_3)\ldots(1-x_n)}{2^n}u \left| \begin{matrix} (a_j,\alpha_j)_{1,n} \\ (b_j,\beta_j)_{1,m} \\ (a_{ji},\alpha_{ji})_{n+1,p_i} \\ (b_{ji},\beta_{ji})_{m+1,q_i} \end{matrix} \right. \right\}$$

$$= 2^{-(\alpha_{11}+\beta_{11}+\alpha_{12}+\beta_{12}+\ldots+\alpha_{1n}+\beta_{1n})} \int_{-1}^{1}\int_{-1}^{1} \ldots n \text{ times} \int_{-1}^{1} (1-x_1)^{\alpha_{11}} (1+x_1)^{\beta_{11}}$$

$$\times (1-x_2)^{\alpha_{12}} (1+x_2)^{\beta_{12}} \ldots (1-x_{1n})^{\alpha_{1n}} (1+x_{1n})^{\beta_{1n}} P_{m_1}{}^{(\alpha_{11},\beta_{11})}(x_1)$$

$$p_{m_2}{}^{(\alpha_{12},\beta_{12})}(x_2)\ldots P_{m_n}{}^{(\alpha_{1n},\beta_{1n})}(x_n)$$

$$\times \left(I\left(\frac{(1-x_1)(1-x_2)\ldots(1-x_n)}{2^n} u \,\middle|\, \begin{matrix} (a_j,\alpha_j)_{1,n} \\ (b_j,\beta_j)_{1,m} \\ (a_{ji},\alpha_{ji})_{n+1,p_i} \\ (b_{ji},\beta_{ji})_{m+1,q_i} \end{matrix} \right) dx_n dx_2 \ldots dx_n \right) \tag{8.16}$$

We substitute the value of I-function from Equations (1.22) and (1.23), on the right-hand side of Equation (8.15) and reverse the order of integral so that it transformed as

$$= 2^{-(\alpha_{11}+\beta_{11}+\alpha_{12}+\beta_{12}+\alpha_{13}+\beta_{13}+\ldots+\alpha_{1n}+\beta_{1n})} \frac{1}{2\pi i} \int_L \phi(\xi) u^{\xi} 2^{-n\xi}$$

$$\int_{-1}^{1}\int_{-1}^{1}\int_{-1}^{1} \ldots n \text{ times} \int_{-1}^{1} (1-x_1)^{\alpha_{11}+\xi} (1+x_1)^{\beta_{11}}$$

$$\times (1-x_2)^{\alpha_{12}+\xi} (1+x_2)^{\beta_{12}} (1-x_3)^{\alpha_{13}+\xi} (1+x_3)^{\beta_{13}} \ldots (1-x_n)^{\alpha_{1n}+\xi} \tag{8.17}$$

$$(1+x_n)^{\beta_{1n}} P_{m_1}^{(\alpha_{11},\beta_{11})}(x_1) P_{m_2}^{(\alpha_{12},\beta_{12})}(x_2)$$

$$P_{m_3}^{(\alpha_{13},\beta_{13})}(x_3)\ldots P_{m_n}^{(\alpha_{1n},\beta_{1n})}(x_n) dx_1 dx_2 dx_3 \ldots dx_n$$

According to Equation (8.8) evaluate Equation (8.16) to obtain

$$2^{(n)} \frac{[\Gamma(\beta_{11}+m_1+1)\Gamma(\beta_{12}+m_{12}+1)\ldots\Gamma(\beta_{1n}+m_n+1)]}{[m_1!m_2!\ldots m_n!]} \frac{1}{2\pi i} \int_L \phi(\xi) u^{\xi}$$

$$\times \frac{[\Gamma(1+\alpha_{11}+\xi)\Gamma(1+\alpha_{12}+\xi)\ldots\Gamma 1+\alpha_{1n}+\xi][\Gamma(m_1-\xi)\Gamma(m_2-\xi)\ldots\Gamma(m_n-\xi)]}{[\Gamma(-\xi)\Gamma(-\xi)\ldots n \text{ times}\Gamma(-\xi)]}$$

$$\times \frac{1}{[\Gamma(\alpha_{11}+\xi+\beta_{11}+m_1+2)\Gamma(\alpha_{12}+\xi+\beta_{12}+m_2+2)\ldots\Gamma(\alpha_{1n}+\xi+\beta_{1n}+m_n+2)]} \tag{8.18}$$

$$J\left\{I\left(\frac{(1-x_1)(1-x_2)(1-x_3)\ldots(1-x_n)}{2^n}u\left|\begin{matrix}(a_j,\alpha_j)_{1,n}\\(b_j,\beta)_{1,m}\\(a_{ji},\alpha_{ji})_{n+1,p_i}\\(b_{ji},\beta_{ji})_{m+1,q_i}\end{matrix}\right.\right)\right\}$$

$$= 2^{(n)} + \frac{\left[\Gamma(\beta_{11}+m_1+1)\Gamma(\beta_{12}+m_2+1)\ldots\Gamma(\beta_{1n}+m_n+1)\right]}{[m_1!m_2!\ldots m_n!]} \tag{8.19}$$

$$\left.\begin{matrix}\left\{(a_j,\alpha_j)_{1,n}\right\};\left\{(a_{ji},\alpha_{ji})_{n+1,p_i}\right\}(0,1)\\\left\{(b_j,\beta_j)_{1,m}\right\};\left\{(b_{ji},\beta_{ji})_{m+1,q_i}\right\}(-1-\alpha_{11}-\beta_{11}-m_1,1)\\(0,1),\ldots,(0,1)\\(-1-\alpha_{12}-\beta_{12}-m_2,1),\ldots,(-1-\alpha_{1n}-\beta_{1n}-m_n,1)\end{matrix}\right)$$

8.2.1 Special Cases

i. If $r=1$, then Equation (8.18) reduces to the corresponding Jacobi transform in the form of Fox's H-function defined in below equation,

$$J\left\{I\left(\frac{(1-x_1)(1-x_2)(1-x_3)\ldots(1-x_n)}{2^n}u\left|\begin{matrix}(a_j,\alpha_j)_{1,n}\\(b_j,\beta_j)_{1,m}\\(a_j,\alpha_j)_{n+1,p}\\(b_j,\beta_j)_{m+1,q}\end{matrix}\right.\right)\right\}$$

$$= 2^{(n)}\frac{\left[\Gamma(\beta_{11}+m_1+1)\Gamma(\beta_{12}+m_2+1)\ldots\Gamma(\beta_{1n}+m_n+1)\right]}{[m_1\ !m_2!\ldots m_n!]}$$

$$\times\left(H^{m+N,n+N}_{p_i+2N,Q_i+2N}\left(\frac{(1-x_1)(1-x_2)(1-x_3)\ldots(1-x_n)}{2^n}u\left|\begin{matrix}(-\alpha_{11}),(-\alpha_{12}),\ldots,(-\alpha_{1n})\\(m_1),(m_1),\ldots(m_n)\end{matrix}\right.\right.\right. \tag{8.20}$$

$$\left.\begin{matrix} \{(a_j,\alpha_j)_{1,n}\};\{(a_j,\alpha_j)_{n+1,p}\}(0,1) \\ \{(b_j,\beta_j)_{1,m}\};\{(b_j,\beta_j)_{m+1,q}\}(-1-\alpha_{11}-\beta_{11}-m_1,1) \\ (0,1),....,(0,1) \\ (-1-\alpha_{12}-\beta_{12}-m_2,1),......,(-1-\alpha_{1n}-\beta_{1n}-m_n,1) \end{matrix}\right) \tag{8.21}$$

ii. If we put again in Equation (1.22) $r=1$, $\alpha_j=\alpha_{ji}=\beta_j=\beta_{ji}=1$ it reduces to the corresponding form of G-function which will be obtained in the form of H- function,

$$= 2^{(n)}\frac{[\Gamma(\beta_{11}+m_1+1)\Gamma(\beta_{12}+m_2+1)\ldots\Gamma(\beta_{1n}+m_n+1)]}{[m_1!m_2!\ldots m_n!]}$$

$$\times\left(G^{m+N,n+N}_{p_i+2N,Q_i+2N}\left(\frac{(1-x_1)(1-x_2)(1-x_3)\ldots(1-x_n)}{2^n}\middle|\begin{matrix}(-\alpha_{11}),(-\alpha_{12}),\ldots,(-\alpha_{1n})\\(m_1),(m_1),\ldots(m_n),\end{matrix}\right.\right.$$

$$\left.\begin{matrix} \{(a_j)_{1,n}\};\{(a_j)_{n+1,p}\}(0,1) \\ \{(b_j)_{1,m}\};\{(b_j)_{m+1,q}\}(-1-\alpha_{11}-\beta_{11}-m_1,1) \\ (0,1),\ldots,(0,1) \\ (-1-\alpha_{12}-\beta_{12}-m_2,1),\ldots,(-1-\alpha_{1n}-\beta_{1n}-m_n,1) \end{matrix}\right) \tag{8.22}$$

iii. If we put $n=1, 2, 3$(one variable) in Equation (8.19), then we obtain the following solutions, which is the Jacobi transform of one variable.

$$2^{-(\xi)}\frac{[\Gamma(\beta_{11}+m_1+1)]}{[m_1!]}$$

$$\left\{I^{m+N,n+N}_{p_i+2N,Q_i+2N}\left(\frac{(1-x_1)}{2}\middle|\begin{matrix}(-\alpha_{11}),\\(m_1),\{(a_j,\alpha_j)_{1,n}\\(b_j,\beta_j)_{1,m}\\(a_{ji},\alpha_{ji})_{n+1,p_i}\\(b_{ji},\beta_{ji})_{m+1,q_i}\ \}(0,1)\\(-1-\alpha_{11}-\beta_{11}-m_1,1)\end{matrix}\right)\right\} \tag{8.23}$$

$$= 2^{-(2\xi)} \frac{\left[\Gamma(\beta_{11}+m_1+1)\Gamma(\beta_{12}+m_2+1)\right]}{(m_1!m_2!)}$$

$$\left\{ I^{m+N,n+N}_{P_i+2N,Q_i+2N} \left(\frac{(1-x_1)(1-x_1)}{2^2} u \left| \begin{matrix} (-\alpha_{11}),(-\alpha_{12}) \\ (m_1),(m_{1z}),,\{(a_j,\alpha_j)_{1,n} \\ (b_j,\beta_j)_{1,m} \\ (a_{jj},\alpha_{ji})_{n+1,p_i} \\ (b_{ji},\beta_{ji})_{m+1,q_i} \\ (-1-\alpha_{11}-\beta_{11}-m_1,1)(-1-\alpha_{12}-\beta_{12}-m_2,1) \end{matrix} \right. \right. \right\}(0,1)(0,1) \tag{8.24}$$

$$= 2^{-(3\xi)} \frac{\left[\Gamma(\beta_{11}+m_1+1)\Gamma(\beta_{12}+m_2+1)\Gamma(\beta_3+m_3+1)\right]}{\left[m_1!m_2!m_3!\right]}$$

$$\left\{ I^{m+N,n+N}_{P_i+2N,Q_i+2N} \left(\frac{(1-x_1)(1-x_2)(1-x_3)}{2^3} u \left| \begin{matrix} (-\alpha_{11}),(-\alpha_{12}),(-\alpha_{13}) \\ (m_1),(m_1),(m_3),\{(a_j,\alpha_j)_{1,n} \\ (b_j,\beta_j)_{1,m} \\ (a_{ji},\alpha_{ji})_{n+1,p_i} \\ (b_{ji},\beta_{ji})_{m+1,q_i} \\ (-1-\alpha_{11}-\beta_{11}-m_1,1)(-1-\alpha_{12}-\beta_{12}-m_2,1) \\ (-1-\alpha_{13}-\beta_{13}-m_3,1) \end{matrix} \right. \right. \right\}(0,1)(0,1),(0,1) \tag{8.25}$$

8.3 APPLICATION TO THE SOLUTION OF GENERALIZED PARABOLIC EQUATION WITH VARIABLE COEFFICIENTS

The above transform in special cases has been used in solving partial differential equation of two and three dimensions governing heat conduction in orthotropic material [91,92]. In this section, we consider a general case of solving a n-dimensional partial differential equation. In this section, we consider a general case of solving an n-dimensional partial differential equation.

$$\sum_{i=1}^{n} \frac{\partial}{\partial x_i}\left[\left(K_i(x_i)\frac{\partial u}{\partial x_i}\right]+Q(x_1+x_2,....x_n)=\lambda_l\frac{\partial u}{\partial t} \tag{8.26}$$

where $U(x_1, x_2, \ldots.., x_n)$ is the n-dimensional field variable which has to be determined. $Q(x_1, x_2, \ldots.., x_n)$ is known 'source' function and λ_i $(i=1, 2, \ldots n)$ are constants while K_i ($i=1, 2, \ldots, n$) are 'conductivity' parameters which may or may not depend on the space and time variables.

$$\frac{\partial}{\partial t} f\left(\begin{array}{c}\alpha_{11},\beta_{11},\alpha_{13},\beta_{13}\\ ;m_1,m_2\ldots m_n\\ \alpha_{12},\beta_{12},\ldots\alpha_{1n},\beta_{1n}\end{array}\right)$$

$$= \frac{\lambda}{c} - \left[m_1\left(m_1+\alpha_{11}+\beta_{11}+1\right)+m_2\left(m_2+\alpha_{12}+\beta_{12}+1\right)+\ldots+m_n\left(m_n+\alpha_{1n}+\beta_{1n}+1\right)\frac{\mu}{\lambda}\right]$$

$$\times f\left(\begin{array}{c}\alpha_{11},\beta_{11},\alpha_{13},\beta_{13}\\ ;m_1,m_2\ldots m_n\\ \alpha_{12},\beta_{12},\ldots\alpha_{1n},\beta_{1n}\end{array}\right) \tag{8.27}$$

$$f_t\left(\begin{array}{c}\alpha_{11},\beta_{11},\alpha_{13},\beta_{13}\\ ;m_1,m_2\ldots m_n\\ \alpha_{12},\beta_{12},\ldots\alpha_{1n},\beta_{1n}\end{array}\right) = A\exp\left[\frac{\mu t}{c} - m_1\left(m_1+\alpha_{11}+\beta_{11}+1\right)\right.$$

$$\left. +m_2\left(m_2+\alpha_{12}+\beta_{12}+1\right)+\ldots+m_n\left(m_n+\alpha_{1n}+\beta_{1n}+1\right)\frac{\lambda t}{c}\right] \tag{8.28}$$

where A is constant of integration applying the initial condition

$$f_0\left(\begin{array}{c}\alpha_{11},\beta_{11},\alpha_{13},\beta_{13}\\ ;m_1,m_2\ldots m_n\\ \alpha_{12},\beta_{12},\ldots\alpha_{1n},\beta_{1n}\end{array}\right) = A$$

$$\alpha_{11} = \frac{c_1+c_2}{2\lambda_1}, \beta_{11} = \frac{c_1-c_2}{2\lambda_1}, \ \alpha_{12} = \frac{c_3+c_4}{2\lambda_2}, \beta_{12} = \frac{c_3-c_4}{2\lambda_2}$$

$$\alpha_{1n} = \frac{c_{n-1}+c_n}{2\lambda_n}, \beta_{1n} = \frac{c_{n-1}-c_n}{2\lambda_n}$$

Finally, with the help of the inversion formula for Jacobi Transform Equation (8.27), we obtain

$$U(x_1, x_2, ...x_n, t) = \sum_{m_1=0}^{\infty}\sum_{m_2=0}^{\infty} ...n - times \frac{1}{\delta_{m1}\delta_{m2}...\delta_{mn}} P_{m_1}^{\alpha_{11},\beta_{11}}(x_1)$$

$$= A\exp\left[\frac{\mu t}{c} - \{m_1(m_1 + \alpha_{11} + \beta_{11} + 1) + m_2(m_2 + \alpha_{12} + \beta_{12} + 1)\right]$$

$$+m_n(m_n + \alpha_{1n} + \beta_{1n} + 1) \times f_t \begin{pmatrix} \alpha_{11}, \beta_{11}, \alpha_{13}, \beta_{13} \\ ; m_1, m_2 ....m_n \\ \alpha_{12}, \beta_{12}, ...\alpha_{1n}, \beta_{1n} \end{pmatrix} \tag{8.29}$$

In heat conduction problem for two variables, Agrawal [92] has defined heat conduction in an orthotropic square with continuous heat generation and variables conductivities in two directions parallel to the sides. The mathematical model for this process is given as,

we take,

$$g(x) = \left\{ I \left(\begin{matrix} \frac{(1-x)^{\rho}}{2} \\ \frac{(1-y)^{\sigma}}{2} \end{matrix} \;w \middle| \begin{matrix} (a_j, \alpha_j)_{1,n} \\ (b_j, \beta_j)_{1,m} \\ (a_{ji}, \alpha_{ji})_{n+1,p_i} \\ (b_{ji}, \beta_{ji})_{m+1,q_i} \end{matrix} \right) \right\} \tag{8.30}$$

and we have

$$f\begin{pmatrix} \alpha, \beta \\ m, n \\ \gamma, \delta \end{pmatrix} = \frac{\Gamma(\delta + n + 1)\Gamma(\beta + m + 1)}{m!\,n!} 2^{\gamma+\beta+\delta} 4^{\alpha+1} \tag{8.31}$$

$$I_{p+3,q+4;r}^{m+2,n+1}\left\{ 2^{\sigma+\rho} z \middle| \begin{matrix} (a_j, \alpha_j)_{1,n}(-\alpha, \rho); \\ (b_j, \beta_j)_{1,m}(n, \rho)(\alpha - \gamma + n, \sigma) \\ (a_{ji}, \alpha_{ji})_{n+1,p_i,}(\alpha - \gamma, \sigma)(0, \rho) \\ (b_{ji}, \beta_{ji})_{m+1,q_i}(-1 - \beta - n - \alpha, \rho) \\ (-1 - \delta - n - \gamma, \sigma) \end{matrix} \right\} \tag{8.32}$$

Temperature distribution in square plate is given by

$$h(x,y,t)=\sum_{m=0}^{\infty}\sum_{n=0}^{\infty}(\delta_m\delta_n)^{-1}P_n^{(\alpha,\beta)}P_m^{(\gamma,\delta)}\frac{\Gamma(\delta+n_l+1)\Gamma(\beta+n_l+1)}{n!\,n!}2^{\gamma+\beta+\delta}4^{\alpha+1}$$

$$\times I_{p+3,q+4;r}^{m+2,n+1}\left\{2^{\sigma+\rho}z\left|\begin{matrix}(a_j,\alpha_j)_{1,n}(-\alpha,\rho);\\(b_j,\beta_j)_{1,m}(n,\rho)(\alpha-\gamma+n,\sigma)\\(a_{ji},\alpha_{ji})_{n+1,p_i,}(\alpha-\gamma,\sigma)(0,\rho)\\(b_{ji},\beta_{ji})_{m+1,q_i}(-1-\beta-n-\alpha,\rho)\\(-1-\delta-n-\gamma,\sigma)\end{matrix}\right.\right\}$$

$$\times\exp\left[\frac{\mu t}{c}-\{m(m+\alpha+\beta+1)+n(n+\gamma+\delta+1)\}\frac{\lambda t}{c}\right]$$

$$\times I_{p+3,q+4;r}^{m+2,n+1}\left\{2^{\sigma+\rho}z\left|\begin{matrix}(a_j,\alpha_j)_{1,n}(-\alpha,\rho);\\(b_j,\beta_j)_{1,m}(n,\rho)(\alpha-\gamma+n,\sigma)\\(a_{ji},\alpha_{ji})_{n+1,p_i,}(\alpha-\gamma,\sigma)(0,\rho)\\(b_{ji},\beta_{ji})_{m+1,q_i}(-1-\beta-n-\alpha,\rho)\\(-1-\delta-n-\gamma,\sigma)\end{matrix}\right.\right\}$$

$$\times\exp\left[\frac{\mu t}{c}-\{m(m+\alpha+\beta+1)+n(n+\gamma+\delta+1)\}\frac{\lambda t}{c}\right]\tag{8.33}$$

8.3.1 Special Cases of Initial Temperature Function

i. putting $r=1$ Equation (8.30) reduces to

$$h(x,y,t)=\sum_{m=0}^{\infty}\sum_{n=0}^{\infty}(\delta_m\delta_n)^{-1}P_n^{(\alpha,\beta)}P_m^{(\lambda,\delta)}\frac{\Gamma(\delta+n_l+1)\Gamma(\beta+n_l+1)}{n!\,n!}2^{\gamma+\beta+\delta}4^{\alpha+1}$$

$$\times H_{p+3,q+4;r}^{m+2,n+1}\left\{2^{\sigma+\rho}z\left|\begin{matrix}(a_j,\alpha_j)_{1,n}(-\alpha,\rho);\\(b_j,\beta_j)_{1,m}(n,\rho)(\alpha-\gamma+n,\sigma)\\(a_j,\alpha_j)_{n+1,p_i,}(\alpha-\gamma,\sigma)(0,\rho)\\(b_j,\beta_j)_{m+1,q_i}(-1-\beta-n_1-\alpha,\rho)\\(-1-\delta-n_1-\gamma,\sigma)\end{matrix}\right.\right\}$$

$$\times\exp\left[\frac{\mu t}{c}-\{m(m+\alpha+\beta+1)+n(n+\gamma+\delta+1)\}\frac{\lambda t}{c}\right]\tag{8.34}$$

If putting $\alpha_j=\beta_j=\sigma_j=\rho_j=1$, then Equation (8.30)

$$h(x,y,t)=\sum_{m=0}^{\infty}\sum_{n=0}^{\infty}(\delta_m\delta_n)^{-1}P_n^{(\alpha,\beta)}P_m^{(\gamma,\delta)}\frac{\Gamma(\delta+n_l+1)\Gamma(\beta+n_l+1)}{n!\,n!}2^{\gamma+\beta+\delta}4^{\alpha+1}$$

$$\times\ G_{p+3,q+4;r}^{m+2,n+1}\left\{2^{\sigma+\rho}z\left|\begin{matrix}(a_j)_{1,n}(-\alpha,);\\(b_j)_{1,m}(n,)(\alpha-\gamma+n_l,\sigma)\\(a_j)_{n+1,p,(\alpha-\gamma,\sigma)}(0,\rho)\\(b_j)_{m+1,q}(-1-\beta-n_l-\alpha,\rho)\\(-1-\delta-n_l-\gamma,\sigma)\end{matrix}\right.\right\}$$

$$\times\ \exp\left[\frac{\mu t}{c}-\{m(m+\alpha+\beta+1)+n(n+\gamma+\delta+1)\}\frac{\lambda t}{c}\right]\tag{8.35}$$

Further specialization of the above results will give the solution in terms of Bessels function and other relevant transcendental functions.

8.3.2 Heat Conduction Problem for Three Variables

Again we consider the heat conduction in an orthotropic cubic solid with continuous heat generalization and variable thermal conductivities in three-dimensional parallel to the sides.

$$\frac{\partial}{\partial x}\left(K_x\frac{\partial u}{\partial x}\right)+\frac{\partial}{\partial y}\left(K_y\frac{\partial u}{\partial y}\right)+\frac{\partial}{\partial z}\left(K_z\frac{\partial u}{\partial z}\right)+Q(x,y,z)=c\frac{\partial u}{\partial t}\tag{8.36}$$

K_x, K_y, and K_z are conductivities along the directions of x, y, z axes Q (x, y, z) is a continuous source of heat situated within the material, c is a constant

$$K_x=\lambda(1-x^2),\ K_y=\lambda(1-y^2),\ K_z=\lambda(1-z^2)\tag{8.37}$$

$$Q(x,y,z,t)=(c_1+c_2x)\frac{\partial u}{\partial x}+(c_3+c_4x)\frac{\partial u}{\partial y}+(c_5+c_6x)\frac{\partial u}{\partial z}+\mu(u)\tag{8.38}$$

Here c_1, c_2, c_3, c_4, c_5, c_6, μ and λ are real constants.

Initial conditions are assumed as

$u(x, y, z)=g(x, y, z)$ for $-1\le x\le 1$, $-1\le y\le 1$, $-1\le z\le 1$

We write

$$J\left\{u(x,y,z,t)\right\}=U_t\begin{pmatrix}\alpha,,\gamma,\xi\\;l,m,n\\\beta,\delta,\eta\end{pmatrix} \tag{8.39}$$

$$J\left\{g(x,y,z,t)\right\}=U_o\begin{pmatrix}\alpha,,\gamma,\xi\\;l,m,n\\\beta,\delta,\eta\end{pmatrix} \tag{8.40}$$

Now substituting the value of K_x, K_y, K_z and $Q(x, y, z, t)$ in Equation (8.32), we get after this taking Jacobi transform of three variables and using Equation (8.38)

$$J\left[c\left\{h(x,y,z)\right\}\right]=-\left[l(l+\alpha+\beta+1)+m(m+\gamma+\delta+1)+n(n+\xi+\eta+1)\right]$$

$$\times f\begin{pmatrix}\alpha,\gamma,\xi & \\ & l,m,n\\ \beta,\delta,\eta & \end{pmatrix} \tag{8.41}$$

$$J\left[\frac{c}{\lambda}\frac{\partial u}{\partial t}\right]=-\left[l(l+\alpha+\beta+1)+m(m+\gamma+\delta+1)+n(n+\xi+\eta+1)-\frac{u}{\lambda}\right]$$

$$U_t\begin{pmatrix}\alpha,\gamma,\xi & \\ & l,m,n\\ \beta,\delta,\eta & \end{pmatrix}$$

or

$$\frac{\partial}{\partial t}U_t\begin{pmatrix}\alpha,\gamma,\xi & \\ & l,m,n\\ \beta,\delta,\eta & \end{pmatrix}$$

$$=\left[-\frac{\lambda}{c}][l(l+\alpha+\beta+1)+m(m+\gamma+\delta+1)+n(n+\xi+\eta+1)-\frac{u}{\lambda}\right]U_t\begin{pmatrix}\alpha,\gamma,\xi & \\ & l,m,n\\ \beta,\delta,\eta & \end{pmatrix} \tag{8.42}$$

$$\alpha = \frac{c_1 + c_2}{2\lambda}, \beta = \frac{c_1 - c_2}{2\lambda}, \gamma = \frac{c_3 + c_4}{2\lambda}, \delta = \frac{c_3 - c_4}{2\lambda}$$

$$\xi = \frac{c_5 + c_6}{2\lambda}, \eta = \frac{c_5 - c_6}{2\lambda}$$

Solution of Equation (8.39) is

$$U_t \begin{pmatrix} \alpha,\gamma,\xi & \\ & l,m,n \\ \beta,\delta,\eta & \end{pmatrix}$$

$$= A \exp\left[\frac{\lambda t}{c}\right]\left[l(l+\alpha+\beta+1)+m(m+\gamma+\delta+1)+n(n+\xi+\eta+1)-\frac{\lambda t}{c}\right] \tag{8.43}$$

$$U_o \begin{pmatrix} \alpha,\gamma,\xi & \\ & l,m,n \\ \beta,\delta,\eta & \end{pmatrix} = A \tag{8.44}$$

Finally, with the help of inversion formula, we obtain

$$u(x,y,z,t) = \sum_{m=0}^{\infty}\sum_{n=0}^{\infty}\sum_{l=0}^{\infty}(\delta_l\delta_m\delta_n)^{-1} P_l^{(\alpha,\beta)}(x) P_m^{(\gamma,\delta)}(y) P_n^{(\xi,\eta)}(z)$$

$$U_o \begin{pmatrix} \alpha,\gamma,\xi & \\ & l,m,n \\ \beta,\delta,\eta & \end{pmatrix} \exp\left[\frac{\lambda t}{c} - \{l(l+\beta+1)+m(m+\gamma+\delta+1)+n(n+\xi+\eta+1)\}\frac{\lambda t}{c}\right] \tag{8.45}$$

with the appropriate condition of the parameters stated earlier.

8.4 I-FUNCTION SOLUTION OF THE HEAT EQUATION

There are different types of initial temperature distribution function $g(x, y, z)$. This function may be algebraic transcendental. The I-function is the most generalized form of the above.

$$I^{m,n}_{Pi,qi;r}\left\{ w\frac{(1-x)^{\sigma}(1-x)^{\rho}(1-x)^{v}}{2} \middle| \begin{matrix} (a_j,e_j)_{1,n} \\ (b_j,f_j)_{1,m} \\ (a_{ji},e_{ji})_{n+1,p_i} \\ (b_{ji},f_{ji})_{m+1,q_i} \end{matrix} \right\}$$

so that we have

$$U_o\left(\begin{matrix}\alpha,\gamma,\xi\\ & ;l,m,n\\ \beta,\delta,\eta\end{matrix}\right)=\frac{8\Gamma(\beta+l+1)\Gamma(\delta+m+1)\Gamma(\eta+n+1)}{l!\,m!n!}$$

$$I^{m+3,n+3}_{p_i+6,q_i+6;r}\left(\begin{matrix}(-\alpha,\rho),(-\gamma,\sigma),(-\xi,v)\\(l,\rho),(m,\sigma),(n,v)\end{matrix}\right.$$

$$\begin{matrix}\{(a_j,e_j);(a_{ji},e_{ji})\}(0,\rho)\\ \{(b_j,f_j);(b_{ji},f_{ji})\}(-1-\alpha-\beta-l,\rho)\end{matrix} \tag{8.46}$$

$$\left.\begin{matrix}(0,\sigma)(0,v)\\(-1-\gamma-\delta-m,\sigma)(-1-\xi-\eta-n,v)\end{matrix}\right)$$

Hence temperature distribution in cubic solid is given,

$$u(x,y,z,t)=8\sum_{m=0}^{\infty}\sum_{n=0}^{\infty}\sum_{l=0}^{\infty}(\delta_l\delta_m\delta_n)^{-1}P_l^{(\alpha,\beta)}(x)P_m^{(\gamma,\delta)}(y)P_n^{(\xi,\eta)}(z)$$

$$\frac{8\Gamma(\beta+l+1)\Gamma(\delta+m+1)\Gamma(\eta+n+1)}{l!\,m!n!}$$

$$\times I^{m+3,n+3}_{P_i+6,q_i+6;r}\left(\begin{matrix}(-\alpha,\rho),(-\gamma,\sigma),(-\xi,v)\\(l,\rho),(m,\sigma),(n,v)\end{matrix}\right.$$

$$\begin{matrix}\{(a_j,e_j);(a_{ji},e_{ji})\}(0,\rho)\\ \{(b_j,\mathrm{f}_j);(b_{ji},f_{ji})\}(-1-\alpha-\beta-l,\rho)\end{matrix}$$

$$\left.\begin{matrix}(0,\sigma)(0,v)\\(-1-\gamma-\delta-m,\sigma)(-1-\xi-\eta-n,v)\end{matrix}\right)$$

$$\times\exp\left[\frac{\lambda t}{c}-\{l(l+\alpha+\beta+1)+m(m+\gamma+\delta+1)+n(n+\xi+\eta+1)\}\frac{\lambda t}{c}\right]$$

provided $\alpha,\beta,\delta,\xi,\gamma,\eta>-1$.

Carslaw and Jaege [93] have suggested many problems on heat diffusion in anisotropic and orthotropic solids. Debnath [94] determines the temperature in an anisotropic rod.

Here, now we consider the diffusion equation for two-dimensional heat flow in orthotropic non-homogeneous matter.

$$\frac{\partial}{\partial x}\left(K_x\frac{\partial u}{\partial x}\right)+\frac{\partial}{\partial y}\left(K_y\frac{\partial u}{\partial y}\right)+Q(x,y,z)=c\frac{\partial u}{\partial t}, \tag{8.47}$$

$$K_x=\lambda\left(1-x^2\right),K_y=\lambda\left(1-y^2\right) \tag{8.48}$$

$$Q(x,y,z,t)=(c_1+c_2x)\frac{\partial u}{\partial x}+(c_3+c_4x)\frac{\partial u}{\partial y}+(c_5+c_6x)\frac{\partial u}{\partial z}+\mu(u) \tag{8.49}$$

where c_1, c_2, c_3, c_4, μ and λ are real constants.

$$J\{u(x,y,t)\}=U_t\begin{pmatrix}\alpha,\beta & \\ & ;l,m \\ \gamma,\delta & \end{pmatrix};\ J\{g(x,y)\}=U_0\begin{pmatrix}\alpha,\beta & \\ & ;l,m \\ \gamma,\delta & \end{pmatrix} \tag{8.50}$$

$$\frac{\partial}{\partial t}U_t\begin{pmatrix}\alpha,\beta & \\ & ;l,m \\ \gamma,\delta & \end{pmatrix}=\left[-\frac{\lambda}{c}\right]\left[\mathrm{m}(m+\alpha+\beta+1)+\mathrm{n}(n+\gamma+\delta+1)-\frac{\mu}{\lambda}\right]\mathrm{U}_t\begin{pmatrix}\alpha,\beta & \\ & ;l,m \\ \gamma,\delta & \end{pmatrix} \tag{8.51}$$

$$\alpha=\frac{c_1+c_2}{2\lambda},\ \beta=-\frac{c_1-c_2}{2\lambda},\gamma=\frac{c_3+c_4}{2\lambda},\delta=-\frac{c_3-c_4}{2\lambda}$$

$$U_t\begin{pmatrix}\alpha,\beta & \\ & ;l,m \\ \gamma,\delta & \end{pmatrix}=A\exp\left[\frac{\mu t}{c}-\left\{m(m+\alpha+\beta+1)+n(n+\gamma+\delta+1)\right\}\frac{\lambda t}{c}\right] \tag{8.52}$$

$$U_o\begin{pmatrix}\alpha,\beta & \\ & ;l,m \\ \gamma,\delta & \end{pmatrix}=A \tag{8.53}$$

Finally, with the help of inversion formula

$$h(x,y)=\sum_{m=0}^{\infty}\sum_{n=0}^{\infty}(\delta_m\delta_l)^{-1}\,P_l^{(\alpha,\beta)}(x)P_m^{(\gamma,\delta)}(y)\times f\begin{pmatrix}\alpha,\beta & \\ & m,n\\ \gamma,\delta & \end{pmatrix} \tag{8.54}$$

We get

$$u(x,y,t)=\sum_{m=0}^{\infty}\sum_{n=0}^{\infty}(\delta_m\delta_l)^{-1}\,P_m^{(\alpha,\beta)}(x)P_n^{(\gamma,\delta)}(y)$$

$$\times U_0\begin{pmatrix}\alpha,\beta & \\ & m,n\\ \gamma,\delta & \end{pmatrix} \tag{8.55}$$

$$=\exp\left[\frac{\mu t}{c}-\{m(m+\alpha+\beta+1)+n(n+\gamma+\delta+1)\}\frac{\lambda t}{c}\right]$$

8.4.1 Particular Cases

We consider two important cases

i. If we put $\gamma=\delta=0$, $\mu=0$ and n by $2n+1$, we get

$$u(x,y,t)=\sum_{m=0}^{\infty}\sum_{n=0}^{\infty}(u_n+3)(\delta_m)^{-1}P_m^{(\gamma,\delta)}(x)P_{2n+1}^{(\gamma,\delta)}(y)$$

$$\times U_0\begin{pmatrix}\alpha,\beta\\ m,2n+1\\ 0,0\end{pmatrix}=\exp\left[\frac{\lambda t}{c}-\{(2n+1)(2n+2)+m(m+\alpha+\beta+1)\}\right] \tag{8.56}$$

This solution is according to the solution given by Sabharwal [95],

ii. If $\alpha=\beta=\gamma=\delta=\mu=0$, m and n replaced by $2m+1$ and $2n+1$ respectively, then

$$\frac{\partial}{\partial x}\left[(1-x^2)\frac{\partial u}{\partial x}\right]+\frac{\partial}{\partial y}\left[(1-y^2)\frac{\partial u}{\partial y}\right]=\frac{c}{\lambda}\frac{\partial u}{\partial t}, \tag{8.57}$$

$$u(x,y,t)=\sum_{m=0}^{\infty}\sum_{n=0}^{\infty}(u_n+3)(u_m+3)\,P_{2m+1}(x)P_{2n+1}(y)$$

$$\times U_0\left(\begin{matrix}0,0\\ & 2m+1,2n+1\\ 0,0\end{matrix}\right) \tag{8.58}$$

$$=\exp\left[-\frac{\lambda t}{c}\{(2m+1)(2m+2)+(2n+1)(2n+1)\}\right]$$

8.5 APPLICATION OF I-FUNCTION IN DIFFUSION PROBLEM

8.5.1 (A)

In this section, we also use the following notation:

$$F\left[\lambda\left(1-y^2\right)^{\mu}\right]={}_U F_Y\left[\begin{matrix}A_u;\\ B_y;\end{matrix}\ \lambda\left(1-y^2\right)^{\mu}\right]$$

$$=\sum_{k=0}^{\infty}\frac{\prod_{j=1}^{U}\left(A_j;k\right)\lambda^k\left(1-y^2\right)^{k\mu}}{\prod_{j=1}^{Y}\left(B_j;k\right)k!} \tag{8.59}$$

We shall make application of the following integral [96]

$$\int_{-1}^{1}\left(1-y^2\right)^{\rho-1}P_N^M(y)F\left[\lambda\left(1-y^2\right)^{\mu}\right]I\left[z\left(1-y^2\right)^{\gamma}\right]dy$$

$$=\frac{\pi 2^M}{\Gamma\left(-\frac{M}{2}+\frac{N}{2}+1\right)\Gamma\left(-\frac{M}{2}-\frac{N}{2}+\frac{1}{2}\right)}F[\lambda]$$

$$\times I^{m,n+2}_{p_i+2,q_i+2:r}\left[z\left|\begin{matrix}(1-\rho-k\mu\pm M/2,\gamma),\ldots; & \ldots\ldots\\ \ldots\ldots; & \ldots\ldots\ldots,(-\rho-k\mu-N/2,\gamma),(1-\rho-k\mu+N/2,\gamma)\end{matrix}\right.\right], \tag{8.60}$$

where μ is a positive integer; γ is a positive number and $|\lambda|<1$,

$$2\mathrm{Re}\left[\rho+\gamma\ \min\left(b_j/\beta_j\right)\right]>|Re(M)|(j=1,2,\ldots,m);$$

$$A=\sum_{j=1}^{p_i}\alpha_{ji}-\sum_{j=1}^{q_i}\beta_{ji}\leq 0;$$

$$B=\sum_{j=1}^{n}\alpha_j-\sum_{j=n+1}^{p_i}\alpha_{ji}+\sum_{j=1}^{m}\beta_j-\sum_{j=m+1}^{q_i}\beta_{ji}>0,\ \forall\ i\in(1,2,\ldots.4);$$

And

$$|\arg z|<\frac{1}{2}B\pi.$$

8.5.2 (B)

In this section as an example of the use of the I-function in Applied Mathematics, we shall consider the problem of determining a function $u(y, t)$, which represents the temperature in a non-homogenous bar with ends at $y=-1$ to $y=+1$ in which the thermal conductivity is proportional to $\left(1-y^2\right)$. If the lateral surface of the bar is insulated, the diffusion equation [97], takes the form:

$$\frac{\partial u}{\partial t}=b\left[\frac{\partial}{\partial y}\left\{\left(1-y^2\right)\frac{\partial u}{\partial y}\right\}-\frac{M^2u}{1-y^2}\right] \tag{8.61}$$
$$=-bN(N+1)u.$$

where $b>0$ and constant. The ends $y=\pm1$ are insulated because the conductivity vanishes there. If $u(y,0)=f(y)$, $-1<y<1$, the solution of Equation (8.61) is given in [98] as follows:

$$u(y,t)=\sum_{M,N=0}^{\infty}C_{N,M}e^{-bN(N+1)t}P_N^M(y), \tag{8.62}$$

where

$$C_{N,M}=\frac{(2N+1)(N-M)!}{2(N+M)!}\int_{-1}^{1}P_N^M(y)f(y)\,dy, \tag{8.63}$$

And

$$f(y)=\sum_{M,N=0}^{\infty}C_{N,M}P_N\sup M(y). \tag{8.64}$$

Now setting

$$f(y)=\left(1-y^2\right)^{\rho-1}F\left[\lambda\left(1-y^2\right)^{\mu}\right]I\left[z\left(1-y^2\right)^{\gamma}\right] \tag{8.65}$$

On combining (8.63) and (8.65) and making the use of integral (8.60), we drive

$$C_{N,M}=\frac{(2N+1)(N-M)!}{2(N+M)!}\times\frac{\pi 2^{M}}{\Gamma\left(-\frac{M}{2}+\frac{N}{2}+1\right)\Gamma\left(-\frac{M}{2}-\frac{N}{2}+\frac{1}{2}\right)}F[\lambda]$$

$$\times I^{m,n+2}_{p_i+2,q_i+2:r}\left[z\left|\begin{matrix}(1-\rho-k\mu\pm M/2,\gamma),....; & .....\\ ....; & ....,(-\rho-k\mu-N/2,\gamma),(1-\rho-k\mu+N/2,\gamma)\end{matrix}\right.\right],$$

(8.66)

Now putting the value of $C_{N,M}$ from (8.66) in (8.62), we get the following solution to the problem (8.61):

$$u(y,t)=\sum_{M,N=0}^{\infty}\frac{(2N+1)(N-M)!}{2(N+M)!}\times\frac{\pi 2^{M}}{\Gamma\left(-\frac{M}{2}+\frac{N}{2}+1\right)\Gamma\left(-\frac{M}{2}-\frac{N}{2}+\frac{1}{2}\right)}F[\lambda]$$

$$\times\ I^{m,n+2}_{p_i+2,q_i+2:r}\left[z\left|\begin{matrix}(1-\rho-k\mu\pm M/2,\gamma),....; & .....\\ ....; & ....,(-\rho-k\mu-N/2,\gamma),(1-\rho-k\mu+N/2,\gamma)\end{matrix}\right.\right]$$

$$\times e^{-bN(N+1)t}P_N^M(y).$$

(8.67)

Provided the conditions stated with (8.60) are satisfied.

8.5.2.1 An Expansion Formula

Making use of (8.65) and (8.66) in (8.64), we drive the following expansion formula

$$\left(1-y^2\right)^{\rho-1}F\left[\lambda\left(1-y^2\right)^{\mu}\right]I\left[z\left(1-y^2\right)^{\gamma}\right]$$

$$=\sum_{M,N=0}^{\infty}\frac{(2N+1)(N-M)!}{2(N+M)!}$$

$$\times\frac{\pi 2^{M}}{\Gamma\left(-\frac{M}{2}+\frac{N}{2}+1\right)\Gamma\left(-\frac{M}{2}-\frac{N}{2}+\frac{1}{2}\right)}F[\lambda]$$

$$\times\ I^{m,n+2}_{p_i+2,q_i+2:r}\left[z\left|\begin{matrix}(1-\rho-k\mu\pm M/2,\gamma),....; & .....\\ ....; & ....,(-\rho-k\mu-N/2,\gamma),(1-\rho-k\mu+N/2,\gamma)\end{matrix}\right.\right]$$

$$\times P_N^M(y),$$

(8.68)

Provided that all the corresponding conditions stated in (8.60) with $\rho\geq 1$.

8.5.2.2 Particular Case

On taking $M=0$ in (8.67), we get the following solution

$$u(y,t)=\frac{\pi}{2}\sum_{N=0}^{\infty}\frac{(2N+1)F[\lambda]e^{-bN(N+1)t}P_N(y)}{\Gamma\left(\frac{N}{2}+1\right)\Gamma\left(-\frac{N}{2}+\frac{1}{2}\right)}$$

$$\times\ I^{m,n+2}_{p_i+2,q_i+2:r}\left[z\left|\begin{matrix}(1-\rho-k\mu,\gamma),(1-\rho-k\mu,\gamma),\ldots; & \ldots\ldots \\ \ldots\ldots; & \ldots\ldots,(-\rho-k\mu-N/2,\gamma),(1-\rho-k\mu+N/2,\gamma)\end{matrix}\right.\right] \tag{8.69}$$

which is a result given by Tiwari and Sharma [96], where the conditions stated in this paper are satisfied, and P_N is the Legendre polynomials.

8.6 APPLICATION RELATED TO COOLING OF A SPHERE

Given a sphere whose initial temperature depends only on the distance from the centre (e.g., a constant initial temperature) and whose boundary is kept at a constant temperature, predict the temperature at any point inside the sphere at a later time. This is the problem, for example, of determining the temperature of the centre of a potato that has been put in a hot oven. The reader might also conjecture that this problem who want to determine the time of an individual's death. Early researchers, notably Kelvin, used this problem to determine the age of the earth based on assumptions about its initial temperature and its temperature today.

This cooling problem may remind the reader of Newton's law of cooling, which is encountered in ordinary differential equations texts. Recall that this law states that the rate at which a body cools is proportional to the difference between its temperature and the temperature of the environment. Quantitatively, if $T=T(t)$ is the temperature of a body and T_e is the temperature of its environment, then $T'(t)=K(T_e-T)$, where K is the constant of proportionality. But the reader should note that this law applies only in the case that the body $g=$has a uniform, homogenous temperature. In the PDE problem we are considering, the temperature may vary radially throughout the body.

In this section, we shall make an application of the following modified form of the integral [99].

$$\int_0^{\pi}(\sin x)^{s-1}\ \sin nx dx=\frac{\pi\sin\frac{n\pi}{2}\Gamma(s)}{2^{s-1}\Gamma\left(\frac{s\pm n+1}{2}\right)} \tag{8.70}$$

where Re(s)> 0.

8.6.1 INTEGRAL

The integral which we need is as follows:

$$\int_0^\pi (\sin x)^{\omega-1} \sin nx \; I^{m,n}_{p_i,q_i;r}\left[z(\sin x)^\lambda\right] dx$$

$$= 2^{1-\omega} \pi \sin\frac{n\pi}{2} I^{m,n+1}_{p_i+1,q_i+2;r} \tag{8.71}$$

$$\times\left[z2^{-\lambda}\left| \begin{matrix} (1-\omega,\lambda),.......... \\ ....,(1/2-\omega/2\pm n/2,\lambda/2) \end{matrix}\right.\right],$$

Provided that $\lambda > 0$, $\omega > 0$, $\mathrm{Re}\left[\omega + 2\lambda \min_{1\le j\le m}\left(b_j / \beta_j\right)\right] > 0$, $|\arg z| < \frac{1}{2}B\pi$;

where B is defined in the introduction section.

8.6.2 Proof

The integral (8.71) can be obtained easily by making use of (8.70) and the definition of the I-function.

For simplicity, let us consider a sphere $\rho = \pi$ whose initial temperature is $T_0=$ constant. We will assume that the boundary is held at zero degree for all time $t > 0$. If u is the temperature, then in general u will depend on three spatial coordinates and time. But a little reflection shows that the temperature will depend only on the distance from the centre of the sphere and on time. Evidently, the temperature u must satisfy the heat equation

$$u_t = k\Delta u,$$

where k is the diffusivity and Δ is the Laplacian operator. It should be clear that a spherical coordinate system ρ,ϕ,θ is more appropriate than a rectangular system, and $u - u(\rho,t)$ Because u does not depend on the angles ϕ and θ, Laplacian takes on a particularly simple form:

$$\Delta = \frac{\partial^2}{\partial \rho^2} + \frac{2}{\rho}\frac{\partial}{\partial \rho}$$

Therefore, we may formulate the model as

$$u_t = k\left(u_{\rho\rho} + \frac{2}{\rho}u_\rho\right), \; 0 \le \rho\langle \pi, \; t\rangle 0. \tag{8.72}$$

$$u(\pi,t) = 0, \; t > 0.$$

$$u(\rho,0) = T_0, \;\; 0 \le \rho < \pi$$

Observe that there is an implied boundary condition at $\rho = 0$, namely that the temperature should remain bounded.

To solve problem (8.72) we assume $u(\rho,t) = u(\rho)g(t)$. Substitute into the PDE and separating variables gives the following solution, which is given in [100].

$$u(\rho,t) = \sum_{n=1}^{\infty} c_n e^{-n^2 kt} \frac{\sin n\rho}{\rho} \tag{8.73}$$

where

$$c_n = \frac{2}{\pi} \int_0^{\pi} T_0 \ \rho \sin n\rho \ \ d\rho \tag{8.74}$$

8.6.3 Solution

The solution of the problem to be obtained is

$$u(\rho,t) = \sum_{n=1}^{\infty} e^{-n^2 kt} \frac{\sin n\rho}{\rho}$$

$$\times 2^{1-\omega} \ \pi \sin \frac{n\pi}{2} I^{m,n+1}_{p_i+1,q_i+2;r} \tag{8.75}$$

$$\left[z2^{-\lambda} \middle| \begin{matrix} (1-\omega,\lambda),.......... \\ ....,(1/2-\omega/2 \pm n/2, \lambda/2) \end{matrix} \right],$$

where all conditions of convergence are the same as in (8.71).

8.6.4 Proof

Choose

$$f(\rho) = u(\rho,0) = T_0 = \rho^{-1} (\sin \rho)^{\omega-1} \ I^{m,n}_{p_i,q_i;r} \left[z (\sin \rho)^{\lambda} \right] \tag{8.76}$$

Now combining (8.76) and (8.74) and making use of the integral (8.71), we drive

$$c_n = 2^{2-\omega} \sin \frac{n\pi}{2} I^{m,n+1}_{p_i+1,q_i+2;r}$$

$$\left[z2^{-\lambda} \middle| \begin{matrix} (1-\omega,\lambda),.......... \\ ....,(1/2-\omega/2 \pm n/2, \lambda/2) \end{matrix} \right] \tag{8.77}$$

Putting the value of c_n form (8.77) in (8.73), we get the required result (8.75).

8.6.5 Special Cases

On specializing the parameters, H-function may be reduced to G-function, Lauricella's functions Legendre functions, Bessel functions, hypergeometric functions, Appell's functions, Kampe de Feriet's functions and several other higher transcendental functions. Therefore the result (8.75) is of the general nature and may reduce to be in different forms, which will be useful in the literature on Applied Mathematics and other branches.

8.7 APPLICATION OF ORTHOGONAL POLYNOMIAL TO NON-LINEAR DIFFERENTIAL EQUATION

8.7.1 (A)

In this section, we have applied the Jacobi polynomials to obtain the linear amplitude-dependent approximate solution of the non-linear differential equation of general type

$$\ddot{x}+\omega I^{k,l}_{p_i,q_i;r}\left[\mu\left(1+{}^{x}\!/_{A}\right)^{m}\right]=NF(t) \tag{8.78}$$

where I-function, defined in introduction. The general initial conditions in which (8.78) can be solved may be taken as $x = A(A-1)$ and $\dot{x} = 0$ at $t = 0$, where $A(A-1)$ is the amplitude of the motion.

Results obtained in [101] have been improved respectively by using Jacobi polynomial approximation for $\sin x$, instead of assuming $\sin x$, equal to x, in the differential equation

$$\ddot{x}=\frac{-g}{l}\sin x+2\omega\cos\lambda\dot{y}, \tag{8.79}$$

$$\ddot{y}=\frac{-g}{l}\sin y-2\omega\cos\lambda\dot{x}, \tag{8.80}$$

and

$$\ddot{x}=\omega^{2}\sin x+\alpha(\dot{x})^{2}, \tag{8.81}$$

where in (8.81), –ve sign is the velocity in the direction of x-axis. In our investigation we shall need the following results:

From [102]:

$$\int_{-1}^{1}(1-x)^{\alpha}(1+x)^{\sigma}P_{n}^{(\alpha,\beta)}(x)\,dx$$
$$=\frac{2^{\alpha+\sigma+1}\Gamma(\sigma+1)\Gamma(\alpha+n+1)\Gamma(\sigma-\beta+1)}{n!\,\Gamma(\sigma-\beta-n+1)\Gamma(\alpha+\sigma+n+2)}, \tag{8.82}$$

where $\mathrm{Re}(\alpha) > -1, \mathrm{Re}(\sigma) > -1$.

8.7.2 Integral

The following integral will be required in our present investigation:

$$\int_{-1}^{1} (1-x)^{\alpha} (1+x)^{\sigma} P_n^{(\alpha,\beta)}(x) I_{p_i,q_i;r}^{k,l}\left[\mu\left(1+{}^{x}\!/_{A}\right)^{-m}\right]$$

$$= \frac{2^{\alpha+\beta+1}\Gamma(1+\alpha+n)}{n!} I_{p_i+2,q_i+2;r}^{k+2,l}\left[\mu 2^{-m}\left|\begin{matrix} \ldots\ldots,(2+\alpha+\beta+n,m),(1-n,m) \\ (1+\beta,m),(1,m),\ldots\ldots\ldots\ldots\ldots\ldots\ldots\ldots \end{matrix}\right.\right] \tag{8.83}$$

provided that m is positive integer $\mathrm{Re}(1+\alpha) > 0$, $\mathrm{Re}(1+\beta+mb_j / \beta_j) > 0, (j = 1,\ldots,k)$

$$\sum_{j=1}^{k}\beta_j - \sum_{j=k+1}^{q_i}\beta_j + \sum_{j=1}^{l}\alpha_j - \sum_{j=l+1}^{p_i}\alpha_j \equiv M > 0,$$

$$\sum_{j=1}^{p_i}\alpha_j - \sum_{j=1}^{q_i}\beta_j = \tau \le 0$$

And

$$\left|\arg \mu 2^{-m}\right| < \frac{1}{2} M\pi.$$

8.7.3 Jacobi Polynomials and Linear Approximation

Jacobi polynomials can be defined as the sets of polynomials orthogonal in the interval (− *A*, *A*) with respect to the weight function $\left(1-{}^{x}\!/_{A}\right)^{\alpha}\left(1+{}^{x}\!/_{A}\right)^{\beta}$. This gives rise to the polynomials $P_n^{(\alpha,\beta)}\left(\frac{x}{A}\right)$. For a function *f*(*x*), which can be expanded in terms of the Jacobi polynomials in the interval (− *A*, *A*), we obtained

$$f(x) = \sum_{n=0}^{\infty} a_n^{(\alpha,\beta)} P_n^{(\alpha,\beta)}\left(\frac{x}{A}\right) \tag{8.84}$$

where the coefficient $a_n^{(\alpha,\beta)}$ are given by

$$a_n^{(\alpha,\beta)} = \frac{\int_{-1}^{1} f(Ax) P_n^{(\alpha,\beta)}(x)(1-x)^{\alpha}(1+x)^{\beta}\,dx}{\int_{-1}^{1}\left[P_n^{(\alpha,\beta)}(x)\right]^2(1-x)^{\alpha}(1+x)^{\beta}\,dx} \tag{8.85}$$

If the series (8.84) is truncated after the second term, we obtain a linear approximation

$$f_*(x) = a_0^{(\alpha,\beta)} P_0^{(\alpha,\beta)}\left(\frac{x}{A}\right) + a_1^{(\alpha,\beta)} P_1^{(\alpha,\beta)}\left(\frac{x}{A}\right) \tag{8.86}$$

where star indicates approximation.

8.7.4 (B)

Here we solve the non-linear differential equation

$$\ddot{x} + f(x) = NF(t) \tag{8.87}$$

where

$$f(x) = \omega I_{pi,qi;r}^{k,l}\left[\mu\left(1+{}^{x}\!/_{A}\right)^{-m}\right] \tag{8.88}$$

by approximating *f*(*x*) in the interval (− *A*, *A*), with the help of linear Jacobi polynomials.

From (8.85), we have

$$\begin{aligned} f_*(x) &= \left(\omega I_{pi,qi;r}^{k,l}\left[\mu\left(1+{}^{x}\!/_{A}\right)^{-m}\right]\right)_* \\ &= a_0^{(\alpha,\beta)} P_0^{(\alpha,\beta)}\left(\frac{x}{A}\right) + a_1^{(\alpha,\beta)} P_1^{(\alpha,\beta)}\left(\frac{x}{A}\right) \qquad (8.87) \\ &= a_0^{(\alpha,\beta)} + a_1^{(\alpha,\beta)}\left[\frac{\alpha-\beta}{2} + \frac{2+\alpha+\beta}{2}\frac{x}{A}\right] \end{aligned}$$

where

$$a_0^{(\alpha,\beta)} = \frac{\int_{-1}^{1} \omega I_{pi,qi;r}^{k,l}\left[\mu(1+x)^{-m}\right] P_0^{(\alpha,\beta)}(x)(1-x)^{\alpha}(1+x)^{\beta}\,dx}{\int_{-1}^{1}\left[P_0^{(\alpha,\beta)}(x)\right]^2 (1-x)^{\alpha}(1+x)^{\beta}\,dx}$$

And

$$a_1^{(\alpha,\beta)} = \frac{\int_{-1}^{1} \omega I_{pi,qi;r}^{k,l}\left[\mu(1+x)^{-m}\right] P_1^{(\alpha,\beta)}(x)(1-x)^{\alpha}(1+x)^{\beta}\,dx}{\int_{-1}^{1}\left[P_1^{(\alpha,\beta)}(x)\right]^2 (1-x)^{\alpha}(1+x)^{\beta}\,dx}$$

Using the result (8.83) and of [103] one finds that

$$a_0^{(\alpha,\beta)} = \omega\frac{\Gamma(2+\alpha+\beta)}{\Gamma(1+\beta)} I_{pi+1,qi+1;r}^{k+1,l}\left[\mu 2^{-m}\left|\begin{matrix} \ldots\ldots,(2+\alpha+\beta,m) \\ (1+\beta,m),\ldots\ldots\ldots\ldots \end{matrix}\right.\right] \tag{8.90}$$

And

$$a_1^{(\alpha,\beta)} = \omega \frac{(\alpha+\beta+3)\Gamma(2+\alpha+\beta)}{\Gamma(2+\beta)} I^{k,l+2}_{p_i+2,q_i+2;r}\left[\mu 2^m \middle| \begin{matrix} \ldots\ldots,(3+\alpha+\beta,m),(0,m) \\ (1+\beta,m),(1,m)\ldots\ldots\ldots\ldots \end{matrix}\right] \tag{8.91}$$

Replacing *f*(*x*) by its approximate $f_*(x)$, Equation (8.87) transforms into

$$\ddot{x} + a_0^{(\alpha,\beta)} + a_1^{(\alpha,\beta)}\left[\frac{\alpha-\beta}{2} + \frac{2+\alpha+\beta}{2}\frac{x}{A}\right] = NF(t) \tag{8.92}$$

Or

$$\ddot{x} + \gamma^2 x = -\frac{(\alpha-\beta)A}{(2+\alpha+\beta)}\left(\gamma^2 - \gamma_1^2\right) + NF(t) \tag{8.93}$$

where

$$\gamma^2 = \frac{(2+\alpha+\beta)}{2A} a_1^{(\alpha,\beta)} \text{ and } \gamma_1^2 = \frac{(2+\alpha+\beta)}{(\beta-\alpha)A} a_0^{(\alpha,\beta)} \tag{8.94}$$

The value of $a_0^{(\alpha,\beta)}$ and $a_1^{(\alpha,\beta)}$ are given by (8.90) and (8.91) respectively.

The approximate general solution of (8.93) is subject to the initial conditions $x = A(A-1)$ and $\dot{x} = 0$ at $t = 0$is given by

$$\begin{aligned} x_* = &\left[A(A-1) + \frac{A(\alpha-\beta)}{(2+\alpha+\beta)}\left(1-\frac{\gamma_1^2}{\gamma^2}\right)\right]\cos\gamma t \\ &- \frac{A(\alpha-\beta)}{(2+\alpha+\beta)}\left(1-\frac{\gamma_1^2}{\gamma^2}\right) \\ &+ \frac{N}{\gamma}\int_0^t F(u)\sin\gamma(t-u)\,du \end{aligned} \tag{8.95}$$

which is the most general approximate solution and on giving different values to *F*(*t*) in (8.87), one can find an approximate general solution corresponding to *F*(*t*). The electric analogue of (8.87) can be given by a circuit consisting of a linear inductor L in series with a non-linear capacitor and a harmonic potential *NF*(*t*). If *x* is the charge separation of the plates of the capacitor, the differential equation of the circuit as introduced by Garde [104] is given by

$$L\ddot{x} + F(x) = NF(t) \tag{8.96}$$

where *f*(*x*) is the potential drop across the non-linear capacitor. Generally, *f*(*x*) is in the form of a curve between voltage and charge known as saturation curve (Figure 8.1).

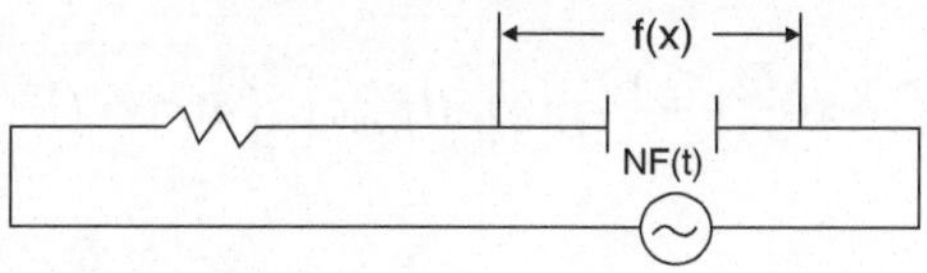

FIGURE 8.1 Resistanceless circuit containing a non-linear capacitor.

8.8 CERTAIN CONDITION PROBLEMS

An insulation-type flux condition at, say, $x = 0$ is a condition on the derivative of the form $u_x(0,t) = 0,\ t > 0$; because the flux is proportional to the temperature gradient (Fourier's heat flow law states flux $= Ku_x$ where K is the conductivity), the insulation condition requires that no heat flow across the boundary $x = 0$.

A radiation condition, on the other hand, is a specification of how heat radiates from the end homogeneous radiation conditions take the form $Ku_x(0,t) + bu(0,t) = 0,\ t > 0$, where b is a constant. If $b > 0$, then the heat flux is negative, which means that heat is flowing from the bar into its surroundings (radiation); if $b < 0$, then the flux is positive, and heat is flowing into the bar (absorption). A typical problem in heat conduction may have a combination of Dirichlet, insulation, and radiation boundary conditions. In other physical contexts, e.g., contaminant transport, biological diffusion, these three types of boundary conditions also play an important role.

8.8.1 Required Result

The following results are required in our present investigation:

From Gradshteyn and Ryzhik [99]:

$$\int_0^l (\sin \pi x / l)^{\omega-1} \cos\left[\left(n+\tfrac{1}{2}\right)\pi x\right]\Big/ l \, dx$$
$$= \frac{l \cos(n/2+1/4)\pi\Gamma(\omega)}{2^{\omega-1}\Gamma\left\{1/2\left(\omega \pm \left(n+\tfrac{1}{2}\right)+1\right)\right\}}, \tag{8.97}$$

where $\operatorname{Re}(\omega) > 0$.

From Rainvile [10], Legendre's duplication formula:

$$\sqrt{\pi}\,\Gamma(2z) = 2^{2z-1}\Gamma(z)\Gamma(z+1/2). \tag{8.98}$$

8.8.2 Integral

The following integral is required in this section:

$$\int_0^l (\sin \pi x / l)^{\omega-1} \cos\left[\left(n+\tfrac{1}{2}\right)\pi x\right] / l\, II^{m,k}_{p_i,q_i;r}\left[z(\sin \pi x / l)^{-2\rho}\right] dx$$

$$\int_0^l (\sin \pi x / l)^{\omega-1} \cos\left[\left(n+\tfrac{1}{2}\right)\pi x\right] / l I^{m,k}_{p_i,q_i;r}\left[z(\sin \pi x / l)^{-2\rho}\right] dx$$

$$= l \Big/ \sqrt{\pi} I^{m+2,k}_{p_i+2,q_i+2;r}\left[z \middle| \begin{matrix} \ldots\ldots.,(1/2+\omega/2 \pm (n/2+1/4),\rho) \\ (\omega/2,\rho),(1/2+\omega/2,\rho),\ldots\ldots\ldots\ldots \end{matrix} \right] \quad (8.99)$$

Provided that $\rho > 0$, $|\arg z| < 1/2B\pi$, where B is given introduction.

The integral (8.99) can be obtained easily in a routine manner by making use of the definition of I-function in introduction, (8.97) and (8.98).

8.8.3 Flux Condition

In this section, we consider a problem on flux condition under boundary conditions.

Consider the diffusion problem

$$\begin{aligned} u_t &= K u_{xx}, \quad 0 < x \langle 1, \ t \rangle 0, \\ u_x(0,t) &= 0, \ u(1,t) = 0, \quad t > 0, \\ u(x,0) &= f(x), \quad 0 < x < 1, \end{aligned} \quad (8.100)$$

where the left end is insulated and the right end is fixed. The solution of (8.100) is given by [105] as follows:

$$u(x,t) = \sum_{n=0}^{\infty} c_n e^{-\lambda_n k t} \cos\frac{(2n+1)\pi x}{2l}, \quad (8.101)$$

where

$$c_n = \frac{2}{l}\int_0^l f(x) \cos\frac{(2n+1)\pi x}{2l} dx, \ n = 0,1,2.... \quad (8.102)$$

8.8.4 Solution of the Problem

The solution of the problem to be obtained is

$$u(x,t) = \sum_{n=0}^{\infty} c_n e^{-\lambda_n k t} \cos\frac{(2n+1)\pi x}{2l} .(2/\sqrt{\pi})$$

$$I^{m+2,k}_{p_i+2,q_i+2;r}\left[z \middle| \begin{matrix} \ldots\ldots.,(1/2+\omega/2 \pm (n/2+1/4),\rho) \\ (\omega/2,\rho),(1/2+\omega/2,\rho),\ldots\ldots\ldots\ldots \end{matrix} \right] \quad (8.103)$$

where all conditions of convergence are same as in the definition of I-function.

8.8.5 Proof

Choose

$$f(x)=(\sin \pi x/l)^{\omega-1}\, I_{p_i,q_i;r}^{m,k}\left[z(\sin \pi x/l)^{-2\rho}\right] \tag{8.104}$$

Now, combining (8.104) and (8.102) and making use of the integral (8.99), we drive

$$c_n=\left(2/\sqrt{\pi}\right) I_{p_i+2,q_i+2;r}^{m+2,k}\left[z\left|\begin{matrix}\ldots\ldots.,(1/2+\omega/2\pm(n/2+1/4),\rho)\\ (\omega/2,\rho),(1/2+\omega/2,\rho),\ldots\ldots\ldots\ldots\end{matrix}\right.\right] \tag{8.105}$$

Putting the value of c_n from (8.105) in (8.101), we get our desired result (8.103).

8.8.6 Special Cases

On specializing the parameters, I-function may be reduced to H-function, G-function, Lauricella's functions Legendre functions, Bessel functions, hypergeometric functions, Appell's functions, Kampede Feriet's functions and several other higher transcendental functions. Therefore the result (8.103) is of general nature and may reduce to be in different forms, which will be useful in the literature on Applied Mathematics and other branches.

8.9 THEORETICAL MODELS FOR RECTANGULAR MICROSTRIP ANTENNA AND I-FUNCTION

In this section, we present a theoretical model T for rectangular microstrip antenna. It is a modal expansion cum line current technique showing good agreement in the frequency range 786–816 MHz with existing measured data. Microstrip antenna is thin conducting patch of any shape fabricated on the top surface of a dielectric slab (of small thickness) placed over infinite conducting ground plane. When electromagnetic power is fed from an oscillator, small radiation occurs and hence behaves as antenna.

In spite of poor radiators, such antennas are used due to advantages like small size, low cost, easy to fabricate and conformity with any surface.

Numerous attempts [106,107] have been made to analyze the electromagnetic properties of rectangular patch, as it is a very popular shape. This chapter attempts to develop new models mentioned above where the dielectric region underneath the patch is treated as cavity with four side H walls, two E walls at the top and bottom and current source (surface on volume depending on feed arrangement) to determine the inner fields by solving Helmholtz's wave equation with appropriate boundary conditions [108,109].

A modal expansion method is used to determine inner fields and line magnetic currents at four edges. From this power losses and stored energy are derived to determine quality factors. These quality factors are exploited to determine Input impedance by the use of complex resonant frequency concept [109].

8.9.1 Theory and Analysis:

Figure 8.2 shows the geometry of rectangular microstrip antenna with edge feeding by microstrip transition time or feedline of width *W*. Parameters *a*, *b*, *h*, etc. are shown with chosen *x*, *y*, *z* axes (Figure 8.3).

Figure 8.4 shows equivalent cavity which is dielectric region underneath the patch surrounded by two *E* walls at top and bottom and four side *H* walls associated with surface current source *J* as shown on side wall containing feed region.

If instead of age feeding coaxial probe feeding is done from outside then equivalent cavity will take form as in Figure 8.4 where equivalent current source will be the same as probe volume.

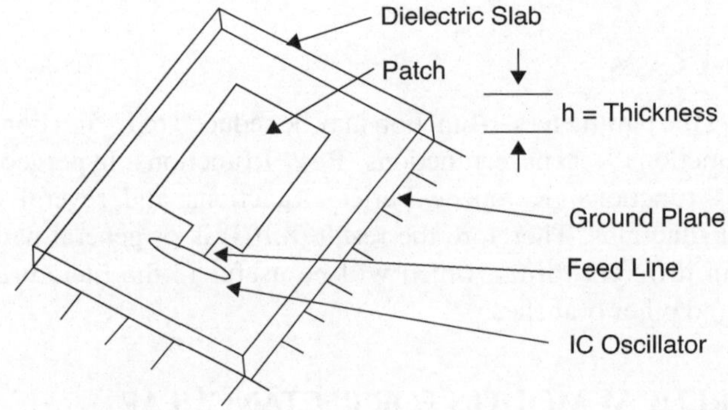

FIGURE 8.2 Geometry of RMA.

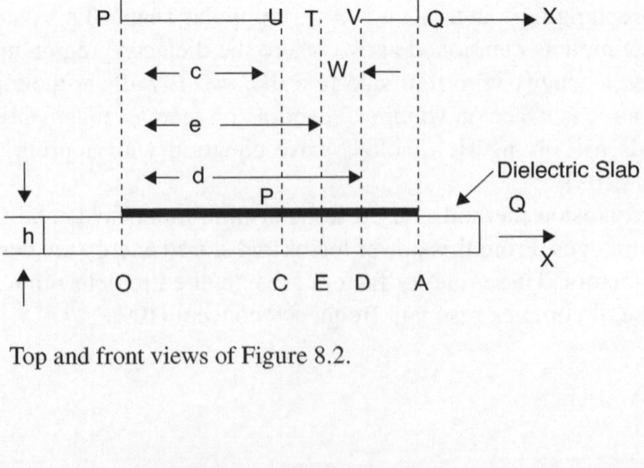

FIGURE 8.3 Top and front views of Figure 8.2.

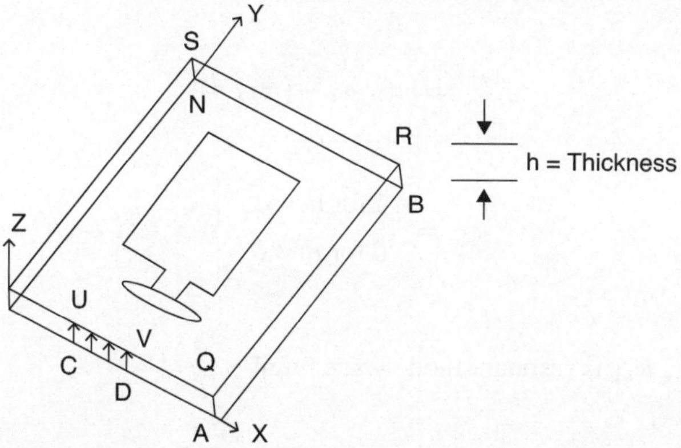

FIGURE 8.4 Equivalently cavity model.

This is a modal expansion cum line current technique where fourline magnetic currents are considered instead of slot radiation. This is shown in

Considering Figures 8.4 and 8.5 and solving source-free Helmholtz's wave equation with appropriate boundary conditions, [108–110] we get

$$E_z = \sum_{m=0}^{\alpha}\sum_{n=0}^{\beta} A_{mn}\cos(m\pi x / a)\cos(n\pi y / b) \tag{8.106}$$

These define resonant dominant mode (*m*, *n*) [109,110].

Now, considering probe source *J* in Figure 8.4 and solving in homogeneous Helmholtz's wave equation A_{mn} is determined as

$$A_{mn} = (c_{om}c_{on} / ab)\int^{a}\int^{b}\left[j\omega\mu_d / \left(K^2 - K_{m,n}^2\right)\right]$$

$$J(x,y)\cos(m\pi x / a)\cos(n\pi y / b)\,dxdy, \tag{8.107}$$

$$x = 0,\ y = 0$$

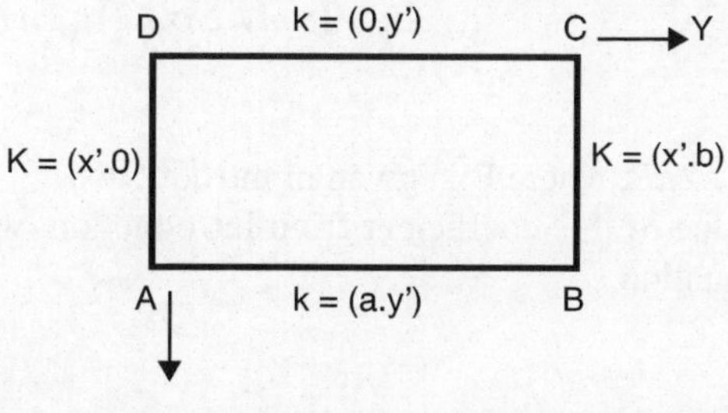

FIGURE 8.5 K current distribution along edges.

where

$$K_{m,n}^2 = (m\pi / a)^2 + (n\pi / b)^2 \tag{8.108}$$

$$c_{om} = \left.\begin{matrix} 1 \text{ for } m = 0 \\ 0 \text{ for } m \neq 0 \end{matrix}\right\} \tag{8.109}$$

Similar for c_{om}, K_{mn} is resonant mode wave numbers.

8.9.2 Solution in terms of I-function

Choose $J(x, y)$ as follows:

$$J(x,y) = I_{p_i,q_i;r}^{k,l}\left[z(\sin \pi x / a)^{\sigma} (\sin \pi y / b)^{\rho} \middle| \begin{matrix} \dots\dots\dots \\ \dots\dots\dots \end{matrix} \right] \tag{8.110}$$

Now from (8.107) and (8.110) and following integral, [20]

$$\int_0^L \{\sin(\pi x / L)\}^{\omega-1} \sin(n\pi x / L)\,dx$$

$$= \frac{L \cos 1/2n\pi\Gamma(\omega)}{2^{\omega-1}\Gamma\{1/2(\omega \pm n + 1)\}} \tag{8.111}$$

where $\operatorname{Re}(\omega) > 0$, we get

$$A_{mn} = (c_{om}c_{on})\left[j_{\omega}\mu_d / (K^2 - K_{m,n}^2)\right] \times \cos(m\pi/2)\cos(n\pi/2)$$

$$\times I_{p_i+2,q_i+4;r}^{k,l+2}\left[z2^{-\sigma}2^{-\rho} \middle| \begin{matrix} (0,\sigma),(0,\rho),\dots\dots\dots \\ \dots\dots\dots,(\pm m/2,\sigma/2),(\pm n/2,\rho/2) \end{matrix} \right] \tag{8.112}$$

$\operatorname{Re}(\omega) > 0$, $|\arg z| < 1/2\pi B$, where B is given in introduction.

Now using the value of this coefficient from introduction, we get the solution of Helmholtz's wave equation.

$$E_z = \sum_{m=0}^{\alpha}\sum_{n=0}^{\beta} \cos(m\pi x / a)\cos(n\pi y / b)\times(c_{om}c_{on})\left[j_{\omega}\mu_d / \left(K^2 - K_{m,n}^2\right)\right]$$

$$\cos(m\pi / 2)\cos(n\pi / 2)$$

$$\times I_{p_i+2,q_i+4;r}^{k,l+2}\left[z2^{-\sigma}2^{-\rho}\left|\begin{matrix}(0,\sigma),(0,\rho),\ldots\ldots\ldots \\ \ldots\ldots\ldots,(\pm m / 2,\sigma / 2),(\pm n / 2,\rho / 2)\end{matrix}\right.\right] \quad (8.113)$$

8.9.3 Conclusion

i. Input impedance is computed for edge feed rectangular microstrip patch using cavity associated two-slot radiator equivalent model (Figure 8.4) in the frequency range 4,100–4,200 MHz.
ii. Input impedance of coaxial fed patch computed using modal expansion cum current line current model (Figures 8.4 and 8.5) in the frequency range of 786–816 MHz.

This theory is much more simple and accurate for performance and design consideration. This will give much future information on the extension of above theory. Also this will employ many applications like satellite communication, aerospace telemetry and radar etc.

8.9.4 Special Cases

It is important to point out that we have performed all calculations in terms of the I-functions, which are more generally functions and whose particular cases are also the hypergeometric functions. On specializing the parameters, I-function may be reduced to H-function, G-function, Lauricella's functions Legendre functions, Bessel functions, hypergeometric functions, Appell's functions, Kampe de Feriet's functions and several other higher transcendental functions. Therefore the result (8.113) is of general nature and may reduce to be in different forms, which will be useful in the literature on Applied Mathematics and other branches.

References

1. Gauss, C.F., 1812, Disquistiones Generales Circaseriem Infinitam, Thesis Gottingen, Published in Ges, Werke, *Gttein*, Vol. II, pp. 437–445, Vol. III, pp. 123–163, 207–229, 446–460.
2. Exton, H., 1978, *Handbook of Hypergeometric Integrals*, John Wiley & Sons, New York.
3. Wallis, J., 1955, *Arithmetical Infinitorum*, Springer, London.
4. Euler, L., 1748, *Introduction in Analysis Infinitorum*, Verlag Birkhäuser, Lausanne, Vol-I.
5. Riemann, B., 1876, *Bernhard Riemann's gesammelte mathematische Werke und wissenschaftlicher Nachlass*. BG Teubner.
6. Kummer, E.E., 1836, Ulredie Hypergeometric Che Reine, *J. Reine Angew. Math.* 15, pp. 39, 127.
7. Thomae, J., 1869, Beiträge zur Theorie der durch die Heinesche Reihe: darstellbaren Functionen, *J. Reine Angew. Math.* 70, pp. 258–281.
8. Barnes, E.W., 1908, A New Development of the Theory of Hypergeometric Functions, *Proc. Lond., Math. Soc.* 6(2), p. 144.
9. Exton, H., 1976, *Multiple Hypergeometric Functions and Applications*, Ellis Horwood, Chichester, England.
10. Rainville, E.D., 1971, *Special Functions*, Chelsea Publication Company, New York.
11. Meijer, C.S., 1946, On the G-function, *Proc. Ned. Acad. Watersch.* 49, pp. 227, 344, 457.
12. Mathai, A.M. and Saxena, R.K., 1973, *Generalized Hypergeometric Function with Application in Statistical and Physical Sciences*, Springer, Verlog, Lecture, Notes No.-343 Heidelberg.
13. Luke, Y.L., 1969, *The Special Function and Their Approximations*, Vol.-I and II, Academic Press, New York.
14. Wright, E.M., 1935, The Asymptotic Expansion of the Generalized Hypergeometric Function, *J. London Math. Soc.* 10, pp. 286–293.
15. Fox, C., 1961, The G and H-Functions as Symmetrical Fourier kernels, *Trans. Am. Math. Soc.*, 98, pp. 395–429.
16. Chandrashekharan, K. and Narasimhan, R., 1962, Functional Equation with Multiple Gamma Factors and the Average Order of Arithmetical Functions, *Ann. Math.* 76(2), pp. 93–136.
17. Bochner, S., 1958, On Reimann's Functional Equation with Multiple Gamma Factors, *Ann. Math.* 67(2), pp. 29–41.
18. Braaksma, B.L.J., 1964, Asymptotic Expansion and Analytic Continuation for a Class of Barnes Integrals, *Camp. Math*, 15, pp. 239–341.
19. Kesarwani, R.N., 1965, A Pair of Unsymmetrical Fourier Kernels, *Trans. Amer. Math. Soc.* 115, pp. 356–365.
20. Saxena, V.P., 1982, A Formal Solution of Certain New Pair of Dual Integral Equations Involving H-Function, *Proc. Nat. Acad. Sci. India.* 52(A), pp. 366–375.
21. Kober, H., 1940, On Fractional Integrals and Derivatives, *Quart. J. Math.* os-11, pp. 193–211.
22. Erdelyi, A., 1950–51, On Some Functional Transformations, Politecnico di Torino, *Rend. Sem. Mat.* 10, p. 217.
23. Saxena, V.P., 1967, On Fractional Integral Operators, *Math. Z.* 96, p. 288.

24. Fox, C., 1965, A Formal Solutions of Certain Dual Integral Equations, *Trans. Am. Math. Soc.* 115, pp. 389–396.
25. Saxena, R.K., 1967, On the Formal Solution of Dual Integral Equations, *Proc. Am. Math. Soc.* 18, pp. 1–8.
26. Kesarwani, R.N., 1967, Fractional Integrals and Dual Integral Equations, *Math. Zeitechv.* 98, p. 83.
27. Dixon, A.L and Ferrar, W.L., 1936, A Class of Discontinuous Integral, *Quart. J. Math.* 7, pp. 81–96.
28. MacRobert, T.M., 1939, I.-Induction Proofs of the Relations between certain Asymptotic Expansions and Corresponding Generalised Hypergeometric Series, *Proc. Roy. Soc. Edinburgh.* 58, pp. 1–13.
29. Slater, J.L., 1966, *Generalized Hypergeometric Function*, Cambridge University Press, Cambridge.
30. Mittag-Leffler, G.M., 1903, Sur la nouvelle fonction Eα (x). *CR Acad. Sci. Paris.* 137(2), pp. 554–558.
31. Saxena, V.P., 2008, *The I-Function*, Anamaya Publishers, New Delhi.
32. Erdelyi, A., 1953, *Higher Transcendental Functions*, Vol. I, McGraw-Hill Book Company, Inc., New York, pp. 1–55.
33. Jat, V., 2016, Analytical Study of Saxena's I-Function and Its Applications, A Thesis Submitted to Barkatullah University, Bhopal.
34. Jain, P., 2016, Some New Properties and Inter-Relations of Saxena's I-Function, A Thesis Submitted to Jiwaji University, Gwalior.
35. Saxena, R.K and Singh, V., 1993, Integration of Generalized H-Function with Respect to their Parameters, *Vijnana Parishad Anusandhan Patrika* 36(1), p. 54.
36. Truesdell, C., 1948, *An Essay toward the Unified Theory of Special Functions*, Princeton University Press, Princeton, NJ.
37. Agarwal, B.M. and Saxena, V.P., 1969, The Solution of a General Integral Equation, *The Mathematics and Student*, XXXVII(1), p. 71.
38. Sneddon, I.N., 1972, *The Use of Integral Transforms*, McGraw-Hill Companies, New York.
39. Gupta, K.C., 1965, On the H-Function, *Am. Soc. Sci.* 179, p. 97.
40. Sharam, O.P., 1967, Relation between Whittaker and Generalized Hankel Transformations, *Proc. Nat. Acad. Sci. India.* 37, p. 97.
41. Agarwal, R.P., 1953, Some Inversion Formulae for the Generalised Hankel Transform, *Bull. Cal. Math. Soc.* 45, p. 69.
42. Goyal, G.K., 1967, Some Relation between Hankel and Meijer Transforms, *Proc. Nat. Acad. Sci. India.* 37, p. 60.
43. Sneddon, I.N., 1966, Mixed Boundary Value Problem in Potential Theory, *J. Frankl. Inst.* 257(2), pp. 121–124.
44. Jat, V. and Saxena, V.P., 2014, Solution of Certain Integral Equation Involving I-function, *Jnanabha.* 4, pp. 43–52.
45. Saxena, V.P., 1971, On a New Class of Kernels, *Jnanabha*, pp. 1–7.
46. Ahmad, A., Jain, R. and Jain, D.K., 2017, Certain Results of (p, q)-analogue of I-Function with (p, q)-Derivative. In *Proceedings of the MSSCID-2017, 20th Annual Conference of VPI*, Jaipur, India, pp. 24–26.
47. Chakrabarti, R. and Jagannathan, R., 1991, A (p, q)-Oscillator Realization of Two-Parameter Quantum Algebras, *J. Phys. A: Math. Gen.* 24, p. L711.
48. Brodimas, G., Jannussis, A. and Mignani, R., 1991, Two Parameter Quantum Groups, Universita di Roma preprint Nr. 820.
49. Arik, M., Demircan, E., Turgut, T., Ekinci, L. and Mungan, M., 1992, Fibonacci Oscillators, *Z. Phys. C Particles Fields.* 55, pp. 89–95.
50. Jackson, F.H., 1910, On q-definite integrals, *Quart. J. Pure Appl. Math.* 41, pp. 193–203.

51. Sadjang, P.N., 2014, On the Fundamental Theorem of (p, q)-Calculus and Some (p, q)-Taylor Formulas, Submitted to Integral Transform. *arXiv e-prints*, arXiv-1309.
52. Sadjang, P.N., 2015, On the (p, q)-Gamma and the (p, q)-Beta Functions. *arXiv preprint*, arXiv:1506.07394.
53. Saxena, R.K., Modi, G.C. and Kalla, S.L., 1983, A Basic Analogue of Fox's H-Function, *Rev. Tec. Ing. Univ., Zulin.* 6, pp. 139–143.
54. Pathak, S. and Jain, R., Some Identities Involving (p, q)-Analogue of Meijer's G-Function, *J. Indian Acad. Math.* 39(2), pp. 137–146.
55. Farooq, A., Jain, R. and Jain, D.K., 2012, q-Integral and Basic Analogue of I Function, *Asian J. Math. Stat.* 5, pp. 99–103.
56. Agrawal, B.M., 1966, A Unified Theory of Special Functions, A Ph.D. thesis, Vikram University, Ujjain.
57. Agrawal, B.M., 1968, Rodrigue's and Schlafli's Formulae for the Solution of F-Equations, *J. Ind. Math. Soc.* XXXXI(3–4), pp. 173–177.
58. Jain, R., Chaturvedi, P.K. and Jain, D.K., 2008, Some Generating Functions of I-Functions Involving Truesdell's F-Equation Technique, *Rajasthan Ganit Parishad.* 21(2).
59. Srivastava, H.M., 1971, A Class of Generating Functions for Generalized Hypergeometric Polynomials, *J. Math. Anal. Appl.* 35(1), 230–235.
60. Jain, P., Gupta, A. and Saxena, V.P, 2016, Multiple Integral Involving I-Function and Bessel-Maitland Functions, *IJMTT.* 39(3), pp. 232–237.
61. Berger, M.J. and Lamkin, J.C., 1958, Sample Calculation of Gamma Ray Penetration into Shelters, Contribution of Sky Shine and Roof Contamination, *J. Res. N.B.S.* 60, pp. 109–116.
62. Siddiqi, R.N., 1989, On a Class of Generalized Elliptic-Type Integrals, *Rev. Bras. Fis.* 19, pp. 137–147.
63. Evans, J.D., Hubbell, J.H. and Evans, V.D., 1993, Exact Series Solution to the Epstein-Hubbell Generalized Elliptic-Type Integral Using Complex Variable Residue Theory, *Appl. Math. Comput.* 53, pp. 173–189.
64. Kalla, S.L., 1984, Results on Generalized Elliptic-Type Integrals, Mathematical Structure-Computational Mathematics-Mathematical Modelling 2, *Publ. house Bulgar. Acad. Sci.*, Sofia, pp. 216–219.
65. Kalla, S.L., 1993, The Hubbell Rectangular Source Integral and Its Generalizations, *Radiat. Phys. Chem.* 41, pp. 775–781.
66. Kalla, S.L. and Al-Saqabi, B., 1986, On a Generalized Elliptic-Type Integral, *Rev. Bras. Fis.* 16, pp. 145–156.
67. Salman, M., 2006, Generalized Elliptic-Type Integrals and Their Representations, *Appl. Math. Comput.* 181(2), pp. 1249–1256.
68. Saxena, R.K., Kalla, S.L. and Hubbell, J.H., 2001, Asymptotic Expansion of a Unified Elliptic-Type Integrals, *Math. Balkanica (N.S.).* 15, pp. 29–38.
69. Srivastava, H.M. and Bromberg, S., 1995, Some Families of Generalized Elliptic-Type Integrals, *Math. Comput. Model.* 21(3), pp. 29–38.
70. Glasser, M.L. and Kalla, S.L., 1989, Recursion Relations for a Class of Generalized Elliptic-Type Integrals, *Rev. Tecn. Fac. Ingr. Univ. Zulia.* 12, pp. 47–50.
71. Al-Saqabi, B.N., 1987, A Generalization of Elliptic-Type Integrals, *Hadronic J.* 10, pp. 331–337.
72. Matera, J., Galue, L. and Kalla, S.L., 2002, Asymptotic Expansions for Some Elliptic-Type Integrals, *J. Rajasthan Acad. Phys. Sci.* 1(2), pp. 71–82.
73. Srivastava, H.M. and Siddiqi, R.N., 1995, A Unified Presentation for Certain Families of Elliptic-Type Integrals Related to Radiation Field Problems, *Radiat. Phys. Chem.* 46, pp. 303–315.
74. Kalla, S.L. and Tuan, V.K., 1996, Asymptotic Formulas for Generalized Elliptic-Type Integrals, *Comput. Math. Appl.* 32, pp. 49–55.

75. Kalla, S.L. and Saxena, R.K., 2000, A New Method for Evaluating Epstein-Hubbell Generalized Elliptic-Type Integrals, *Int. J. Appl. Math.* 2, pp. 732–742.
76. Saxena, R.K. and Pathan, A.M., 2003, Asymptotic Formulas for Unified Elliptic-Type Integrals, *Demonstratio Math.* 36(3), pp. 579–589.
77. Chaurasia, V.B.L. and Pandey, S.C., 2008, Unified Elliptic-Type Integrals, *Integral Transforms Spec. Funct.* 15(4), pp. 359–368.
78. Saxena, R.K. and Kalla, S.L., 2004, Asymptotic Formulas for Unified Elliptic-Type Integrals, *Integral Transforms Spec. Funct.* 15(4), pp. 359–368.
79. Mohammed, W.Ch., 1997, Bilinear and Bilateral Generating Functions of Generalized Polynomials, *J. Aust. Math. Soc. Ser. B.* 39, pp. 257–270.
80. Saran, S., 1972, Theorems on Bilinear Generating Functions, *Indian J. Appl. Math.* 3, pp. 12–20.
81. Srivastava, H.M. and Yeh, Y.N., 2000, Certain Theorem on Bilinear and Bilateral Generating Functions, *Anziam J.* 43, pp. 567–574.
82. Chaursaia, V.B.L. and Meghwal, R.C., 2012, Unified Presentation of Certain Families of Elliptic-Type Integrals Related to Euler Integrals and Generating Functions, *Tamkang J. Math.* 43(4), pp. 507–519.
83. Srivastava, H.M., 1967, Generalized Newman Expansions Involving Hypergeometric Functions, *Proc. Camb. Phil. Soc.* 63, pp. 425–429.
84. Appell, P. and Kampe De Feriet, J., 1926, *Functions hypergeometriques et hyperspheriques, polyno, es d'Hermite*, Gauthier-Villers, Paris.
85. Bromwich, T.J.I.A., 1965, *An Introduction to the Theory of Infinite Series* (Vol. 335), American Mathematical Society, Providence, RI.
86. Erdelyi, A., 1954, *Project: Tables of integrals Transforms*, Vol. I, McGraw-Hill, New York.
87. Jain, R. and Jain, D., 1995, Saxena's I-Function and Heat Conduction in Rod under Specific Boundary Conditions, *J. Indian Acad. Math.* 17(1), pp. 50–54.
88. Jain, R. and Singh, D., 2012, An Integro-Differential Equation of Volterra Type with Sumudu Transform, *Math. Aeterna.* 2(6), pp. 541–547.
89. Erdelyi, A., Magnus, W., Oberhettinger, F. and Tricomi, F.G., 1954, *Tables of Integral Transform*, Vol. 1, Vol. 2, McGraw-Hill, New York.
90. Saxena, V.P., 1969, A Study of Transform Calculus with Technical Applications, A Ph.D Thesis Submitted to Vikram University, 7, pp. 184–202.
91. Vaishya, G.D., 2007, Saxena's I-Function and Its Applications, A Ph.D Thesis Submitted to Jiwaji University, 3, pp. 78–104.
92. Agarwal, L., 2002, Recent Advances in I-Function and Its Applications, A Ph.D Thesis Submitted to Jiwaji University, 5, pp. 199–222.
93. Carslaw, H.S. and Jaeger, J.C., 1959, *Conduction of Heat in Solids*, Second Edition, Oxford University Press, Oxford.
94. Debnath, L., 1993, On Jacobi Transform, *Bull. Cal. Math. Soc.* 55, p. 113.
95. Sabharwal, K.C, 1996, Heat Conduction in a Non-homogenous Anisotropic Slab, *Indian J. Pure & App. Phys.* 4, p. 282.
96. Tiwari, I.P. and Sharma, C.K., 1993, Evaluation of a Definite Integral by Using E-Operator and Its Application in Heat Conduction, *Journal of M.A.C.T.* 26, pp. 1–6.
97. Churchill, R.V., 1963, *Fourier Series and Boundary Value Problems*, McGraw-Hill, New York.
98. Paliwal, D.D., 1992, Application of Associated Legendre Polynomials in Diffusion Problem, *Journal of M.A.C.T.*, 25, pp. 45–47.
99. Gradshteys, I.S. and Ryzhik, I.M., 1980, *Table of Integrals, Series and Products*, Academic Press, New York, p. 372.
100. Logan, J.D., 2004, *Partial Differential Equations*, Springer, New York.

101. Pipes, L.A., 1958, *Applied Mathematics for Engineers and Physicists*, 2nd Ed. McGraw-Hill, New York.
102. Erdelyi, A., 1953, *Tables of Integral Transform*, Vol. II, McGraw-Hill, New York.
103. Horn, J., 1931, Hypergeometrische Funkionen zweir varanderliche, *Math. Ann.* 105, pp. 381–407.
104. Garde, R.M., 1967, Applications of Jacobi Polynomials to Non-linear Oscillations–1, Free Oscillations, *Proc. Nat. Acad. Sci. India* 37, pp. 119–120.
105. Demeryd, A.G., 1997, A Theoretical Investigation of Rectangular Microstrip Antennas, Rome Air Development Centre, *Tech. Rep. TR77-206* AP-26(4), pp. 532–535. DOI: 10.1109/TAP.1978.1141890.
106. Despandey, M. D, and Baily, M.G., 1982, Input Impedance of Microstrip Antenna, *IEEE Trans. AP.* 30(4), pp. 645–650. DOI: 10.1109/TAP.1982.1142863.
107. Munson, R.E., 1974, Conformal Microstrip Antennas and Microstrip Phased Array, *IEEE Trans. AP* 22, pp. 74–77.
108. Jordon, E.C. and Balrnann, K.G., 1974, *Electromagnetic Waves and Radiating Systems*, Prentice-Hall, Inc., Englewood Cliffs.
109. Collins, R.E., 2000, *Foundations for Microwave Engineering*, 2nd Edition, Wiley-IEEE Press, New York.
110. Prasad, S., 1993, A Line Magnetic Current Method for the Analysis of Rectangular Microstrip Antenna, ME thesis, BIT (Mesra).

Index

Note: *Italic* page numerals refer to figures.